Air Transport Management

Air Transport Management: An International Perspective provides in-depth instruction in the diverse and dynamic area of commercial air transport management. The 2nd edition has been extensively revised and updated to reflect the latest developments in the sector.

The textbook includes both introductory reference material and more advanced content so as to provide a solid foundation in the core principles and practices of air transport management. This 2nd edition includes a new chapter on airline regulation and deregulation and new dedicated chapters focusing on aviation safety and aviation security. Four new contributors bring additional insights and expertise to the book.

The 2nd edition retains many of the key features of the 1st edition, including:

- A clearly structured topic-based approach that provides information on key air transport management issues including: aviation law, economics; airport and airline management; finance; environmental impacts, human resource management; and marketing;
- Chapters authored by leading air transport academics and practitioners worldwide that provide an international perspective;
- Learning objectives and key points that provide a framework for learning;
- Boxed case studies and examples in each chapter;
- Keyword definitions and stop and think boxes to prompt reflection and aid understanding of key terms and concepts.

Designed for undergraduate and postgraduate students studying aviation and business management degree programmes and industry practitioners seeking to expand their knowledge base, the book provides a single point of reference to the key legal, regulatory, strategic and operational concepts and processes that shape the form and function of the world's commercial air transport industry.

Lucy Budd is Professor of Air Transport Management and Programme Director of the MSc in Air Transport Management in Leicester Castle Business School at De Montfort University, Leicester. She has extensive experience of teaching air transport management at both undergraduate and postgraduate levels and has published extensively in the area of airline and airport operations.

Stephen Ison is Professor of Air Transport Policy in Leicester Castle Business School at De Montfort University, Leicester. He has taught air transport programmes for over 20 years and has published extensively in the area of aviation business and economics.

Air Transport Management

An International Perspective

Second Edition

Edited by
Lucy Budd and Stephen Ison

Routledge
Taylor & Francis Group

LONDON AND NEW YORK

Second edition published 2020
by Routledge
2 Park Square, Milton Park, Abingdon, Oxon, OX14 4RN

and by Routledge
52 Vanderbilt Avenue, New York, NY 10017

Routledge is an imprint of the Taylor & Francis Group, an informa business

First edition published by Routledge 2016

British Library Cataloguing-in-Publication Data
A catalogue record for this book is available from the British Library

Library of Congress Cataloging-in-Publication Data
A catalog record has been requested for this book

ISBN: 978-0-367-28056-7 (hbk)
ISBN: 978-0-367-28057-4 (pbk)
ISBN: 978-0-429-29944-5 (ebk)

Typeset in Minion Pro
by Apex CoVantage, LLC

Contents

17 Air cargo and logistics

Martin Dresner and Li Zou

18 Environmental impacts and mitigation

Thomas Budd

19 Human resource management and industrial relations

Geraint Harvey and Peter Turnbull

Figures

Tables

Contributors

Ronald Bartsch is Professor of Aviation Law at the University of the South Pacific and Visiting Professor at the Australian National University and lecturer in aviation law at the University of New South Wales for the past two decades. Ron is Chairman and founding Director of AvLaw, a highly respected international consultancy firm. He was the former Head of Safety and Regulatory Compliance at Qantas Airways and Manager of Air Transport Operations with Australia's Civil Aviation Safety Authority (CASA). Ron was admitted as a barrister to the High Court of Australia in 1993, is a presiding member of Australia's Administrative Appeals Tribunal as an aviation specialist and holds a current Airline Transport Pilot Licence (ATPL). He is also a Director of Regional Express Holdings Limited (REX) and Chair of the Board of REX Safety and Risk Management Committee. In 2019, he was appointed as Senior Aviation Consultant to Clyde & Co's Global Drones Strategy Group. He is author of numerous publications including the bestselling *International Aviation Law: A Practical Guide, Aviation Law in Australia* (5th edition) and *Drones in Society* (2016).

Lucy Budd is Professor of Air Transport Management and Programme Director of the MSc in Air Transport Management in Leicester Castle Business School at De Montfort University in Leicester. She has published extensively in the area of airline and airport operations and has a particular interest in exploring human interactions with, and interventions affecting, the global air transport system. She sits on the international editorial boards of the *Journal of Transport Geography*, the *Journal of Transport Policy* and the *Journal of Air Transport Management* and is a co-editor of *Research in Transportation Business and Management*.

Thomas Budd is a Lecturer in Airport Planning and Management in the Centre for Air Transport Management at Cranfield University, UK, and Course Director for the MSc in Airport Planning and Management. His teaching and research activities focus on issues of air transport environmental sustainability and resilience. He has published his research in a range of leading academic peer-review journals, presented at numerous international conferences, and sits as an advisor on London Heathrow Airport's Centre of Excellence for Sustainability Think Tank. He is also the Academic Lead for Cranfield's Passenger Experience Laboratory, a dedicated research facility within Cranfield's Digital Aviation

Research Technology Centre (DARTeC) examining how disruptive technologies can be leveraged to create more seamless, connected and sustainable air passenger journeys.

Martin Dresner has served on the faculty of the University of Maryland since 1988, where he is currently Professor in the Robert H. Smith School of Business's Logistics, Business and Public Policy Department. His research focuses on two broad areas: air transport policy and logistics management. Professionally, Martin is President of the Air Transport Research Society and is a member of the Scientific Committee of the World Conference on Transportation Research. He serves as associate editor for the *Journal of Business Logistics* and *Decision Sciences*.

David Gillen is Professor of Transportation and Logistics in the Sauder School of Business and Director, Centre for Transportation Studies, University of British Columbia, Canada. He is former Editor of the *Journal of Transport Economics and Policy* and Associate Editor of *Transportation Research E: Logistics & Transportation Review*. His published research has examined the linkages between governance structures, ownership and regulation, measuring performance of transportation infrastructure, evolving strategies and business models in airlines and airports, and studying the role of transportation in the supply chain and the linkage to the economy.

Anne Graham is Professor of Air Transport and Tourism Management at the University of Westminster, UK. She has been involved in air transport teaching and research for over 30 years. She has a particular interest in airport management, economics and regulation, and air travel demand and tourism. She was recently the Editor-in-Chief of the *Journal of Air Transport Management*. Anne has published extensively in the field of air transport and tourism, and her book *Managing Airports: An International Perspective* is now in its 5th edition (2018). Other recent co-authored or edited books include *Air Transport: A Tourism Perspective* (2019), *Destination London: The Expansion of the Visitor Economy* (2019), *The Routledge Companion to Air Transport Management* (2018), *Airport Finance and Investment in the Global Economy* (2016), *Aviation Economics* (2016) and *Airport Marketing* (2013).

Nigel Halpern is Professor of Air Transport and Tourism Management in the Department of Marketing at Kristiania University College in Norway. Nigel has previously worked as Associate Professor in Transport and Logistics at Molde University College, Principal Lecturer and Subject Group Director in Aviation Management and Operations at London Metropolitan University, Chief Instructor at PGL Travel, Policy Adviser at the UK Department for Transport, and Airline Licensing Executive at the UK Civil Aviation Authority. His research is focused largely on marketing. He has authored and edited books with Anne Graham on *Airport Marketing* (Routledge, 2013) and *The Routledge Companion to Air Transport Management* (Routledge, 2018).

Geraint Harvey is Professor of People and Organisation at the School of Management, Swansea University. His research focuses on the changing nature of work. In particular he is interested in the implications for managers, workers and trade unions of the increasing use both of precarious employment contracts and of the replacement of employment contracts by commercial contracts. Whereas the industrial focus of his research extends beyond civil aviation, much of his previous and current research focuses on the airline industry. He has conducted research for the European Commission, the European Transport Workers' Federation and International Transport Workers' Federation, and the International Labour Organisation.

Peter Hind is Chief Executive of RDC and a visiting lecturer at City University, London, where he teaches sustainable aviation, and at Loughborough University, where he teaches airline business strategy. He has worked in the air transport industry for over 25 years, starting at British Midland Airways as a pricing, prorate and revenue management specialist. He spent the following ten years as a consultant at RDC Aviation Economics, undertaking a range of strategic advisory assignments for airline and airport clients worldwide. As Chief Executive at RDC, he leads the company's technology and data strategy.

Gieri Hinnen obtained a PhD in Management from the University of St Gallen (HSG), Switzerland. He has a background in International Affairs and Business Administration; he studied at the University of St Gallen (HSG), the London School of Economics (LSE) and Schulich School of Business. His research investigates innovation and organisational behaviour, focusing on sustainable mobility. Besides his research activities, Gieri works for Swiss International Air Lines, where he holds the position of Head of Labour Relations & HR Steering.

Stephen Ison is Professor of Air Transport Policy in Leicester Castle Business School at De Montfort University, Leicester. He has published extensively in the areas of low-cost airlines, airport policy and airport ground access and is the author/co-author/editor of 18 books. Stephen is a member of the Scientific Committee of the World Conference on Transport Research, Editor of the *Journal of Research in Transportation Business and Management* (Elsevier), and Book Series Editor of *Transport and Sustainability* (Emerald).

Joe Kelly is a Fellow of the Institute of Chartered Accountants and holds an MBA from Aston University. He was Deputy Chief Executive and before that Finance Director at Birmingham Airport between 1998 and 2015. Prior to that he was a Finance Director of companies in the manufacturing, distribution, retail and service sectors. He was also a Member of the Institute of Credit Management, the Chartered Institute of Marketing, the Institute of Data Processing and the Pensions Management Institute. A past President of the Chamber of Commerce and Governor of Solihull College, he is currently Deputy Chairman of Birmingham City University and acts as Pension Trustee for several schemes.

Gareth Kitching currently works as Head of Product R&D for RDC Aviation, a leading aviation software company having previously worked at Manchester Airport Group as Head of Strategic Insight. During his career, Gareth has gained expert knowledge in airport strategy, air service route development, airport traffic forecasting, airline fare and yield analysis, and the economic impact of air services, among other areas. His main interest is in the understanding of key drivers behind air service developments, from both demand and supply perspectives. He has provided analysis and assessment on some of the world's largest aviation projects.

Erik Linden is a research associate and doctoral candidate at the University of St. Gallen in Switzerland. His research focuses on long-term planning, with a special focus on strategy planning, strategy processes, business models, business plans, future research, strategic marketing as well as aviation and transportation management. He studied business management at the University of St. Gallen and has a professional business background in transportation management, business management, and public transport. Also, he is currently managing director of the Swiss Aerospace Cluster and lecturer at various universities and universities of applied sciences.

Gui Lohmann is Associate Professor in Aviation Management in the School of Engineering and Built Environment at Griffith University, Australia. His research expertise includes airline business models, passengers' travel patterns and behaviours, and aviation and tourism. Gui is also an enthusiastic educator and currently the Deputy Head of School, Learning and Teaching. He has consulted to several airports in Australia, including Adelaide, Brisbane and Gold Coast. Throughout his international career, Gui has worked at universities in Brazil, New Zealand and the US.

Stephen J Maher is a lecturer at the University of Exeter. His work involves the development of the mixed integer programming solver SCIP, primarily focused on general frameworks for decomposition methods. Stephen completed his PhD at the University of New South Wales, Australia. He has experience with robust optimisation approaches to aviation applications and the development of novel solution algorithms to solve airline recovery problems. Stephen was awarded the Anna Valicek medal from the Airline Group of IFORS for his paper 'The Recoverable Robust Tail Assignment Problem'. His PhD thesis was awarded the dissertation prize from the Aviation Applications Section of INFORMS.

Rico Merkert is Chair in Transport and Supply Chain Management and Deputy Director of the Institute of Transport and Logistics Studies at The University of Sydney. He has taught and researched at a number of high-profile institutions such as Cranfield University and Haas Business School, University of California, Berkeley; and is Visiting Professor at the University of Johannesburg. Rico is Editor-in-Chief of the *Journal of Air Transport Management*, Associate Editor of the *Journal of Transport and Supply Chain Management* and an appointed member of three US Transportation Research Board (TRB) standing committees. He has been involved in a number of projects on performance measurement and management (organisation, finance and M&A), benchmarking, value and supply chain analysis, digitisation, strategy and policy for a range of clients such as the European Commission, Transport for NSW and a number of major airlines. Most of his recent projects, such as the iMOVE CRC project, focus on the smart (technology/data supported) and efficient management of various elements of the aviation value chain both in the global and regional context.

Richard de Neufville is Professor of Engineering Systems at the Massachusetts Institute of Technology (MIT), Institute for Data Systems and Society. He co-authored the texts *Airport Systems: Planning, Design, and Management* (McGraw-Hill, 2nd edition 2013, with MIT Professor Odoni) and *Flexibility in Engineering Design* (MIT Press, 2011, with Cambridge Professor Scholtes). His aviation awards include the US FAA prize for

Excellence in Aviation Education, the US Transportation Research Board McKelvey Prize in Aviation, and the Horonjeff Prize in Airport Planning from the American Society of Civil Engineers. He has consulted on airport design on every inhabited continent; major recent engagements involve airports at Boston, Mumbai, Paris, Qatar, and Singapore.

Mohammed Quddus is a Professor of Intelligent Transport Systems (ITS) at Loughborough University, UK. He holds a PhD in Intelligent Transport Systems (ITS) from Imperial College, London, and is a Fellow of the UK Higher Education Academy. Currently Mohammed serves as an Associate Editor of *Transportation Research Part C: Emerging Technologies*. His primary research areas include transport safety, autonomous transport systems and transport modelling. Since 2006, he has been teaching aviation safety and quantitative analysis in aviation to undergraduate students at Loughborough University.

Andrew Timmis is a lecturer in Transport in the School of Architecture, Building and Civil Engineering at Loughborough University. His research interest concerns transitions to a more sustainable aviation system with particular focus on the management of environmental noise and understanding the environmental impacts of aerospace supply chains. He obtained his PhD in Management from Sheffield University Management School as part of the Logistics and Supply Chain Management research group.

David Trembaczowski-Ryder is Head of Aviation Security at Airports Council International – Europe. He has spent his whole career in aviation, defence and security. He is a retired Royal Air Force officer with significant experience in NATO and EU strategic policy development. David's last appointment was with the UK Permanent Representation to the EU dealing with European Security and Defence Policy, and prior to that he was a seconded national expert in the EU Council General Secretariat. Before Brussels he was the Station Commander of RAF Gibraltar, where he was responsible for all aspects of airfield operations. In February 2010 he joined ACI EUROPE, where he covers the whole range of aviation security issues affecting Europe's airports, including preparing proposals to counter new and emerging threats. He is directly involved in the drafting of EU legislation. David holds a BSc (Hons) in Economics and is a graduate of the RAF Staff College and the European Security and Defence College.

Samantha Lucy Trimby is a Solicitor at Professionals Australia. She was admitted as a lawyer in the Supreme Court of New South Wales in October 2015. Samantha tutors in Aviation Law and Environmental Law at Swinburne University of Technology. She has drafted and edited aviation law courses for the University of New South Wales, Swinburne University of Technology and the Australian National University. Samantha has edited and drafted material for publishers such as Thomson Reuters and Ashgate Publishing, including *International Aviation Law: A Practical Guide and Aviation Law in Australia* (4th edition). She has also conducted research for the Space Law Guide Card for Halsbury's Laws of Australia.

Peter Turnbull is Professor of Management in the new School of Management at the University of Bristol, UK. He was previously an Academic Fellow at the International Labour Organization (ILO) and has produced several reports on the civil aviation industry for the ILO, European Commission, International Transport Workers' Federation (ITF),

European Transport Workers' Federation (ETF) and the European Cockpit Association (ECA). His current research focuses on business strategies and atypical forms of employment in the European single aviation market. Peter is an Academic Fellow of the Chartered Institute of Personnel and Development (CIPD) and a member of the Advisory, Conciliation and Arbitration Service (ACAS) arbitration panel.

Randall Whyte lectures in aviation and behavioural economics at Griffith University, Australia. He has a long-standing interest in, and detailed knowledge of, the air transport industry, and he has conducted extensive research into the growth of low cost carriers and carriers-within-carriers, with a particular focus on the Australian market, as well as the potential for long-haul low cost carriers.

Andreas Wittmer is Head of International Networks at University of St Gallen and Director of the Centre for Aviation Competence at the University of St Gallen, Switzerland. He holds teaching positions at several universities in Switzerland and internationally. His research interests cover consumer behaviour, marketing and service management in transport, tourism and especially air transport. Andreas has published research papers, book chapters and three books. He is a member of several editorial journals and conference boards. Furthermore, he is Vice President of the Swiss Aerospace Cluster, Vice President of the Foundation for Aviation Competence, Vice President of the Aviation Research Center Switzerland and a freelance aircraft accident investigator at the Swiss Aircraft Accident Investigation Bureau.

Cheng-Lung Wu is an Associate Professor at the School of Aviation, UNSW Sydney. His research interests include airline/airport operations, airline scheduling, passenger behaviour modelling, airport retail development and data analytics. Cheng-Lung provides consultancy to airlines and airports on scheduling, operations, retail business and data modelling. He is a certified IATA trainer in the area of airports and ground operations and has delivered IATA training courses to senior industry managers since 2010. He also provides commentary on TV and magazines regarding aviation development.

Li Zou is Professor of Marketing and Supply Chain Management in David B. O'Maley College of Business at Embry-Riddle Aeronautical University, Daytona Beach, US. Her research into the economics of air cargo, airline alliances and aviation logistics has been published in leading academic journals, and she currently teaches international business and air cargo logistics management to students at Embry-Riddle.

Preface to the 2nd edition

It is with great pleasure that we have been able to edit the 2nd edition of *Air Transport Management: An International Perspective*. We have received many positive comments and favourable reviews of the 1st edition of the book from students, academics and practitioners, and we have sought to incorporate many of their suggestions regarding content, structure and presentation to further enhance the 2nd edition. We have assembled the combined skills, expertise and insight of leading academics and industry experts from around the world. The time and effort they have invested in their contributions for this 2nd edition is reflected in the quality of their outputs and their commitment to sharing their knowledge with the next generation of air transport managers. We have made every effort to ensure that this 2nd edition informs, inspires and challenges your understandings of this most dynamic and interesting of subjects.

Lucy Budd and Stephen Ison
Leicester, UK, 2020

Editor acknowledgements

All book projects involve intense collaboration between multiple individuals and organisations. We consider ourselves fortunate to have assembled the expertise of leading aviation scholars and practitioners from around the world. We would like to thank all the authors for their time, effort and positive response to our editorial requests.

We would like to thank and acknowledge Ian Humphreys and Graham Francis (Chapter 6), John Jackson (Chapter 11) and David Pritchard (Chapter 14), who authored these chapters in the first edition.

We would also like to thank: Nena Adrienne (Manchester Airport), Camille Burban (FlyBe), Rachel Burbidge (Eurocontrol), Nick Collard (retired), Tim Coombs (Aviation Economics), Lynnette Dray (UCL), David Gillen (University of British Columbia), Anne Graham (University of Westminster), Robert Mayer (Cranfield University), Faye McCarthy (Leicester University), Frankie O'Connell (University of Surrey), Romano Pagliari (Cranfield University), Aneil Patel (ACI-North America), Jon Proudlove (NATS), René Puls (University of St Gallen), Daniel Rodriguez-Jimenez (IAG World Cargo), Mohammed Raifath (De Montfort University), Monika Simonaityte (Stansted Airport), Mazen Tarawiyeh (Loughborough University), Saskia Vulturius (Emirates) and David Warnock-Smith (Bucks New University), for undertaking such considered reviews, providing meticulous and detailed comments on individual chapters, and for generously sharing their time and expertise with us. Nicolette Formosa (Loughborough University) expertly drew Figures 1.2, 2.1, and 2.2 for us. We would like to thank Tony Budd for proof reading the text.

We would like to acknowledge the valuable feedback provided by past and present students at Loughborough University and De Montfort University and the support of James Gore at Bristol Airport for supplying the LAGS graphic which appears in Chapter 14, and Megan Westmoreland at HIAL for supplying the Twin Otter photo which appears in Chapter 21.

At Routledge, Guy Loft has provided continued good humour, support and encouragement while Matthew Ranscombe has offered unwavering support and advice. At Apex, Kate Fornadel provided expert production support.

Acronyms

AAIB	Air Accidents Investigation Branch (UK)
A-CDM	Airport Collaborative Decision Making
ACI	Airports Council International
ACMI	aircraft crew maintenance and insurance (lease)
AFUA	advanced flexible use of airspace
AIS	aeronautical information service
AMSL	above mean sea level
ANSP	Air Navigation Service Provider
APD	Air Passenger Duty
API	Advanced Passenger Information
APU	auxiliary power unit
AR	aircraft routing
ASA	air service agreement
ASK	available seat kilometre
ASM	airspace management
ASQ	airport service quality
ATC	air traffic control
ATFM	air traffic flow management
ATM	air traffic movement *or* air traffic management (depending on context)
ATS	air traffic services
AWB	air waybill
BA	British Airways
CAA	Civil Aviation Authority (UK)
CAS	controlled airspace
CASK	cost per available seat kilometre
CCO	continuous climb operation
CDO	continuous descent operation
CFIT	controlled flight into terrain
CFR	Code of Federal Regulations
CNS	communication, navigation and surveillance
CO_2	carbon dioxide

COMAC	Commercial Aircraft Corporation of China
CPI	consumer price index
CRM	customer relationship marketing (or management)
CRS	computer reservation system
CUSS	common-use self-service (check-in kiosk)
CUTE	common-use terminal equipment
CWC	carrier-within-a-carrier
DBS	distance based separation
EAS	Essential Air Service
EASA	European Aviation Safety Agency
EBIT	earnings before interest and tax
EBITDAR	earnings before interest, tax, depreciation, amortisation and rent
EC	European Commission
ECAA	European Common Aviation Area
EEA	European Economic Area
EFB	electronic flight bag
EIA	environmental impact assessment
EMS	environmental management system
ETBS	enhanced time-based separation
ETOPS	Extended Range Twin-engine Operational Performance Standards
ETS	Emissions Trading System
EU	European Union
FAA	Federal Aviation Administration (US)
FAL	final assembly line
FAM	fleet assignment model
FDI	foreign direct investment
FEGP	fixed electrical ground power
FFP	frequent flyer programme
FIATA	International Federation of Freight Forwarders Associations
FIFO	fly-in fly-out
FIR	flight information region
FL	flight level
FOD	foreign object debris
FSNC	full service network carrier
FTK	freight tonne kilometre
GDP	gross domestic product
GDS	Global Distribution System
GSIE	Global Safety Information Exchange
IAG	International Airlines Group
IATA	International Air Transport Association
ICAO	International Civil Aviation Organization
ICT	information communication technology
IFR	Instrument Flight Rules
ILS	instrument landing system

IMC	instrument meteorological conditions
IPCC	Intergovernmental Panel on Climate Change
KPI	key performance indicator
LAGs	Liquids, Aerosols and Gels
LCC	low cost carrier
LEDS	liquid explosive detection systems
LOC-I	loss of control in-flight
LOS	level of service
MARS	Multi Aircraft Ramp System
MCT	minimum connection time
MNJV	metal neutral joint venture
MRO	maintenance, repair and overhaul
NNI	Noise and Number Index
NO_x	nitrous oxides
NPR	noise preferential route
NTSB	National Transportation Safety Board (US)
OD	origin and destination
OEM	original equipment manufacturer
OTA	online travel agent
PSO	Public Service Obligation
PSR	primary surveillance radar
RAAP	Regional Aviation Access Programme (Australia)
RAI	Remote Aerodrome Inspection (Australia)
RAIF	Remote Aviation Infrastructure Fund (Australia)
RASK	revenue per available seat kilometre
RASP	Remote Aerodrome Safety Programme (Australia)
RASS	Remote Air Services Subsidy (Australia)
RDF	Route Development Fund
RET	rapid exit taxiway
RF	radiative forcing
ROCE	return on capital employed
RPI	retail price index
RPK	revenue passenger kilometre
RSOO	Regional Safety Oversight Organisations
RTK	revenue tonne kilometre
RVSM	reduced vertical separation minima
SARPs	Standards and Recommended Practices
SCM	Swiss Cheese Model
SDR	Special Drawing Rights
SEL	sound exposure level
SEM	security management system
SES	Single European Sky
SIDS	small island developing states
SMS	safety management system

SSR	secondary surveillance radar
STAMP	Systems-Theoretic Accident Model and Processes
STOL	Short Take-Off and Landing
TA	tail assignment
TAWS	terrain awareness and warning system
TBS	time based separation
UAV	unmanned aerial vehicle
UHF	ultra-high frequency
UIR	upper flight information region
ULCC	ultra-low cost carrier
ULD	unit load device
UN	United Nations
VFR	Visual Flight Rules *or* visiting friends and relatives (depending on context)
VHF	very high frequency
VMC	visual meteorological conditions
VOR	VHF omnidirectional range (beacon)
WLU	work load unit
WTO	World Trade Organization

✈ Introduction to air transport management

Lucy Budd and Stephen Ison

Air transport is rarely out of the news. Whether it is a story about the inconvenience caused by a pilots' strike or airline failure, community concerns about a proposed expansion of an airport, changes to existing flightpaths or the introduction of a new security regulation or aircraft safety directive, the appetite for aviation news appears insatiable. Indeed, the industry is complex, interconnected and dynamic. The purpose of this textbook is to introduce students of air transport management to the key legal, regulatory, commercial, operational, environmental and social dimensions that underpin the industry.

Commercial air transport

Commercial air transport describes the scheduled or non-scheduled aerial conveyance of passengers, cargo or mail in exchange for revenue. Commercial air transport is a mode of public transport which is distinct from both military aviation, which concerns the use of specialised aircraft by countries for defence and national security, and general aviation, which refers to the recreational, agricultural and/or instructional use of aircraft which are not available for public use and which are not flown in exchange for remuneration. Despite this apparently straightforward classification and definition, the commercial air transport industry includes a wide range of business approaches and operational practices, incorporating everything from long-range widebody aircraft flown by leading airline operators between major airports to small single-engine aircraft that serve some of the most remote regions on Earth.

Air transport management

Every day, over 100,000 flights transport almost 12 million passengers and around USD$18 billion worth of goods, equivalent to 4.3 billion passengers and 38 million scheduled flights a year, on 48,500 routes (ICAO, 2019). Managing this level of global activity in accordance with strict international regulations concerning safety, security, consumer affairs, finance, auditing, environmental performance and professional competence is inherently complex. It requires the combined and coordinated actions of 10.2 million direct employees and billions of dollars' worth of assets and infrastructure to ensure that people, goods and information are safely, efficiently, cost-effectively and profitably transported around the world by air at agreed standards of service.

Air transport management describes the business processes and functions that are deployed by companies involved at all stages of the air service delivery chain to achieve corporate objectives with the optimum use of available resources. It requires detailed consideration of forecasting, planning, procurement, staffing, supervision and operational coordination from the initial conception, design and manufacture of new aircraft and cabin service products to the collation of post-flight customer feedback and evaluation of financial performance.

To meet the industry's growing requirements for highly skilled labour, dedicated degree programmes in air transport management have been established to equip students with the practical knowledge they need to successfully manage air transport operations and services worldwide. Students studying such specialised aviation curricula require both subject-specific knowledge of aircraft and airport operations and a broader understanding of economics, law, regulation, finance, geography and international relations. Being cognisant of the diverse social, economic, political, physical and regulatory environments in which air transport operates not only enriches the learning experience and the value of the qualification(s) students obtain but also enhances employment prospects and provides a valuable mechanism through which theoretical constructs can be applied to real-world challenges.

Structure and organisation of the 2nd edition

This book has been designed to be accessible, informative and structured for ease of use. Each of the 21 chapters addresses a different but intrinsically interrelated aspect of commercial air transport management.

The first three chapters establish the legal, economic and regulatory context for air transport operations worldwide.

- ➤Chapter 1 details the legal environment in which air transport operates and documents the development of key international regulations pertaining to air service agreements and consumer rights.

- Air transport is a derived demand. This means that consumer demand for flight arises from a need for passengers, freight or mail to be somewhere else, not because the actual journey is desired for its own sake. ➤Chapter 2 examines the economics of air transport which underpins air transport management.

- An exploration of the motivation for, and consequences arising as a result of, policies of deregulation and liberalisation which have stimulated increased competition and led to the development of alternative airline business models is provided in ➤Chapter 3.

This is followed by four chapters which detail the complexities of the design, management and operation of airports and airport infrastructure.

- The provision and utilisation of runway, terminal and ground access infrastructure are vital to safe and efficient operations, and the relationship between airlines and airports is integral to the effective management of these scarce resources and to mitigating their adverse environmental effects. ➤Chapter 4 focuses on airfield design, ➤Chapter 5 examines airport systems planning and ➤Chapter 6 covers airport management and performance. ➤Chapter 7 examines the complex relationship that exists between airports and airline operators.

Chapters 8–14 concern airlines and cover business models, pricing, passenger segmentation, flight scheduling, finance, safety and security.

- Selecting the most appropriate airline business model for the market, taking into account the competitive environment and prevailing market conditions, is vital and is discussed in ➤Chapter 8.

- The airline product is a single-use consumer good. Once an aircraft is airborne, empty seats cannot be sold, so airlines have to carefully manage demand and their yields to efficiently utilise their inventory, cover their costs, and return a profit. Pricing and revenue management is examined in detail in ➤Chapter 9.

- Airlines need to ensure that they attract and retain the custom of passengers. The important process of passenger segmentation and airline loyalty programmes are covered in ➤Chapter 10.

- One of the ways in which airlines can attract and retain custom is via efficient scheduling and ensuring flight times meet consumer needs. This is a complex process which involves considerations of aircraft utilisation and crew resourcing in line with basic economic principles. The scheduling of aircraft and crew is discussed in ➤Chapter 11.

- Aircraft procurement and financing are all prerequisites of efficient airline operations, and different business models lend themselves to particular approaches. ➤Chapter 12 details the importance of finance to air transport management.

- Ensuring the safety of passengers, crew and third parties through the adoption of company-wide safety management systems that comply with international protocols and regulations is essential to ensuring continued air transport provision. ➤Chapter 13 details the models and processes that are utilised.

- In recognition that safety can be compromised by malicious activity, ➤Chapter 14 details the importance of security in ensuring the smooth and safe operation of air transport.

The final seven chapters relate to the air transport industry as a whole and deal with issues of airspace management, aircraft manufacturing, air cargo, environmental impacts, human resource management, air transport marketing and operating air services in remote regions.

- Issues of airspace organisation and management are detailed in ➤Chapter 15.

- No airline could operate without suitable aircraft, and ➤Chapter 16 focuses on issues of aircraft design and manufacturing.

- While the transport of passengers represents a major component of global air transport operations, air cargo services perform a vital economic function by transporting high value low weight goods quickly around the world. ➤Chapter 17 covers this dynamic area of air transport activity.

- Air transport operations create a range of negative environmental externalities. Managing and mitigating the effects of these externalities is important given scientific evidence about aviation's impact on the global climate and its local impact in terms of human health and wellbeing. These issues are addressed in ➤Chapter 18.

- The cyclical and seasonal variations in demand make the management of human resources inherently challenging. Given the central role of employees in air transport, the effective management of human resources is vital to the aviation sector's continued reputation and operations. As such, ➤Chapter 19 covers this increasingly important area in detail.

- Marketing the airline product to attract and retain customers and engender brand loyalty can be achieved using a variety of online and offline channels. These are discussed in detail in ➤Chapter 20.

- Although much of the focus of air transport activity is on the major airports, the final chapter, ➤Chapter 21, concerns the operation of air services in remote regions and examines the challenges associated with this type of operation.

Each chapter includes:

- a series of detailed **learning objectives** which convey the chapter's key content and purpose;

- **keyword** definitions which are provided in the margin the first time they appear; the location of keywords is also highlighted in blue in the index;

- carefully selected **case studies** and **examples** to aid understanding and demonstrate the practical application of key concepts;

- **cross-references** to other chapters, where appropriate, to emphasise the interconnected nature of the topics;

- strategically placed 'Stop and think' boxes, which are highlighted by an exclamation mark, invite readers to pause and reflect on key topics and test their understanding of important issues.

Each chapter concludes with:

- a series of **key points**;

- **references** and **suggestions for further reading** for those who wish to read more widely around a particular subject.

Key features of the 2nd edition include the following:

- Each chapter has been updated to reflect the changes that have occurred in the industry since the first edition was published. Updated case studies and data reflect the latest developments in the industry.
- A new chapter on airline regulation and deregulation has been added.
- A dedicated chapter on aviation safety and a new chapter on aviation security are included.
- The ICT chapter from the first edition has been subsumed within the marketing chapter.
- The addition of four new contributors from industry and academia who enhance the overall knowledge base.

Reference

ICAO. (2019). *Aviation benefits report*. Available at: www.icao.int/sustainability/Documents/AVIATION-BENEFITS-2019-web.pdf

CHAPTER 1

Aviation law and regulation

Ronald Bartsch and Samantha Lucy Trimby

LEARNING OBJECTIVES

- To identify the difference between air law and aviation law.

- To appreciate the importance of national sovereignty over airspace.

- To understand the reasons for, and implications arising from, the Chicago Convention 1944.

- To recognise the freedoms of the air.

- To describe the scope and purpose of the Warsaw and Montreal conventions.

- To determine the extent to which aviation law and regulation shape contemporary air transport operations and management.

1.0 Introduction

It is the freedom and agility by which air transport operations can readily transcend previously restrictive geographic and political boundaries that differentiates flying from other modes of transport. To harness this freedom, aviation regulation provides the requisite authority, responsibility and sanctions. The regulation of aviation is as fundamental and important to the industry as civil order is to modern society.

Irrespective of the particular discipline being studied, it is always desirable to have an understanding of the background and development of how and why the subject matter evolved. In respect to the study of international aviation law, this understanding is essential. No other field of human endeavour or branch of law is as harmonised as the international aviation regulatory regime, and that has occurred only because of the close cooperation of sovereign states collaborating for national and international objectives. The following sections describe the development of the very first aviation laws and explain the motivations and objectives of governments in respect to how aviation was to be regulated. The branch of law that has developed essentially remains unchanged from its humble beginnings because of the unique features and requirements of the international aviation industry.

Almost since its inception, commercial aviation has been subject to stringent legal and regulatory control. This was required for reasons of national security, defence, consumer protection, national economic interest and the protection of life and property. There are early recorded instances involving ballooning accidents in which damage to personal property occurred and the courts were required to pass judgement. One such accident occurred in New York in 1822 (Case Study 1.1).

CASE STUDY 1.1

GUILLE V *SWAN*, 19 JOHNS 381 (NY SUP CT, 1822), SUPREME COURT OF NEW YORK

Mr Guille, the defendant balloonist, landed his balloon in the vicinity of the plaintiff Mr Swan's garden. When the defendant descended, he was in a dangerous situation and asked for assistance from a person who was working in Swan's field. The event attracted the attention of hundreds of local residents who, in all the excitement, broke through Swan's fences and spoiled the plaintiff's vegetables and flowers. The damage caused to the balloon was minimal, totalling approximately US$15, whereas the damage resulting from the stampede was in the order of US$90. The court held that the defendant was liable for all damages that occurred to the premises as the defendant should have anticipated that his descent and landing would most likely have attracted such a large crowd.

Sovereign state: a country with a defined territory that administers its own government and is not subject to, or dependent on, another country.

Only six weeks after the commencement of the first regular international passenger air service, 26 states signed the *Convention Relating to the Regulation of Aerial Navigation* in Paris on 13 October 1919. The Paris Convention 1919 (as it became known) saw the beginning of international air law by confirming, virtually at the beginning of airline operations, the desire of governments throughout the world to systematically control aviation.

Today, the *Convention on International Civil Aviation* (Chicago Convention 1944), which updated and replaced the Paris Convention 1919, has been ratified by more than 190 **sovereign states**. These countries have agreed, under international air law, to be bound by the technical and operational standards developed by the International Civil Aviation Organization (ICAO) and which are detailed in the 19 Annexes.

1.1 Air law

Air law: that branch of law governing the aeronautical uses of airspace.

Throughout the world there has been considerable debate in relation to the formation of a universally agreed definition for the terms 'air law', 'aeronautical law' and 'aviation law'. Sometimes the terms are even used interchangeably. With respect to the terms 'air law' and 'law of the air', if they were to apply to the literal or common meaning of the word 'air' as the medium or the atmosphere, then this would include all the law associated with the use of the air, including radio and satellite transmissions. In the main, **air law**, as it applies to aviation, has a far narrower interpretation and is generally considered to be law which governs the aeronautical uses of airspace (Milde, 2016).

An alternative definition, and one which has received considerable support, is 'that body of rules governing the use of airspace and its benefits for aviation, the general public and the nations of the world' (Diederiks-Verschoor 2006, p. 1). This definition significantly expands the scope of activities to which air law applies.

Not that there is anything fundamentally irreconcilable with the second definition; however, to deviate so substantially from the subject matter of the first potentially creates confusion and ambiguity as to its meaning and usage. Throughout this chapter, air law will be considered as originally defined as 'that branch of law governing the aeronautical uses of airspace'.

1.2 Aviation law

Aviation law is a broader term than air (aeronautical) law and has been defined as 'that branch of law that comprises rules and practices which have been created, modified or developed to apply to aviation activities'. Aviation law is to air law what maritime law is to the law of the sea. To assist with the clarity of expression and reduce the potential for problems to arise in the application of these terms, the previous definitions will respectively apply to the terms 'air law' and 'aviation law'.

Aviation law therefore encompasses the regulation of the business aspects of airlines and general aviation activities. Consequently, aspects of insurance law, commercial law and competition law all form part of aviation law. Security and environmental regulations applicable to aviation activities are also within the scope of aviation law. Also included within the domain of aviation law is the regulatory oversight of aviation activities by regulators and government agencies.

> **Aviation law:** that branch of law that comprises rules and practices which have been created, modified or developed to apply to aviation activities.

Stop and think

What are the main differences between air law and aviation law?

1.3 International air law

International law is that body of legal rules that apply between sovereign states and such entities that have been granted international personality. Within the aviation community, the concept of international personality extends to organisations including ICAO, which is a division of the United Nations. International conventions (e.g. the Chicago Convention 1944 with regard to ICAO) detail and confer international personality upon these respective organisations.

As there is no sovereign international authority with the power to enforce decisions or even compel individual states to follow rules, international law has often been considered as not being a 'true law'. In aviation, however, because of the extensive and important role of international institutions such as ICAO and IATA (International Air Transport Association) and the proliferation of **bilateral air service agreements** between nations, including the

> **Bilateral air service agreement:** an agreement which two nations sign to allow international commercial air transport services to occur between their territories.

Conflict of laws: the laws of different countries, on the subject matter to be decided, are in opposition to each other or that laws of the same country are contradictory.

almost universal ratification of international conventions concerning international civil aviation, the existence of an international law would be difficult to refute.

The branch of international air law that determines the rules between contracting states and other international personalities is known as 'public international air law'. The Paris Convention 1919 and the Chicago Convention 1944 are true charters of public international air law. This term contrasts with the law relating to private disputes in which one of the parties may be of another state. This is the realm of 'private international air law' or **conflict of laws**.

International air law is essentially a combination of both public and private international air law. It has been suggested that its principal purpose is to provide a system of regulation for international civil aviation and to eliminate conflicts or inconsistencies in domestic air law.

Stop and think

Consider why international air law is required.

1.4 International convention law

Convention law is the major source of international air law and is constituted by multilateral and bilateral agreements between sovereign states. To provide a further insight into the application and importance of both public and private international air law to the aviation industry, three major international conventions will be examined; but first it is important to highlight the importance of the concept of sovereignty as it applies to airspace.

1.5 Sovereignty of territorial airspace

Sovereignty: the authority of a state to govern itself.

In international aviation, the concept of **sovereignty** is the cornerstone upon which virtually all air law is founded. At the Paris Convention 1919, 26 Allied and Associated nations had to decide whether this new mode of transport was to follow the predominantly unregulated nature of international maritime operations, or whether governments would choose to regulate this new technology. The First World War had brought about the realisation of both the importance of aviation and its potential danger to states and their citizens in threatening their sovereignty.

It was, therefore, not surprising that the first Article of the Paris Convention 1919 stated:

The High Contracting Parties recognise that every Power has complete and exclusive sovereignty over the air space above its territory.

This proclamation addressed the debate of whether airspace was 'free', as it is with the high seas, or whether it was part of the subjacent state or territory. The decision to follow the latter path was almost unanimous. While the Paris Convention 1919 clearly asserted that exclusive or absolute sovereignty extends to the airspace above the territory of the state, issues were raised as to what constitutes the vertical and horizontal territorial limits of each state.

In respect to vertical limits, customary law, based on an ancient Roman principle, had long recognised that absolute sovereignty of the state over its territorial airspace extended to an unlimited height. The Roman principle was based on an old maxim, *cujus est solum ejus usque ad coelum*, translated to mean 'whose is the soil, his is also that which is up to the sky'. Although international treaties have since modified this position in asserting that '[no] national appropriation by claim of sovereignty' can prevent overflight rights of satellites in outer space (space beyond the navigable airspace), no precise definition of outer space is provided (➤Chapter 15). The *Treaty on Principles Governing the Activities of States in the Exploration and Use of Outer Space, Including the Moon and Other Celestial Bodies* (Outer Space Treaty) (1967) does not provide a precise definition of outer space either.

Once again, in respect to horizontal or lateral limits of sovereignty, international treaties have clarified the situation. Article 2 of the Chicago Convention 1944 states:

> For the purposes of this Convention the territory of a State shall be deemed to be the land areas and the territorial waters adjacent thereto under the sovereignty, **suzerainty**, protection or mandate of such State.

Suzerainty: the situation in which a dominant state controls the foreign relations of another state but allows it sovereign authority in internal affairs.

The *United Nations Convention on the Law of the Sea* (UNCLOS) (1982) defines the limits to which sovereignty of the coastal state may apply to the airspace above the territorial waters or sea.

It is important to realise that the Paris Convention 1919 did not create the principle of exclusive air sovereignty but rather *recognised* it. Furthermore, the principle extends to all nations, irrespective of whether a particular state has signed or ratified the convention.

Subsequent conventions in Madrid in 1926 and Havana in 1928 achieved little by way of advancement in international air law. Significantly, however, the Havana (Pan-American) Convention 1928 was the first multilateral convention which challenged the principle of absolute sovereignty and was signed by the United States, Mexico and 14 South American states.

The principle of absolute sovereignty was again challenged with the Chicago Convention 1944, but ultimately the status quo prevailed. The Chicago Convention 1944 recognised and confirmed the principle that every state has complete and exclusive sovereignty over the airspace above its territory. The territory of a state for the purposes of the Chicago Convention 1944 was deemed the land areas and the territorial waters adjacent to them under the sovereignty, suzerainty, protection or mandate of the state. The question of the vertical extent of the airspace above a state's territory remains undetermined. However, the view that rights in airspace extend to a height without any limit has been firmly rejected (➤Chapter 15). Apart from the right of overflight by satellites in outer space, the concept of sovereignty remains the basis upon which both the structure and proliferation of bilateral air service agreements continue. This chapter now examines what is the most important international treaty in aviation.

1.6 Chicago Convention 1944

As in the aftermath of the First World War, the positive contributions of aviation during times of peace were realised following the improved performance and capabilities of aircraft during the Second World War. By the end of the Second World War, advances in aircraft design and technology had culminated in the development of the first jet engine. Following preliminary discussions initiated by the British government in early 1944, the United States called for an international conference in Chicago in November 1944. It was the intention of the US and Allied nations to establish post-war civil aviation arrangements and institutions and, in particular, the US sought to promote the freedom of international exchange by removing the restrictions to international air travel imposed by absolute air sovereignty. The conference was attended by most of the established nations of the world, including Britain, the US and Australia.

The general aims of the conference, in terms of promoting international air transportation, were:

- *Economic.* These included the promotion of freedom of airspace to nations and airlines; procedures for determining airfares, frequencies, schedules and capacities; and arrangements for simplifying customs procedures and standardising visas and other documentation.

- *Technical.* These were concerned with establishing international standards with respect to a variety of technical standards, including the licensing of pilots and mechanics, registering and certifying the airworthiness of aircraft, and the planning and development of navigational aids.

The resulting Chicago Convention, which was signed on 7 December 1944, applies only to civil aircraft and not to state aircraft. However, Annex 13 of the Convention implies that states are expected to apply its provisions domestically, while Annex 17 of the Convention was amended following the events of 11 September 2001 to 'require' states to implement certain security standards domestically, except where it is impracticable to do so. ICAO was also swift to promulgate international security standards to address the deficiencies apparent with the international aviation system.

1.7 Freedoms of the air

As with the Paris Convention 1919, the Chicago Convention 1944 restated and reinforced the principle of absolute air sovereignty. Consequently, air transit and traffic rights between contracting states required specific agreement. The US advocated complete freedom of the air for commercial air transportation, while Britain, supported by Australia and New Zealand,

Freedoms of the air: freedom to cross the territory of another country and conduct commercial services to other countries.

proposed varying degrees of international regulation. A Canadian proposal for **freedoms of the air** was documented as the *International Air Transport Agreement* (see Figure 1.1). Only 20 states signed the agreement at Chicago, including the US, but not all subsequently ratified it. Only five freedoms were discussed at Chicago. Four other freedoms, although not officially recognised by the Chicago Convention 1944 or granted in bilateral air service agreements, are referred to and taken into account in bilateral air service agreements (Figure 1.2).

Although most of the delegates at Chicago agreed that some degree of regulatory control was desirable, and indeed necessary for the cooperative development of international civil aviation, there was no general consensus apart from agreement of the first two freedoms. It was hoped that the other freedoms might be settled on a multilateral basis, but that was not

Figure 1.1   Chicago Convention 1944

practicable as the more powerful nations stood to gain more through negotiating bilateral arrangements. As the free market approach was not acceptable and multilateral approaches were not practicable, the only other way to secure international air travel consensus was by way of individual bilateral air service agreements that were reciprocally negotiated between two national governments.

Under the Chicago Convention 1944, all scheduled international air services (that either pass through the airspace of more than one state, carry passengers, mail or cargo or service two or more destinations in accordance with a published timetable) must acquire prior permission before flying into or over foreign territories. To fill the gap with regard to scheduled international air services, most states, including Australia, Britain and the US, signed the *International Air Services Transit (Two Freedoms) Agreement* (Transit Agreement). This agreement has proved to be extremely effective in terms of simplifying overflight rights and practical when diplomatic tensions arise between contracting states. In practice, although ICAO is authorised to resolve disputes arising from the Transit Agreement, this power is rarely invoked.

It is at the contracting state's discretion whether to adhere to the Transit Agreement. Bilateral agreements can, and usually do, include terms exchanging these two freedoms. This is an alternative arrangement for overflight rights where one or both states are not party to the multilateral agreement. The Transit Agreement does not specifically require contracting states to obtain a permit prior to exercising transit or non-traffic stopovers. In practice, irrespective of how overflight rights have been established, the filing of flight plans for

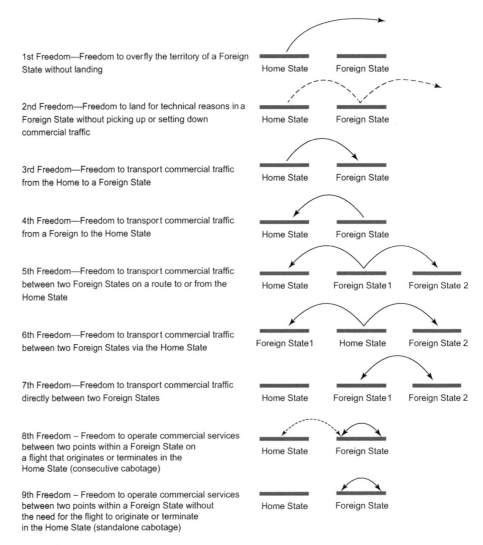

1st Freedom—Freedom to overfly the territory of a Foreign State without landing

2nd Freedom—Freedom to land for technical reasons in a Foreign State without picking up or setting down commercial traffic

3rd Freedom—Freedom to transport commercial traffic from the Home to a Foreign State

4th Freedom—Freedom to transport commercial traffic from a Foreign to the Home State

5th Freedom—Freedom to transport commercial traffic between two Foreign States on a route to or from the Home State

6th Freedom—Freedom to transport commercial traffic between two Foreign States via the Home State

7th Freedom—Freedom to transport commercial traffic directly between two Foreign States

8th Freedom – Freedom to operate commercial services between two points within a Foreign State on a flight that originates or terminates in the Home State (consecutive cabotage)

9th Freedom – Freedom to operate commercial services between two points within a Foreign State without the need for the flight to originate or terminate in the Home State (standalone cabotage)

Figure 1.2 The nine freedoms of the air

operational purposes is usually all that is required to provide the requisite safety, technical and security information.

> ## Stop and think
>
> Consider the extent to which airspace sovereignty influences the pattern of air transport provision worldwide.

1.8 The International Civil Aviation Organization (ICAO)

The most important contribution of the Chicago Convention 1944 was the agreement over technical matters and the groundwork which led to the establishment of ICAO. ICAO is the most important international organisation in the area of public international air law.

Article 44 of the Chicago Convention 1944 describes the purpose of ICAO:

> To develop the principles and techniques of international air navigation and foster the planning and development of international air transport so as to insure the safe and orderly growth of international civil aviation throughout the world.

On 6 June 1945, the required 26th state, including each of the 20 states elected to the ICAO Council, had accepted the *Interim Agreement on International Civil Aviation*. The 'Provisional' ICAO or PICAO came into effect as planned within six months of the signing of the Chicago Convention 1944. It was agreed by Member States that PICAO would remain in operation until the permanent forum, ICAO, came into force within the three-year limit prescribed in the convention.

ICAO provides the structure for the achievement of international cooperation and coordination in civil aviation. Through a variety of mechanisms, ICAO works to uphold the principles underlying the Chicago Convention 1944. It develops and adopts internationally agreed standards and procedures for the regulation of civil aviation, coordinates the provision of air navigation facilities on a regional and worldwide basis, collates and publishes information on international civil aviation, and acts as the medium by which aviation law develops at an international level.

Apart from technical matters, ICAO has also been instrumental in providing the organisational structure for the determination of less contentious economic arrangements. ICAO has addressed matters such as customs procedures and visa requirements and also assumed responsibility for collecting statistical data for international civil aviation, including information on safety-related issues, most notably incident and accident statistics (➤Chapter 13).

The international specifications for civil aviation appear in 19 Annexes to the Chicago Convention 1944. Each Annex addresses a particular subject. The specifications are divided into two categories, namely, Standards and Recommended Practices, although they are collectively, and most commonly, referred to as 'SARPs' (➤Chapter 4). Today, the 190 signatories to the Chicago Convention 1944 are obliged to comply with the extensive and comprehensive technical, safety, operational, security and environmental provisions as set out in the SARPs.

Stop and think

Outline the principal functions of ICAO.

!

1.9 Warsaw Convention 1929

International carriage by air is predominantly governed by international conventions. These international conventions were established as a result of the development in the air transport industry and were aimed at addressing conflict of law problems commonly associated with international carriage.

The first true instrument of private international air law was the *Convention for the Unification of Certain Rules Relating to International Carriage by Air* (Warsaw Convention 1929). It adopted a uniform set of rules governing international carriage by air, deals with the rights of passengers and owners or consignors of cargo and provides for internationally accepted limits on a carrier's liability for death, injury or damage.

Prior to the establishment of the Warsaw system, there were no uniform rules of law concerning international carriage by air. The problems inherent in international air travel often relate to matters concerning conflicts of law. The rights of passengers and owners of cargo, most of which had been previously stated in the contract of carriage, would vary from country to country and in accordance with each country's domestic law. Similarly, the liabilities of the carriers would vary enormously. The Warsaw Convention 1929 represented the first uniform international effort to implement universal laws relating to international air carriage, especially in respect of carriers' liability.

The implementation of internationally accepted limits on a carrier's liability for death, injury or damage was also a driving force which ultimately led to the Warsaw Convention 1929. At that time airlines were predominantly state-owned and particularly supportive of the introduction of known limits on liability. Arguments advanced in favour of liability limits included:

- protection of a developing and financially vulnerable aviation industry;

- distribution of potentially large risks;

- practicality of carriers being able to fully insure against liabilities;

- standardised and readily quantifiable damages awards;

- allowing passengers to take out their own insurance policies;

- reducing litigation against airlines and facilitating settlement of disputes.

The objectives of the Warsaw Convention 1929 were achieved for approximately two decades after implementation, but its effectiveness and support have been gradually eroded. In an attempt to retain its effectiveness, the Warsaw Convention 1929 was updated several times by way of amendments. As a consequence of the US (and other countries) not having adopted all of the subsequent amendments to the Warsaw Convention 1929, a non-uniform international system of liability of carriers governing international air carriage emerged, thereby frustrating the most fundamental objective as to why it was originally created. Moreover, the terminology and language used in the Warsaw Convention 1929 (and amending protocols) had become outdated and was the source of much ambiguity and dispute.

It is important to realise that the Warsaw Convention 1929 applies only to international carriage. Case Study 1.2 illustrates the importance of determining whether or not a particular flight is deemed to be international carriage and thereby limiting the liability of the carrier.

CASE STUDY 1.2

STRATIS V *EASTERN AIR LINES LIMITED* 682 F 2D 406 (1982)

Mr Stratis was a crewman on a Greek tanker. He was booked to fly on Delta Air Lines from Baton Rouge to New Orleans, then on Eastern Air Lines from New Orleans to New York and finally on Olympic Airways from New York to Athens. He had paid for all travel prior to departure but was issued only with tickets for the domestic **sectors**, having arranged to collect his international ticket in New York. The Eastern Air Lines flight crashed on approach to New York, and Mr Stratis became a quadriplegic. The court awarded him US$6.5 million in damages, which was apportioned as follows: 60 per cent for the negligence of the hospital who had treated him, and 40 per cent against Eastern Air Lines and the US government for the negligence of their pilots and air traffic controllers respectively. On appeal, the US Circuit Court of Appeals accepted Eastern Air Lines' defence that their liability was limited under the Warsaw Convention 1929 because it was international carriage. It was held that the contemplation of the parties was for international carriage, and the fact that the domestic tickets were annotated to that effect was a significant factor.

On 4 November 2003, the *Convention for the Unification of Certain Rules for International Carriage by Air* (Montreal Convention 1999) came into force and replaced, for those states which have ratified it, the Warsaw Convention 1929. This convention has fundamentally changed this area. The new liability rules were developed during an International Air Law Conference called by ICAO to modernise the Warsaw Convention. Since it took effect, the importance of the Warsaw Convention 1929 and subsequent amendments to the development of international air law cannot be overstated.

1.10 Montreal Convention 1999

The development of various international conventions relevant to international air carriage, in the context of a maturing commercial aviation industry, gave rise to a complex system of international treaties, many of which have now become unwieldy and outdated. Although the Montreal Convention 1999 consolidates the many amendments to the Warsaw Convention 1929, it is an entirely new treaty that unifies and replaces the system of liability established by the Warsaw Convention 1929 and its subsequent amendments.

Prior to the introduction of the Montreal Convention 1999, compensation limits remained generally low, in line with the early philosophies aimed at supporting a fledgling air transport industry. The industry, despite present-day challenges, has developed significantly in respect of its commercial stability and relative safety standards. Commercial arrangements such as intercarrier, **codeshares** and airline alliance agreements and the complex nature of international trade have led to practices never envisaged by the drafters of earlier conventions, such as electronic documentation in place of traditional paper tickets and **air waybills (AWBs)**.

Sector: (also known as a 'route') a single flight that connects an origin and destination airport (an OD pair).

Codeshare: a reciprocal agreement between two or more airlines that enables a flight that is operated by one carrier (e.g. ABC123) to be marketed by another using its own code and flight number (ZYX987). The carrier operating the flight (ABC) is the operating carrier, while the airline marketing the flight (ZYX) is the marketing carrier. The resulting flight is identified by the shared codes (in this case ABC123/ ZYX987).

Air waybill (AWB): a type of bill that serves as a receipt of goods by an airline (carrier) and as a contract of carriage between the shipper and the carrier. It includes conditions of carriage that define (among other terms and conditions) the carrier's limits of liability and claims procedures. Further, it provides a description of the goods and the applicable charges.

The Montreal Convention 1999 establishes an alternative carriage by air regime for determining the liability of air carriers for injury or death of a passenger, loss or damage to luggage or cargo and damage caused by or delay in the transport of passengers, luggage or cargo which occurs during the course of international carriage. The Montreal Convention 1999 resulted from the need to modernise the Warsaw Convention 1929. Despite such modernisation, the Montreal Convention 1999 does not address certain areas adequately. For example, one area of contention is in cases regarding the liabilities of airlines and their obligations to passengers in the event of death and/or injury.

The legal ramifications of the accidents involving MH370, which disappeared over the Indian Ocean in March 2014, and MH17, which was shot down over eastern Ukraine in July 2014, both of which resulted in the deaths of all persons aboard, were profound. A key issue was liability under the Montreal Convention of airline operators to *non*-passengers who have suffered enormously as a result of the disasters. This has included a range of psychological and financial losses including pure mental harm (nervous shock) and loss of dependency. There is a significant amount of case law for passenger claims dealing with liability for injuries, but essentially none for non-passengers making claims relating to 'death' in Article 17 of the Montreal Convention – and not 'bodily injury'.

With regards to post-traumatic stress disorder injury related claims, this contention stems from the wording of Article 17 which states:

> The carrier is liable for damage sustained in case of death or bodily injury of a passenger upon condition only that the accident which caused the death or injury took place on board the aircraft or in the course of any of the operations of embarking or disembarking.

The Montreal Convention's reference to an airline's liability where a passenger has sustained 'bodily injury' has caused considerable debate particularly in Australia. The debate centres around what the meaning of the term 'bodily injury' is and the scope of that term (see Case Study 1.3).

CASE STUDY 1.3

PEL-AIR AVIATION PTY LTD V CASEY [2017] NSWCA 32

Ms Karen Casey was a nurse employed by Care Flight. Ms Casey travelled on a small aircraft to Samoa to evacuate a patient and her husband to Melbourne. The aircraft was scheduled to refuel at Norfolk Island on the return journey but bad weather prevented the pilot landing, as a result of which he ditched the aircraft in the sea. All six of the passengers and crew on board were rescued after spending a period of time in the water. The experience proved terrifying for Ms Casey and, as a result, she suffered significant physical injuries and post-traumatic stress disorder (PTSD) amongst other disorders. Ms Casey brought District Court proceedings against Pel-Air, claiming damages. After the proceedings were transferred to the Supreme Court of NSW, judgement was entered in favour of Ms Casey to the sum of $4,877,604. Pel-Air then appealed the matter to the NSW Court of Appeal. One of the main issues on appeal was whether the primary judge erred in concluding that Ms Casey's PTSD constituted a 'bodily injury', as the term is used in the 1999 Montreal Convention relating to International Carriage by Air (the Montreal Convention).

The NSW Court of Appeal held that the primary judge had erred in concluding that Ms Casey's PTSD constituted a 'bodily injury'. The Court of Appeal found that while 'bodily injury' does not exclude consideration of damage to a person's brain, there must be evidence of actual physical damage to the brain. The court held that there was no evidence that Ms Casey's PTSD resulted from actual physical damage to her brain.

The decision of the NSW Court of Appeal has realigned Australia's position with the international position, namely reaffirming that the term 'bodily' purposefully distinguishes bodily injury from psychological injury.

Overall, the Montreal Convention 1999 has sought to address the problems that developed in the Warsaw system by substantially raising carriers' liability limits, presenting the liability framework in a single consistent convention and updating the language and terminology used.

The Montreal Convention 1999 distinguishes between international and domestic carriage. The convention applies to international carriage only. Domestic travel is treated as 'other carriage', to which the *Civil Aviation (Carriers' Liability) Act 1959* (Cth) applies.

The convention lists five guiding principles agreed by its contracting parties as a preamble to its substantive provisions; namely, the Montreal Convention 1999:

- recognises the significant contribution of the Warsaw Convention 1929 (as amended) to the harmonisation of private international air law;

- recognises the need to modernise and consolidate the Warsaw Convention 1929 (as amended);

- recognises the importance of ensuring protection of consumer interests in international air transport and the need for equitable compensation based upon the principle of restitution;

- reaffirms the desirability of the orderly development of international air operations and the smooth flow of passengers, baggage and cargo in accordance with the Chicago Convention 1944;

- promotes collective state action for further harmonisation and codification of certain rules governing international air carriage through a new convention as the most adequate means of balancing interests.

Interesting issues have arisen throughout the ratification and implementation process of the Montreal Convention 1999. Legislation in operation in the European Union (EU) has taken a broad interpretation of the provisions of the convention. EC Regulation 261/2004 in particular has attracted much attention. Argued to be overly focused on 'passenger protection', the regulation imposes obligations on carriers to assist passengers in the event of delay, including those situations where the events giving rise to a delay are beyond the control of the carrier, such as when the delay is caused by adverse weather conditions or air traffic disruptions.

This regulation proved to be particularly controversial because it has the potential to affect foreign, non-European carriers and also since it appears to contravene those provisions of the Montreal Convention 1999 which provide carriers with a defence in circumstances in which delay is beyond their control. Further, such issues will likely continue to arise as additional implementing legislation is introduced by various state parties, who are likely to look to the example set by the EU as a point of guidance and comparison – most notably the 'Passenger Bill of Rights' enacted by the State of New York in the US and later overturned by the US courts.

1.11 Differences between the Warsaw and Montreal Conventions

It is important to note that carriage under the Warsaw system does not cease to be legally binding because of the entry into force of the Montreal Convention 1999. The Warsaw Convention 1929 still applies to round trips departing from a state which is not a member of the Montreal Convention 1999 and to one-way flights between two states where either has adhered to the Montreal Convention 1999. The convention applies to all international air carriage in which the country of departure and the country of destination have both adopted it.

Special Drawing Rights (SDRs): an international monetary reserve currency created by the International Monetary Fund.

Through its Chapter III, the Montreal Convention 1999 establishes a new two-tiered scheme to govern passenger compensation. The first tier, which operates up to 100,000 **Special Drawing Rights (SDRs)**, imposes strict liability upon the carrier. The carrier's liability under the first tier can be reduced only by the demonstrated contributory negligence of the passenger. Liability under the second tier is unlimited if damages are proven in excess of 100,000 SDR but can be avoided by the carrier proving that the damage was not caused by its negligence or was caused solely by the negligence or other wrongful act or omission of a third party.

The Montreal Convention 1999 applies only if the parties agree to its application to transportation between two locations (the destination may be changed during the flight or the flight may be a round trip). This rule excludes pilot training and test flights. Although carriage occurs in these examples, it does not occur pursuant to a contract of carriage. Therefore, the Montreal Convention 1999 is excluded by the absence of a contract and not by the absence of carriage. It follows that carriage does not need to be defined according to the parties' subjective intentions. Over 100 states have adopted the Montreal Convention 1999, and accordingly, the importance and application of the Warsaw Convention 1929 has, and will continue to be, significantly reduced.

!

Stop and think

How do the Warsaw and Montreal Conventions differ?

1.12 International carriage by air

As the Montreal Convention 1999 applies to international carriage only, it is imperative, in the first instance, to determine whether or not a particular flight is domestic or international. The leading authority on this issue is *Stratis v Eastern Air Lines Limited* (see Case Study 1.2). International carriage under the Montreal Convention 1999 includes baggage (luggage) and cargo. In the case of cargo, Article 4 of the Montreal Convention 1999 requires that every carrier of cargo has the right to require the consignor to generate an air consignment note, called an 'air waybill'. Every consignor has the right to require the carrier to accept this document.

The question arises as to whether the Montreal Convention 1999 provides an exclusive right of action in respect of claims arising from international air transportation. This question was discussed in the context of the Warsaw Convention 1929 in *Sidhu v British Airways plc (Scotland)* (see Case Study 1.4).

CASE STUDY 1.4

SIDHU V BRITISH AIRWAYS PLC (SCOTLAND) [1997] AC 430

The claimants in this case had been travelling from London on BA Flight 149 to Kuala Lumpur via Kuwait in August 1990. The flight landed in Kuwait immediately after Iraq had invaded. While the aircraft refuelled, the airport was seized by Iraqi troops. The passengers and crew were taken by force to Baghdad and detained for approximately one month. The claimants alleged the airline knew of, or ought to have known of, the dangerous situation between Iraq and Kuwait and the possibility of imminent invasion. Damages were claimed in respect of both physical and psychological injuries. The passengers brought their claim at common law, arguing that the Warsaw Convention 1929 did not prevent or extinguish their rights at common law. The House of Lords dismissed the claimant's arguments and held that the objects and structure of the Warsaw Convention 1929 supported its interpretation as a uniform international code that could be applied by all the high contracting parties without reference to their own domestic law.

In delivering the judgement of the UK House of Lords, Lord Hope stated that the structure of Article 17 and Article 24 of the Warsaw Convention 1929 required the carrier to surrender its freedom to exclude or limit its liability on one hand, while restricting the passenger in the claims which can be brought in any action for damages on the other. He stated:

> The idea that an action of damages may be brought by a passenger against the carrier outside the Convention in the cases covered by art 17 seems to be entirely contrary to the system in which these two articles were designed to create.
>
> (*Sidhu v British Airways plc (Scotland)* [1997] AC 430 at 447)

Lord Hope concluded that while the Warsaw Convention 1929 did not purport to deal with all matters relating to contracts of international carriage by air, it was intended to be uniform and exclusive of any resort to the rules of domestic law in the areas dealt with by its terms.

1.13 European regulations

Regulation (EC) No 216/2008 of 20/02/2008 (the Regulation) established the common rules in the field of civil aviation in Europe and created a European Aviation Safety Agency (EASA), repealing the earlier 2002 and 2004 regulations. The 2008 Regulation expanded EASA's responsibilities and extended their role to air operations, pilots' licenses and the safety of non-EU Member State aircraft. The primary purpose of the law is to ensure a high protection of European citizens in civil aviation through the adoption of common safety rules and to introduce measures ensuring products, persons and organisations in the Community comply with such rules. The economic purpose is also intertwined, as the Regulation prescribes that safety for European citizens should contribute to facilitating 'the free movement of goods, persons and organisations in the internal market' (the ninth freedom).

The Regulation applies to the 'design, production, maintenance and operation of aeronautical products, parts and appliances, as well as personnel and organisations involved in the design, production and maintenance of such products, parts and appliances' and also the personnel and organisations involved in operating aircraft. However, the Regulation does not apply when these products and personnel are engaged in military, customs, police or similar services.

EASA has become critical to EU's aviation safety strategy and operates with both the European Commission and Member States to achieve uniform application of laws. It is pan-European, with the involvement of other non-EU countries including Iceland, Norway, Switzerland and Liechtenstein. EASA prepares draft rules for consideration in the EU legislative processes and consults with a range of organisations across the industry.

The European Regulations are essentially separated into three components:

- regulations and implementing regulations (hard law);

- acceptable means of compliance (AMC) and guidance material (GM) (soft law);

- certification specifications (soft law).

The 'hard law' is the law that must be complied with, and it includes the areas listed in Figure 1.3. The other soft laws offer guidance: they do not have to be complied with exactly but aid compliance and implementation of the Regulations.

EASA has not restricted its operations to Europe. In March 2017, EASA and ICAO jointly hosted the Global Forum on Regional Safety Oversight Organizations (RSOOs) in Switzerland where progress was made on initiatives to improve the worldwide safety recognition, efficiency and cooperation (➤Chapter 13). EASA therefore also takes an advisory role in strengthening regional initiatives for the benefit of aviation safety on an international basis. RSOOs are essentially regional collaboration mechanisms that are created to fulfil states' obligations under Article 37 of the Chicago Convention, which provides that each State undertakes 'to collaborate in securing the highest practicable degree of uniformity in regulations, standards, procedures and organisation in relation to aircraft, personnel,

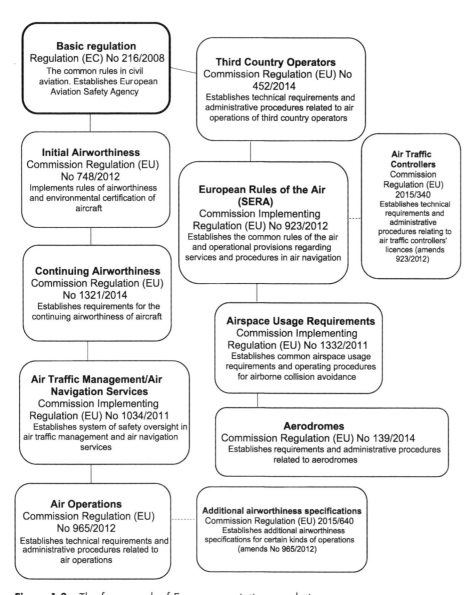

Figure 1.3 The framework of European aviation regulations

airports, airways, and auxiliary services in all matters in which such uniformity will facilitate and improve air navigation'. ICAO believes that when such regional aviation safety organisations are implemented effectively, 'regional arrangements can help States overcome specific challenges, such as a lack of resources and qualified technical personnel, while helping to harmonise regulatory systems, reduce duplicative functions, create economies of scale, and ensure long-term sustainability and self-sufficiency'. EASA is often cited as a RSOO that works well.

1.14 US regulations

US aviation regulations are enshrined in the Federal Aviation Act 1958, which delegates authority to the Federal Aviation Administration (FAA) to oversee and regulate safety in the airline industry and the use of US airspace, including both civil and military aviation. It is unlike EU regulation as EU Regulations do not apply to military aircraft. The Federal Aviation Regulations (FARs) are regulations created by the FAA to govern all US aviation activities, and are part of Title 14 of the Code of Federal Regulations (CFR); 'Aeronautics and Space'. The FAA, like EASA, set airworthiness standards and establish the requirements for aircraft registration. They also provide rules for airspace and air traffic and a range of other matters.

1.15 Globalisation of aviation

The extent to which there has been an adoption of international treaties such as the Chicago Convention 1944 is unique to aviation. This particular treaty not only *influences* all aviation activities – that is, international, domestic and, to an increasing degree, military – but to a large and increasing extent *dictates* all operational, technical, safety and security standards within the industry. The study of international air law is important, not just to attain a more complete picture of aviation but also to provide a clear understanding of the legal basis upon which *all* aviation law is founded.

As an industry, what makes aviation unique can be explained in terms of both its development and how it is regulated. These two aspects of aviation, although quite distinct, are in fact highly interrelated and, to a large extent, account for why there is a greater degree of international harmonisation of aviation legislation than with any other industry.

From the outset, aviation activities have been subject to strict regulatory control. Soon after the first hot-air balloon assents by the Montgolfier brothers in 1783, the Paris police required flight permits to protect the safety of persons and property on the ground. The trend of international harmonisation towards universal conformity of aviation activities is not only increasing but is doing so at an ever-increasing rate. The catalyst for this was the First World War, and the trend has continued to be fuelled by major worldwide events which include: the Second World War; international terrorism; government economic rationalisation; airline strategic alliances; pandemics and epidemics; customer loyalty (frequent flyer) programmes; codesharing; global reservation systems; highly dynamic oil prices; proliferation of low cost carriers; internet ticketing; the global financial crisis; and increased government liberalisation towards more 'open skies' policies. Unlike any other mode of transportation, air transport is not restricted by political and geographical boundaries. The internationalisation of aviation activities, and the legal processes that have supported and assisted this development, commenced with the invention of aircraft.

Key points

- Air travel is an inherently international mode of transport, but this gives rise to potential conflict of laws.

- Air transport is regulated at an international and national level for reasons of national security, defence, safety, consumer protection and competition.

- Air law is a branch of international law that governs use of and access to airspace.

- Airspace is sovereign territory, and airlines have to seek permission to enter and overfly foreign territory.

- The Chicago Convention 1944 led to the formation of ICAO and resulted in international accord on Standards and Recommended Practices (SARPs).

- The Warsaw Convention 1929 established private air law and defined carrier liability.

- The Montreal Convention 1999 replaced and updated the Warsaw Convention.

References and further reading

Bartsch, R. I. C. (2018). *International aviation law.* 2nd edn. London: Routledge.

Diederiks-Verschoor, I. H. Ph. (2006). *An introduction to air law.* 8th edn. The Netherlands: Kluwer Law International.

Matte, N. (1981). *Treatise on air-aeronautical law.* Montreal, Canada: Institute and Centre of Air and Space Law.

Mendes de Leon, P. M. J. (2017). *Introduction to air law.* 10th edn. Deventer: Wolters Kluwer.

Milde, M. (2016). International Air Law and ICAO. *Essential air and space law series*, Volume 18. 3rd edn. Montreal, Canada: Eleven International Publishing.

CHAPTER 2

Aviation economics and forecasting

David Gillen

LEARNING OBJECTIVES

- To identify the factors that affect the demand for air travel.

- To understand the concept of price elasticity as it relates to air transport demand.

- To recognise the key features of airline supply: joint production, perishability and overcapacity.

- To understand the term 'deregulation' and its impact on airlines.

- To appreciate how government policy can change the structure of airline markets and service delivery.

- To examine the factors that affect aggregate air travel demand and understand how to forecast aggregate and origin and destination (OD) demand.

2.0 Introduction

This chapter will provide an explanation of the fundamental features of aviation economics. This includes a description of the aviation supply chain and its characteristics, such as the importance of network design, and the fact that airline seats cannot be stored, and that aviation services are offered by a variety of companies with different business models. The chapter also explains the factors that affect the demand for aviation services and how aggregate air travel has grown considerably as a result of deregulation. It also examines demand forecasting in aggregate and between origins and destinations.

2.1 The aviation value chain

Airlines deliver a service to customers, passengers and shippers by moving people and goods between places. Airlines are one component of the aviation value chain, which begins with airframe, engine and aircraft component manufacturers, infrastructure and service providers, and ends with the distribution of passengers and freight (see Figure 2.1).

This chapter will focus primarily on passenger airlines in the value chain. Both organisations that are input providers as well as organisations that distribute an airline's product will affect how the airlines produce and deliver their service and structure their business models. For example, the high costs of distribution through travel agents and computer reservation systems (CRSs) have resulted in airlines moving to internet distribution. The internet has resulted in lower distribution costs but has also allowed fare transparency and comparison shopping, which have changed the way airlines compete and set fares. The subsequent development of Global Distribution Systems (GDSs) was shaped by the initial foray into the new technology of internet distribution.

Experience good: a good or service where its characteristics (price, quality) can be difficult to observe without consuming it.

Airline service prior to deregulation in the US in 1978 was an **experience good** with an emphasis on quality, as price was not a strategic focus. With regulated high fares, only a small proportion of the population flew. This meant that the market structure had little, if any, effect on airfares. The emphasis of regulators was on supply availability. However, deregulation changed every aspect of the industry, as new competition through market entry led to lower prices owing to significant improvements in efficiency, and new business models led to a focus on costs. Airline service was commoditised, and the key demand parameter was price.

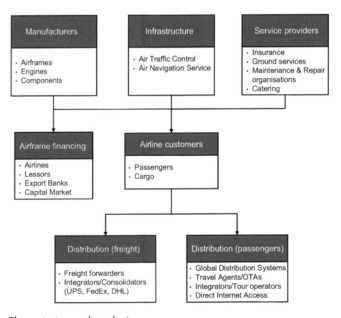

Figure 2.1 The aviation value chain

Source: Adapted from Tretheway and Markhvida (2014)

The delivery of the airline product also affected network design as the business model of traditional **full service network carriers (FSNCs)** moved to **hub-and-spoke** networks, while **low cost carriers (LCCs)** developed a **point-to-point** system (➤Chapters 8 and 9).

The evolution of the airline industry has been shaped by the growth in international aviation. Unlike domestic markets that had been deregulated, international aviation markets were regulated by bilateral air service agreements. These agreements controlled prices, market entry, flight frequency and seat capacity (➤Chapter 1). The development of international airline alliances was affected significantly by constraints imposed by bilateral agreements as airlines sought access to **beyond and before traffic.** SkyTeam, oneworld and Star Alliance have developed into major transnational alliances (➤Chapter 10) and most FSNCs have become a member of an alliance. Even non-aligned carriers, including Virgin Atlantic, Emirates and Etihad, have entered into strategic partnerships with overseas airlines to extend their geographic presence and market reach.

Over time, these alliances have developed into **metal neutral joint ventures (MNJVs)** which enjoy significant market power. At the same time as alliances were gathering more members and decreasing the number of independent competitors, there was increasing consolidation in North America and Europe, and there were constraints on the number of carriers in China, the fastest-growing aviation market.

Another segment of the industry that was dynamic was the low cost carrier (LCC) market; and this was true in every market: the US, Canada, Australia, Asia, Europe, Latin America and the Middle East (➤Chapter 8). The traditional LCC business model has undergone three main developments. First, LCCs morphed into carriers that moved away from a strict focus on costs and low price. For example, UK-based easyJet focused on passengers who valued access to major airports and higher flight frequency; while in Canada, WestJet started a regional airline (Encore) and entered codeshare agreements with other carriers. A second shift was the emergence of a class of LCCs known as ultra-low cost carriers (ULCCs), which concentrated on minimising costs, no-frills service and revenue from add-ons and relied on market stimulation; Wizz Air and Ryanair are examples in Europe, while Allegiant, Frontier and Spirit are US examples. The most recent shift has been LCCs flying longer-haul services and flying internationally; AirAsia, for example, started flying long-haul routes with **wide-body aircraft**.

> ## Stop and think
>
> Outline what is meant by the aviation value chain, and detail its importance to airline operators.

2.2 Airline markets: demand

Air travel demand is driven by aggregate economic activity: trade, gross domestic product (GDP) growth, urbanisation and growth in emerging economies. Demand is also affected by the growth in connectivity provided by more seats, more frequency and more destinations. The substantial increase in global trade as a result of the reduction in trade barriers and tariffs under the World Trade Organization (WTO) has resulted in a significant increase in global air travel.

Full service network carrier (FSNC): also known as a 'legacy carrier', an airline that offers high levels of in-flight service and connectivity, attracts a range of passenger segments to its network of short- and long-haul routes and flies a variety of aircraft types.

Hub-and-spoke: a network in which passengers are transported between two locations via an intermediate (hub) airport.

Low cost carrier (LCC): an airline that adopts a rigorous cost-minimisation strategy to keep its costs and fares low.

Point-to-point: a network in which each airport is directly connected to other airports.

Beyond and before traffic: passengers who travel from an origin that is not a departure gateway, through the gateway, to a destination that is beyond the gateway.

Metal neutral joint venture (MNJV): an alliance in which members are indifferent to who operates the 'metal' (aircraft) when they jointly market services.
Wide-body aircraft: an aircraft with two aisles.

Derived demand: a demand which is a consequence of the demand for something else; e.g. for [air] transportation, the demand arises from a desire for passengers or cargo to be somewhere else, not because the trip between origin and destination is desired.

The key characteristics of demand for airline services are that it is a **derived demand** and there are a range of market segments with differing degrees of price and service quality sensitivity. As a derived demand, it is also affected by the prices of complementary products or services. The movement of people and cargo between two points is based on the desire to be at a destination. If this involves a leisure destination, the price of hotels and leisure activities will affect the demand for airline services.

Demand is affected by a number of factors, including price and income, service quality (reliability), passenger demographics, the frequency and timing of flights and the price and availability of alternative modes of transport. Demographic factors are also important, such as population, age distribution and cultural ties between cities. The demand for a specific airline will be affected by relative prices and service quality but also by amenities such as food, entertainment options and loyalty programmes; people may choose one particular airline in order to accumulate credits that will offer 'free' flights in the future (►Chapter 10).

Figure 2.2 illustrates the different market segments. Each segment will have its price and service quality sensitivity depending on a number of factors. For example, in the leisure market, whether long- or short-haul, there will be differences between true leisure (holiday) and VFR (visiting friends and relatives) traffic and whether these travellers are retired or not.

The degree to which demand is responsive to price and service quality will vary across these segments. Studies of air travel demand provide a range of estimates for these elasticities. There is a distinction between short- and long-haul passengers, domestic and international, and business and leisure. Generally business travellers have a low-price, high-service quality and a low **income elasticity** of demand. Leisure travellers, on the other hand, have a broader range of sensitivity (elasticity) estimates. The range depends on whether travellers are holidaymakers or VFR and whether they are retired or still working.

!

Stop and think

Detail what is meant by derived demand, and outline the factors that impact on demand for air services.

Income elasticity: a measure of how the demand for a good or service will change when the income of the individual changes.

Own-frequency elasticity of demand: the elasticity of demand with respect to an increase in flight frequency.

Table 2.1 summarises route/market elasticity estimates for different trip types, sector lengths (short hauls and long hauls) and markets (domestic and international). The values reported here are average values including being averaged across geographic regions. If these price elasticities were calculated for a specific geographic region, one would expect some differences. The differences arise because price sensitivity will depend on the transport options available (mainland Europe has more rail services than North America or the UK), the average income, the emergence of a LCC segment and the degree of urbanisation, among other factors. Little formal work has been undertaken on measures of other key factors that affect demand such as frequency, destinations and service quality. Fu, Oum and Yan (2014) have produced a model that provides estimates of the **own-frequency elasticity of demand** for air trips with respect to an increase in flight frequency. For short-haul trips (< 500 km) the frequency elasticity is 0.39, and for long-haul trips (> 500 km) it is 0.42. These values indicate markets are responsive to increased frequency, but there are two issues. If airline X

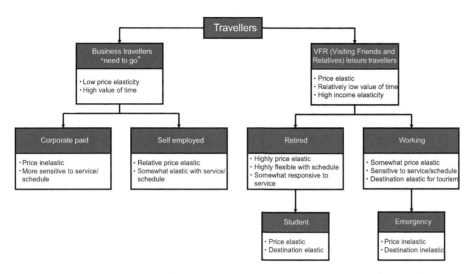

Figure 2.2 Market segments for air travel demand

increases frequency relative to airline Y, X will 'steal' high-time-value passengers from Y; however, if all carriers increase frequency, the market will grow. These additional passenger trips must come from somewhere. They would come from other modes, and some would be induced trips (see Example 2.1 for examples of demand and pricing).

Stop and think

Explain what is meant by income elasticity of demand and why it is important to the air transport industry.

Table 2.1 Summary of route/market elasticity values

Category	Elasticity values	
	All studies	
	Short haul	Long haul
Intra North America	−1.54	−1.40
Intra Europe	−1.96	−1.96
Intra Asia	−1.46	−1.33
Intra Sub-Sahara Africa	−0.92	−0.84
Intra South America	−1.93	−1.75
Transatlantic (North America–Europe)	−1.85	−1.68
Trans-Pacific (North America–Asia)	−0.92	−0.84
Europe–Asia	−1.39	−1.26

Source: IATA Report, Estimating Air Travel Demand Elasticities, Final Report 2008

Example 2.1

Air travel demand, market segmentation and differential pricing

In the era of regulation (up to 1978 in the US and the mid-1990s in Europe and the UK), airlines were allowed to have two fares, first class and unrestricted economy (also called coach), and these fares were typically set on a mileage-based formula. Deregulation resulted in a revolution in pricing where airlines based their prices not on costs but on the value of the service to customers; cost was the level below which prices should not fall. **Differential pricing** was first introduced into different **origin and destination (OD)** markets based on their demand characteristics. So, unlike mileage-based fares, short-haul fares could exceed long-haul fares and fares on one segment of a two-segment flight could exceed the fare for both segments.

Airlines were able to gather significant amounts of information from their passengers as to how they responded to different prices, services and competition. This led to the introduction of 'revenue management'. The basic idea was that since the flight was scheduled to go and that the majority of costs were fixed costs, the airline would set prices in order to maximise the revenue from the flight. The airline was able to set different prices for different passengers based on when they booked, where they were flying, the amount of competition and other factors.

Revenue management used differential prices based on passengers' willingness to pay; airlines charge different prices based on demand sensitivity. The different prices could apply to different demand segments in the same OD market as well as across markets. The airline's ability to engage in price discrimination was based on three factors: the existence of different willingness to pay across customers, a way of identifying the different customer segments and a means of preventing customers engaging in **arbitrage** between low and high fares. Revenue management also controlled the number of seats that were allocated to any given fare class. The numbers of seats would be adjusted depending on whether the observed demand was more or less than what had historically occurred. If current demand was above the historic booking curve, low fare seats were moved to higher fare categories, for example.

Airlines also introduced service-based pricing through product differentiation, whereby different fares were based on the quality of the product. Higher-quality products – a larger sleeper seat in business class versus a narrower seat in economy, for example – are more costly for the airline, and higher prices reflect in part these higher costs.

Airlines are able to identify different levels of willingness to pay by experimenting with different fare structures and levels. They collect large amounts of data daily across the markets they serve. Airlines are therefore able to estimate demand and identify own and **cross-price elasticities of demand**. They have been able to maintain the range of fares and minimise any fare arbitrage (diversion) by placing restrictions on different fares, such as Saturday night stay overs, minimum length of return trips,

Differential pricing: charging different customers different prices for the same service.

Origin and destination (OD): the start and end points of a passenger's itinerary, regardless of their actual routing.

Arbitrage: the buying of an asset at a high or low price and reselling it at a low or high price respectively.

Cross-price elasticity of demand: the change in the quantity demand of one good or service when the price of another good or service changes. If the cross-price elasticity is positive, the two goods or services are substitutes, and if it is negative, the two are complements.

and purchase date restrictions prior to a flight departure date (e.g. by purchasing a ticket three weeks in advance of travel).

The natural evolution of differential pricing and demand segmentation has been the unbundling of the airline product so separate charges are levied for on-board food, checked luggage, early boarding and more legroom – so-called à la carte pricing. There has been mixed reaction to this type of pricing depending in part on what the base fare for the airline service is.

2.3 Airline markets: supply

Airlines supply products which are produced jointly are perishable and generally in excess supply. On any given flight between A and B, FSNCs may have multiple classes (first, business and economy) on the same aircraft. This results in common costs that must be allocated across classes in an efficient way to ensure maximum profits. Another form of **joint production** is that many markets can be served by a single flight. For example, on a flight from Boston (BOS) to Chicago (ORD), there can be passengers who arrived in Boston from overseas, others may be flying through Chicago to another destination and still others may be flying BOS–ORD. FSNCs provide air transportation over a network, so any one flight will have a combination of origin and destination (OD) and connecting traffic. These airlines generally deliver services through a hub-and-spoke network with the hub serving to redistribute connecting passengers among flights. In many cases, 70 per cent or more of passengers on a given flight through a hub will be connecting. Demand is an OD trip, which can be accomplished using non-stop or connecting routings or paths; but supply is via a network, where a flight segment can be supplying seats to many OD markets.

> Joint production: a situation in which two products or services are created at the same time in the production process.

LCCs offer services through a point-to-point network with ODs connected directly. However, as LCCs have developed, there have been concentrations of traffic, such as Southwest Airlines at Phoenix (US), WestJet at Calgary (Canada) and Jetstar at Sydney (Australia). These are not hubs in the way FSNCs use hubs, although some LCCs will sell a through ticket with a connection; a FSNC hub strategy coordinates incoming and outgoing flights, and LCCs do not do this.

The product is also perishable since a flight leaving with an empty seat means the seat is wasted as it cannot be placed in inventory. The perishability of seats has led to an emphasis on yield management, whereby airlines carefully manage their seat inventory to ensure seats are reserved for late-booking high-yield passengers and lower-priced seats are sold to more price-elastic passengers.

A third important feature of airline supply is that capacity tends to be oversupplied, meaning utilisation is rarely 100 per cent (the **load factor** of passengers to seats is < 100 per cent). Load factors vary regionally and in domestic versus international markets. The International Air Transport Association (IATA) reported in 2018 that the average load factors have increased from 80.5 per cent in 2016 to 81.7 per cent in 2018. Furthermore, they note that aircraft are being flown more intensively, with an average of 74 aircraft departing somewhere in the world each minute of 2018. The variation across international markets in 2013 and 2018 appears in Table 2.2.

> Load factor: the ratio of the number of passengers to the number of seats on a flight, also sometimes measured by ratio of revenue passenger kilometres (RPK) to available seat kilometres.

Table 2.2 Variation in load factors, international markets, 2013 and 2018

Airline	Load factor %	
	2013	2018
Asia-Pacific airlines	77.7	82.6
European airlines	81.0	88.9
North American airlines	82.8	87.2
Middle Eastern airlines	77.3	80.7
Latin American airlines	79.2	81.4
African airlines	69.0	78.2

Source: IATA (2013, 2018)

Spilled: when a passenger wants to book a particular flight but cannot get a ticket, this passenger is 'spilled'. This represents a potential loss of revenue for the airline and dissatisfaction and inconvenience for the passenger.

S-curve effect: an observation from airline markets whereby airlines that offer more flights (frequencies) obtain proportionately more market share.

Capacity discipline: the practice of airlines restricting the capacity they bring to a market.

Supply commonly exceeds demand in the airline industry. If flights operated at 100 per cent load factor, many customers willing to pay for a flight would not be able to purchase a ticket on a desired flight. As a result, a passenger may be '**spilled**'. An airline may have the passenger spilled onto another of its flights, but the passenger may move to another airline.

There are a number of reasons carriers would seek to increase seats in a market. First, traditionally, scheduling frequency disproportionately increased revenues. This was due to a claimed **S-curve effect**. Airlines use frequency as a strategic competitive variable. The airline with the higher frequency is more likely to offer passengers a flight close to their preferred departure time. This is considered to be a higher quality of service, which translates into a greater proportion of travellers and a greater market share. The widespread belief in the S-curve effect has led to excess flight scheduling.

Recently, what has been observed in both North America as well as the EU is what has been termed '**capacity discipline**' (Wittman, 2013). The increase in average load factor in both jurisdictions reflects a change in airline strategy from one of pursuing market share to one of pursuing profits. Traditionally, seat growth increases more than proportionately compared to GDP growth, but over the last four or five years it has been less than proportionate. This change in behaviour is a result of airline consolidation in the US market and the increase in the number of metal neutral joint ventures. Both factors have resulted in a decrease in competition in many markets.

!

Stop and think

Define joint production, and provide examples of where it might occur in the air transport industry.

A second reason for supplying more seats is that the addition of a new network point geometrically increases the product lines (city-pairs) of an airline. For example, if the number of network points connected to a hub increases from 9 to 14 points (5 added points), the potential additional city-pairs rise from 45 to 105. An approximate 50 per cent increase in points served increases the number of markets served by 133 per cent. Airlines recognise that network reach affects whether passengers choose their airline.

A third reason for increasing supply is that aircraft represent a significant fixed cost, and the carrier incurs these fixed ownership costs regardless of how much the aircraft is used. It may be more sensible to operate the aircraft even if it covers the incremental flying costs and makes some contribution to fixed costs. Another reason is that a route may be of strategic value to the airline. Many network industries, besides airlines, have high fixed costs and produce excess capacity: hotels, telecommunications firms, TV and radio broadcasters, energy firms, railways and logistics companies are examples. These industries, including airlines, can produce differentiated products, and there is ample room for competition to exist and fixed costs to be spread over increasing output(s). The high fixed-cost nature of airline services is not a sufficient condition to regulate the industry.

Stop and think

!

Outline why the airline product might be oversupplied, and identify the factors that influence the oversupply.

2.4 The evolution from economic regulation to deregulation

Aviation developed in the 1920s with the introduction of commercial scheduled passenger and freight services; there was a greater emphasis on passenger service. Many airlines have come and gone over the last 100 years, but some airlines started back in the 1920s are still in existence; KLM started in 1919, Qantas in 1920, Finnair in 1923, Lufthansa began in 1926, United Airlines in 1926 and Delta in 1928 are examples. The majority of countries for political reasons and trade established a national flag carrier to provide important links between countries.

Following World War II, governments realised they needed to establish a set of rules as to how airlines could access each other's airspace. Access was deemed the responsibility of a government's department of trade, and gaining access to a foreign nation was something only governments could do, not airlines. The national governments established a set of freedoms of the air that governed access by airlines to another country's sovereign territory. These agreements ranged from highly restrictive to highly liberal (➤Chapter 1).

The movement towards airline deregulation for domestic markets faced two important questions.

1   Are there consumer benefits in allowing airlines to engage in price and service competition?

2   In the case of national airlines, should these airlines be privatised?

The US was without a national carrier and thus faced only the question of deregulation. It was also an economy that had sympathy with a market-based solution for allocating goods, services and resources. It is therefore not surprising that the US was the first place to consider and introduce airline deregulation; it did so in 1978.

Economists played an important role in providing the economic support for liberalising airline markets. In the US the Civil Aeronautics Board (CAB), a federal body, regulated airline commerce in all cases where an airline flew a route that crossed state lines. However, two states, Texas and California, were large enough to have intra-state services. Studies of airfares in California that compared intra-state fares with inter-state fares found the latter to be significantly higher. Hence there appeared to be consumer benefits by allowing greater commercial freedom in commercial schedule aviation services. Following a set of hearings before the US government in 1976–77, the US made the decision to allow airlines to set their own fares, chose their own routes and determine their own capacities on those routes. Importantly, the US deregulated scheduled commercial aviation almost overnight, there was no phased-in approach to the change in market regime that has occurred elsewhere, including Europe and Canada. The outcome of the phased approach was to simply entrench the power of the incumbent national airlines; this outcome occurred in every country with a national airline which sought protection from greater market competition.

International markets have been liberalised at a much slower pace than domestic airline markets. The international air transport sector has grown under a complex regime of regulations since the conclusion of the Chicago Convention 1944 (➤Chapter 1). Lack of agreement at that time on how the market for air services should be regulated led to the growth of bilateral agreements between countries. The US favoured open skies with no control on tariffs or capacity and a maximum exchange of rights, including fifth freedoms (➤Chapter 1). The UK and other European countries were more protectionist. The two divergent views could not be reconciled and no multilateral agreement on traffic rights, tariff control and capacity was reached. The most important outcome of the Chicago Convention 1944 was that it provided a framework for the orderly development of international air transport. It also agreed on the first and second freedoms. The key institutions that emerged from the Chicago Convention were: first, air service agreements (ASAs) for the exchange of traffic rights (the ASAs were matters for negotiation between states, not carriers); second, the tariff-fixing machinery of IATA; and third, the control of capacity and frequencies by inter-airline agreement. All agreements that emerged were highly protectionist or 'predetermined'. The ASAs are trade agreements between governments, not airlines, and contain administrative (soft) and economic (hard) provisions. The soft provisions cover taxation, exemption from duties on imported aircraft parts, airport charges and transfer of funds from ticket sales from abroad. The hard provisions cover pricing and capacity limits.

In 1946, the UK and US negotiated an ASA for travel between the two countries. It became known as the Bermuda Agreement and was more liberal than agreements emerging from the Chicago Convention. The two 'liberal' features of these agreements were that fifth freedoms (➤Chapter 1) were more widely available and there were no controls on capacity or frequency.

The Bermuda II Agreement was signed in 1977, also between the UK and US. A renegotiation of the 1946 Bermuda Agreement, it allowed four airlines to operate direct flights from London Heathrow to the US and barred any other carriers from operating such flights. The designated airlines were: British Airways, Virgin Atlantic (added in 1991),

American Airlines (which replaced Pan American World Airways in 1991) and United Airlines (which replaced Trans World Airlines in 1991). Other airports in Britain, including London Gatwick Airport, were restricted to other carriers.

The ASAs came under pressure in the early 1970s as nearly 30 per cent of transatlantic traffic was flying on charter carriers, a segment that had developed as a result of the high fares and capacity restrictions inherent in the bilateral ASA arrangements. The rapid deregulation of the US air transport market from 1978 and the domestic market deregulation in numerous other countries soon thereafter gave an impetus for international reform of both cargo and passenger air services. Considerable progress has been made since that time in liberalising international air transport. Some of the changes have come through renegotiation of bilateral agreements to remove many barriers to competition. The US open skies policy reflected a new approach to international markets. They were successful in negotiating with one country, which led to adjacent countries also seeking a similar arrangement. This was successful in Europe, where individual countries still negotiated ASAs. From the early 1990s, it allowed the US and many trading partners to sign a liberal template bilateral accord, which has led to a common framework of agreements. The US open skies policy is a clear example of bilateral liberalisation, with 114 agreements having been signed to date. The EU created the European Common Aviation Area (ECAA) in 2006, which included 26 member countries. This agreement essentially removed the need for bilateral agreements between EU Member States.

There have been a number of liberal regional air trade agreements which have open skies features. Canada signed its first open skies agreement with the US in 1995, and this agreement was re-negotiated and made even more open in 2005. Australia and New Zealand also have a liberal accord, particularly across the Tasman Sea, with a single aviation market accord signed in 1996 and further liberalised in 2000. The EU has taken additional steps, which focus on liberalisation within the European Economic Area (EEA), although individual Member States and the EU have also concluded aviation agreements with countries outside the EU.

In 2008, the EU and US signed an open skies (first stage) agreement that provided significant liberalisation for air services and included the entire EEA. In this first stage, both the commercial agreement, as well as the legal framework for cooperation, had to be negotiated. In the second stage, the legal issues become more contentious: night flight bans in the EU, symmetric traffic rights, foreign ownership and control, US homeland security and EU-style slot coordination. Rising marginal costs and declining incremental benefits are likely to be the outcome.

The rise of globalisation, which is the growth of greater interaction and integration of companies and economies, has been both a cause and an effect of air transportation deregulation. The lower costs and greater access through the growth of new routes has facilitated globalisation by lowering the transaction costs of access. Advances in information technology also stimulated globalisation, including the use of information technology used by airlines to manage their fleets, network and schedules.

Stop and think !

To what extent have airline business models been affected by deregulation, and how has the liberalisation of international air services changed the pattern of service provision?

2.5 Airline profit, yield and unit costs

Airlines' units of sale differ from their units of output. Unlike an automobile manufacturer which produces and sells cars and trucks, airlines produce seat kilometres but sell passenger kilometres. Passenger traffic is measured as a passenger travelling a unit distance, which takes account of the spatial nature of demand. Passenger traffic is measured as revenue passenger kilometres (RPK); however, care must be taken in distinguishing one passenger travelling 1,000 km and ten passengers travelling 100 km: both would be measured as 1,000 RPK.

Airlines produce a product termed '**available seat kilometres (ASKs)**', which again reflects the spatial nature of their output; the relative proportions of seats and distance mentioned with respect to RPK is also true for ASKs. The utilisation of capacity is termed load factor and is measured as RPK/ASK; for a given OD this would be passengers/seats.

Two measures of airline revenue performance are yield, which is revenue divided by RPK, or the average fare paid per kilometre, and revenue per available seat kilometre (RASK), which is total operating revenue divided by ASK; RASK is also equal to yield times load factor. Cost per available seat kilometre (CASK) is a measure of cost performance and is measured as total operating costs per ASK; CASK is also referred to as unit cost. Airline costs are composed of salaries and benefits, purchased materials (which includes fuel), purchased services and landing fees, rentals, depreciation and other (➤Chapter 12). Airline operating profit is a simple calculation of revenues (including ancillary revenues) less costs, or RPK × Yield – ASK × unit cost. The objective is not to maximise revenues or to minimise costs, but to maximise the difference between them.

Available seat kilometre (ASK): the product of the number of seats and route distance and a measure of available capacity.

!

Stop and think

What are the issues associated with using ASKs and RPKs as measures of airline output?

2.6 Alliances

Alliances are a common feature of the airline industry (➤Chapter 10). After US domestic deregulation, FSNCs formed alliances with regional air carriers to feed passengers into the FSNC domestic hubs (Gillen, Morrison & Stewart, 2002). The formation of international airline alliances was a direct outcome of the failure to have significant and broad liberalisation of international air travel markets. There were three factors at work. First, many countries prohibit or limit foreign ownership of domestic airlines (➤Chapter 12). Therefore, mergers were not possible, and alliances were a means of achieving many of the benefits a merger would bring. Second, there were (and are) restrictions on **cabotage** rights, which are the rights of a foreign airline to operate between two or more airports in a domestic market. Third, increasingly passengers are originating in before hub markets and going to beyond hub markets, and alliances provide a means of accessing these points. By forming alliances, two (or more) carriers can increase service frequency and the number of accessible

Cabotage: the right of a foreign airline to operate domestically between two points in another country.

destinations. This increased connectivity provides a higher quality of service through increased accessibility and improves load factors.

The motivation to join alliances arises from marketing cooperation, cost synergies and increases in market power. Marketing benefits include broader **frequent flyer programmes (FFPs)**, codeshare agreements and lounge access (➤Chapter 10). Cost synergies stem from shared airport facilities, joint scheduling, reciprocal sales agreements and increased purchasing power. Alliances do have anticompetitive effects; these include higher fares, reduced capacity and a lack of access to alternative alliances.

The range of types and degrees of cooperation is illustrated in Figure 2.3. On the axis the amount of integration is compared to the amount of cooperation needed. In the lower left portion of the figure, the gains in revenues are relatively high compared to costs, but as integration increases, more cooperation is required and hence more costs. Toward the top right portion of the diagram the marginal costs are increasing, while the marginal benefits, in terms of increases in revenue, may be increasing or growing slowly. In this portion of the alliance integration picture, the gains stem from increased market power rather than simply cost savings through greater efficiencies. It extends from simple interlining to codesharing to co-investments and metal neutral joint ventures. The forms of limited cooperation are essentially marketing agreements and tend to have a high pay-off in the short to medium term. As expanded cooperation occurs, airlines co-invest in coordinating schedules, sharing facilities and cooperative pricing. Because this latter activity involves agreeing on prices, an

Frequent flyer programme (FFP): a membership scheme with different status levels that enables passengers to collect mileage points for flights and related purchases which can be redeemed for free flights, upgrades or discounts on selected retail and travel products.

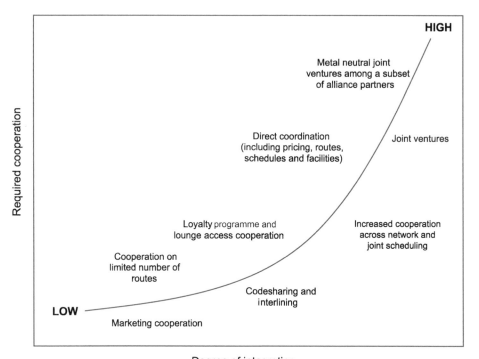

Figure 2.3 Range of alliance cooperation

activity that is generally not allowed under national competition law, it requires the approval of antitrust authorities. The alliances have convinced authorities that, despite a reduction in competition, the synergies and cost savings as well as joint fare determination result in net consumer benefits. These close ties among alliance partners have evolved into joint ventures where there is a high degree of coordination and some co-investment. In many cases, these joint ventures have become metal neutral joint ventures, which means the alliance acts as one airline in setting prices, capacity and scheduling, and alliance members involved in the joint venture share revenues and profits (➤Chapters 8 and 10) irrespective of who does the flying.

Based on the work by Bilotkach (2019), 85 per cent of passengers who flew transatlantic in 2016 were on one of the three alliances (Star, SkyTeam or oneworld). The dominance of the three alliances was significantly enhanced by mergers that occurred in the US (Continental merged with United, US Airways merged with American and Virgin Atlantic has a joint venture with Delta). These actions removed what were smaller but effective competitors from the market.

However, as Table 2.3 indicates, the alliances have been losing market share to LCCs who are providing increased amounts of seat capacity. Initially wide-body aircraft such as the Boeing 787-8 and 787-9, flown primarily by Norwegian, provided the growth in LCC capacity. However, more recently narrow-body aircraft including the Airbus 321neo, 737 NG and 737 MAX (until its grounding in Spring 2019) were supplying a large number of

Table 2.3 North Atlantic alliance and LCC market shares (seats)

2018 growth in seat capacity %		**Narrow-body aircraft capacity %**
LCC	60.0	19.0 (2016)
Alliances	3.8	26.7 (2017)
Total market	6.6	31.7 (2018)
		LCCs have 37.7% of narrow-body aircraft seats
Seat share %		**Seat share by alliance 2018 %**
Immunised joint ventures		STAR 27.2
79.8 (2015)		oneworld 20.5
73.7 (2017)		SkyTeam 20.0
72.3 (2018)		
LCCs 7.7 (2018)		

Seat growth in 2018 %	
Immunised joint venture	4.6
Non-joint venture airlines	12.2

Narrow-body Aircraft capacity %
19.0 (2016)
26.7 (2017)
31.7 (2018)
LCCs have 37.7% of narrow-body aircraft seats

Source: Based on CAPA (2018)

seats and all of these were from non-alliance airlines, mostly LCCs. The evolution of seat growth between metal neutral joint-venture alliances and LCCs is illustrated in Table 2.3.

2.7 Demand and demand forecasting

In Section 2.2, demand segments were defined according to two key parameters: price and time sensitivity. These two variables are key in forecasting the demand in OD markets and for measuring the respective elasticities in these markets. However, the aggregate demand for air travel is established by linking air travel to macroeconomic variables.

The demand for air travel is a derived demand. It should therefore be influenced by general economic conditions. Aggregate travel demand models will relate total or interregional demand to measures of GDP, as well as fuel prices, and include dummy variables for negative traffic events, such as SARS, the 11 September attacks in New York City, wars and traffic-generating events such as the Olympic Games. These models will generally yield a parameter estimate on the GDP variable which is greater than 1. This means that air passenger travel grows faster than the growth in GDP; Oum, Fu and Zhang (2009) estimate this parameter to be 1.58, which implies for each 1 per cent growth in GDP, global air traffic will grow by 1.58 per cent. Gillen (2009) estimates a forecast model that disaggregates GDP into trade (imports and exports) and investment and finds trade and connectivity have the largest impacts on aggregate air travel demand.

The problem with many macro air travel demand-forecasting models is that they exclude key market variables that will influence aggregate air travel. However, over time, as markets have liberalised, fares have decreased, new routes have developed and connectivity has grown. The liberalisation of air service agreements has led to the expansion in markets, but it also leads to more efficient continental and international networks that further stimulate traffic growth. The indirect efficiency effect would reinforce the direct effect of liberalisation on opening markets. The degree to which this would occur depends on the extent of liberalisation and the way it is done. Another key factor is the growth in global trade, which has led to increased globalisation of manufacturing and assembly. This growth in world trade has subsequently led to a growth in air travel.

Another consideration is the way industries and services that make up the economy have changed. High technology industries and the financial services sector are aviation intensive. Rapid growth in these sectors leads to even more rapid growth in air travel than would be expected with growth in GDP.

To sum up, the five fundamental drivers of long-term international air passenger growth are: GDP growth, political disruption, cost changes (such as fuel costs), service quality changes and trade growth. Political disruption would include terrorism, geopolitical tension and protectionism. While protectionism reduces trade growth, it also appears in the form of reductions in foreign direct investment. Foreign ownership of 'strategic assets' such as ports, energy and airlines are either up for review or prohibited. Such constraints increase capital costs and reduce trade in the long term. Political disruption and friction also increase costs in the form of security and regulation. These costs make shippers and service providers worse off and lessen trade and air travel. Cost changes, particularly fuel costs, are a long-term threat. In the past, growth in real fuel costs was zero or negative. In the future, this will not

---

be the case as the real cost of energy will go up and environmental taxes will become a permanent feature. In the past, cost reductions provided a 0.7 per cent stimulus to passenger growth (Swan, 2008). It is unlikely this will continue, and even advances in engine and fuel technology will not fully offset the costs of raw materials and taxes.

Quality changes have occurred in the network over the last two to three decades. International networks reorganised with gateway hubs and airline alliances. This increased accessibility and stimulated traffic growth. A significant quality change was the growth in new markets; old markets did not simply get bigger, but more routes were opened and frequencies grew. Both of these outcomes stimulated traffic growth by 1 per cent or more. In the future, the network will not improve due to higher costs, hence bigger aircraft and less frequency; frequencies were a significant stimulus to traffic growth in the past. As trade growth slows, frequencies decline, fewer routes are added, some abandonments may occur and underserved cities continue to be underserved. All of this adds up to a negative net effect on past forecast traffic growth.

The slowing of trade growth over the longer term will also reduce the previous growth forecasts. As important will be the restructuring of trade as merchandise trade (trade in goods) falls and trade in services grows. In the past, trade growth was double that of GDP growth and added 1–2 per cent to forecast air traffic growth. In the short term, with recession and trade reductions, traffic growth will also be negative. In the longer term, increased protectionism, a failure to reduce tariffs and increased costs from security and regulatory barriers will mean zero stimuli from the trend in the future.

Market OD demand forecasts rely much more on route-specific values such as airfares, the prices of alternative modes, the total trip time, and the income and population of the origins and destinations. A conventional demand-forecasting model for economy air travel would be represented as:

$$D_E = M \cdot q_E^\delta \cdot q_B^\tau \cdot p_E^\alpha \cdot p_B^\beta \cdot t^\gamma \cdot \epsilon$$

Own-price elasticity of demand: a measure of how demand for a good or service responds to a change in the price of that good or service.

Revenue management: the process of maximising revenue from every flight.

where D_E is demand for economy travel (the OD subscripts are excluded for ease of presentation); M is a market size variable such as population; q_E and q_B are measures of service quality in economy and business class respectively; p_E and p_B are fares in economy and business; t is a trip time variable; and ϵ stands for economy. The parameters $\alpha, \beta, \gamma, \delta$ and τ can be interpreted as elasticities with this functional specification. Thus, δ is the own-service quality elasticity of demand, α is the **own-price elasticity of demand** and β is the cross-price elasticity of demand for economy travel with respect to the price of business. These types of demand models are used for forecasting in **revenue management** systems. Airlines and aircraft manufacturers are constantly getting new data that they will use to re-estimate their models to update the parameters. Example 2.2 provides details of this.

Example 2.2

Forecasting market growth versus growth in the number of markets

The Airbus A380 seats up to 850 people (most are configured to seat approximately 525) and was developed by Airbus in the belief that large city-pair markets were going to get larger and airport capacity was going to become scarcer. Airbus believes in the hub-and-spoke network model and predicts airlines will continue to fly small aircraft into big hubs to fill large aircraft. The A380 has a range of 15,200 km.

The Boeing 787–9, which seats up to 280 passengers, was developed by Boeing because it saw more growth in the number of markets and routes rather than in the size of markets. Boeing does not believe the hub-and-spoke model will dominate. Boeing's research showed that since 1990 the number of city pairs greater than 4,800 km apart has doubled, frequencies have doubled and ASK has doubled. Average aircraft size has been declining. The 787 has a range of 14,000 km.

Airbus forecast sales of 1,200 A380s when it launched the aircraft in 2000. As of September 2019, Airbus had delivered 239 A380s out of a total order book of 290. Airbus announced in February 2019 it would cease production of the A380 in 2021 after it had satisfied the firm orders for the aircraft.

Boeing forecast sales of 1,300 787 aircraft in early 2000. As of September 2019, Boeing had 1,450 orders for the B787 family and had delivered 894 aircraft. Boeing has three models of 787; 787–8, 787–9 and 787–10. The 787 has two unique features, an all-composite fuselage and a completely new supply chain. Boeing's forecast was based on five key factors: deregulation drives airlines to offer more routes and more frequencies using smaller aircraft; there are diminishing cost savings with aircraft size; there are cost savings to airlines by avoiding intermediate stops; route networks evolve into highly connected networks; and, as average income increases, people's value of time increases.

Stop and think

What are the challenges of forecasting demand, and why are forecasts used despite their limitations?

!

Key points

- Airlines are just one (albeit very important) component of the aviation value chain which starts with aircraft and aircraft engine manufacturers and ends in the delivery of an air service to passengers and freight demanders.

- Airlines deliver services to passengers, goods and shippers by enabling people and freight to move from A to B.

- Air travel is a derived demand, which means it is affected by the health of the global economy, GDP growth and levels of disposable income among consumers.

- Multiple air travel market segments exist, and each one exhibits different price elasticities of demand.

- Airlines practice price discrimination and differential pricing to maximise their flight revenue and overall profits.

- Airlines supply products that have three distinguishing features: they are produced jointly, they are perishable and they are generally oversupplied because demand is variable and airlines provide capacity to serve expected demand.

- Domestic deregulation and international liberalisation have changed the nature of airline service provision. The emergence and rapid expansion of LCCs has put increased emphasis on cost and price.

- Demand forecasting is complex and has to consider a range of variables. OD demand forecasts, which provide estimates of own- and cross-price and service quality elasticities, are used in airline revenue management.

References and further reading

Bilotkach, V. (2019). Airline partnerships, antitrust immunity, and joint ventures: What we know and what I think we would like to know. *Review of Industrial Organization*, 54, pp. 37–60.

CAPA. (2018). *North Atlantic aviation market: LCCs grow market share* (April). Available at: https://centreforaviation.com/analysis/reports/north-atlantic-aviation-market-lccs-grow-market-share-410928

Fu, X., Oum, T. and Yan, J. (2014). An analysis of travel demand in Japan's intercity market: Empirical estimation and policy simulation. *Journal of Transport Economics and Policy*, 48(1), pp. 97–113.

Gillen, D. (2009). *Trends and developments in inter-urban passenger transport: International air passenger transport in the future* (Paper prepared for OECD/ITF Eighteenth International Symposium on Transport Economics and Policy, Madrid, Spain, 16–18 November 2009), in OECD, The Future for Interurban Passenger Transport: Bringing Citizens Closer Together, May 2010. Available at: http://dx.doi.org/10.1787/9789282102688-4-en

Gillen, D., Morrison W. and Stewart, C. (2002) *Air travel demand elasticities: Concepts, issues and measurement* (Report to Department of Finance, Canada). Available at: www.fin.gc.ca/consultresp/airtravel/airtravstdy_-eng.asp

IATA. (2008). *Estimating air travel demand elasticities, final report 2008*.

IATA. (2013). *International air transport association annual review 2013*. Available at: www.iata.org/publications/economics/Pages/index.aspx

IATA. (2018). *Passenger load factor hits 28-year high*. Available at: www.airlines.iata.org/news/passenger-load-factor-hits-28-year-high, 19 October 2018

Oum, T., Fu, X. and Zhang, A. (2009). *Air transport liberalization and its impact on airline competition and air passenger traffic*, final report, OECD International Transport Forum, Leipzig, Germany. Available at: www.internationaltransportforum.org/Pub/pdf/09FP04.pdf

Swan, W. (2008). *Forecasting Asia air travel with open skies*, Mimeo Seabury Group, Seattle. Available at: www.sauder.ubc.ca/Faculty/Research_Centres/Centre_for_Transportation_Studies/William_Swan_Publications

Tretheway, M. and Markhvida, K. (2014). The aviation value chain: Economic returns and policy issues, *Journal of Air Transport Management*, 41, pp. 3–16.

Wittman, M. (2013). *New horizons in US airline capacity management: From rationalization to 'capacity discipline'*. Cambridge, MA: MIT International Center for Air Transportation.

CHAPTER 3

✈ Airline regulation and deregulation

Stephen Ison and Lucy Budd

LEARNING OBJECTIVES

- To understand why airlines were regulated.

- To appreciate the motivation for deregulating the airline market.

- To understand the significance of the 1978 US Airline Deregulation Act on air service provision in the United States.

- To examine processes of airline deregulation and liberalisation in Europe and other world markets.

- To explain how economic deregulation led to new consumer and environmental regulations.

- To explore future trends in air service deregulation worldwide.

3.0 Introduction

This chapter examines why the airline industry was historically heavily regulated, explores the reasons for deregulation and discusses the implications of deregulation and liberalisation for airlines and consumers. Under normal market conditions, the frequency, regularity and patterns of connectivity that are offered by transport providers are a function of consumer demand (➤Chapter 2). In the absence of demand or falling demand, services are either not offered or are suspended or permanently withdrawn from the market.

3.1 Early regulation

In the early years of aviation, flying was dangerous and expensive. Forced landings were common, and there was a risk that serious injury or death could result from an accident. Consequently, national governments felt compelled to regulate airline operations through legally enforceable national legislation that focused on safety in order to protect passengers. Safety, however, had a financial implication for operators, as maintaining aircraft and training pilots to the required standards were expensive and many pioneering independent operators went out of business. As a result, government involvement in the sector expanded in the 1920s and 1930s to include state ownership of airlines and the introduction of new regulations (➤Chapter 1). These included regulations governing the airworthiness of aircraft, pilot licensing and accident investigation. As airlines symbolised a country's economic and political power, a regulated market was deemed important.

The Chicago Convention

The 1944 Chicago Convention (➤Chapter 1) introduced new Standards and Recommended Practices (SARPs). These SARPs, which countries should (or should endeavour) to meet, were an attempt to promote international unity and harmonisation in the provision and safe operation of air services. The Convention also sought to facilitate the economic development of international services by establishing a series of freedoms of the air.

These international regulations were enshrined in national law by ICAO Member States and enforced by each individual country's Competent Authority. This authority was usually a division of the national government which was given special responsibility for aviation safety oversight and economic regulation. By the mid-1940s, a high level of regulatory oversight and intervention in the airline sector was considered necessary in order to:

- safeguard the economic interests of the country by ensuring the quality and provision of air services, by preventing potentially damaging competition between service providers and by denying other nations unrestricted access to their domestic markets (➤Chapter 1);

- protect public and passenger safety by ensuring that aircraft were airworthy and were being operated and maintained by qualified pilots and engineers (➤Chapter 13);

- secure the country's borders (for defence and national security reasons) by permitting only authorised foreign aircraft to enter its sovereign airspace (➤Chapter 15).

The economic regulatory environment intended to promote the development of each nation's civil air transport industry by protecting state-owned airline companies from competition and preventing new operators from entering the market. The economic arguments in favour of regulation were that the high cost of market entry meant that, in most instances, only one carrier could profit from providing an air service and so strict regulations would prevent damaging competition from occurring between service providers. The aim was to bring stability and guaranteed continuity of service to the market.

This effectively created a situation in which state-owned airlines were protected from market forces by national governments who set airfares, seat capacity, service frequency and issued route licences. Every new route, cargo rate and passenger airfare had to be approved by a government committee. It was argued that this level of regulation would ensure the orderly development of international air services that were in the national interest and also guarantee that domestic services would be provided to remote areas where services otherwise could not be justified on commercial grounds (➤Chapter 21).

As a consequence of this regulatory regime, which protected the airline operators from external competition, the global airline industry from the mid-1940s to the early 1980s was characterised by economic inefficiency and high prices which ensured flying, whether domestically or internationally, remained beyond the financial means of most people. Although the highly regulated environment prevented the airlines from making any more than a modest profit, it protected them from competition, guarded them against financial loss regardless of economic conditions, and eliminated the possibility of failure. Yet, it had the consequence of inhibiting the growth of markets for aviation services.

In addition to controlling which airports and routes could be served and the airfares that could be charged, national governments also held authority over market entry and exit. To prevent service duplication and protect existing carriers from competition, approval for new airlines to enter the market was rarely granted. As well as fostering inefficiency and preventing price competition, the regulated system was also bureaucratic and labour intensive. Any proposed change to airfares, route entitlements or market entry required complex approval. Under these conditions of price regulation, non-price competition occurred in the form of increased flight frequencies. However, this led to an oversupply of seats and costs which rose to meet fares rather than fares which fell to meet costs.

The introduction of new high-capacity wide-body jet aircraft during the 1960s and 1970s exacerbated these oversupply issues and load factors (a measure of the proportion of seats sold on a flight) fell to as low as 50% (which meant on any given flight half of the seats were empty). To the extent that the regulations allowed, airlines began to adopt a rudimentary form of yield management by offering last minute discount fares and promotional rates to try to stimulate demand from passengers who otherwise would not have flown. In common with other transport modes, aircraft seats are a perishable commodity or a single use consumer good, and once a flight departs with empty seats on board, these potential revenue streams are lost and can never be recovered. Airlines were thus keen to maximise the revenue generating potential of each and every service by heavily discounting unsold seats close to departure (➤Chapter 9).

Stop and think

!

What are the potential advantages to passengers and countries of regulating the airline sector?

3.2 Moves towards deregulation

Although economic regulation allowed a limited number of airlines to operate, it kept ticket prices artificially high and created a situation in which airfares did not reflect the actual costs of providing the service. The economic and political arguments favouring deregulation started to appear during the early 1970s. It was argued that economic regulations governing market access, frequency, capacity, and fares were unnecessarily restricting competition and preventing consumers from enjoying the benefits of a more liberalised and competitive market.

It was argued that allowing airline operators greater freedom of market entry and exit would improve consumer welfare as carriers could adjust airfares to reflect changes in their production costs. It was also believed that the potential for monopolistic behaviour would be reduced by the introduction of a more contestable market. Critics of deregulation countered that unrestricted competition would lead to a reduction in service quality and safety standards, non-profitable but socially essential air services to remote regions would be abandoned to the detriment of communities who lived there (➤Chapter 21), and the welfare of airline employees would be damaged (➤Chapter 19). Nevertheless, the arguments favouring deregulation gained political support, and from 1978 the US domestic air freight market was liberalised and discount and promotional fares were approved. This had two effects:

- Freight operators could charge a premium for express delivery.

- Passenger airlines could market a greater range of fare products, including heavily discounted 'saver' fares.

3.3 The 1978 US Airline Deregulation Act

On 24 October 1978 the US Congress passed a law that enabled US-based airlines to set their own fares and routes. The Airline Deregulation Act sought to encourage the development of a competitive US domestic aviation market in which innovation was permitted and airlines were able to determine the variety, quality and price of air services. Restrictions on market entry and exit were removed and new operators were permitted to enter into competition with existing carriers.

The 1978 Act had a number of effects:

- It increased market access and enabled new operators to commence services.

- It granted fifth freedom flying rights (see ➤Chapter 1) to US airlines.

- It removed capacity constraints.

- It lifted restrictions on the routes airlines could fly within the US.

- It revoked US government authority to regulate fares.

The removal of price controls meant that airlines could price fares according to route-specific operating conditions rather than averaging costs out over their whole network. This, however, had the consequence that some routes were no longer commercially viable and the state had to intervene and provide a subsidy to ensure continuation of these services, which were commercially not viable but socially and politically important (➤Chapter 21).

3.4 The impact of US deregulation

Deregulation resulted in a number of immediate spatial, economic and operational effects. In a bid to increase revenues, reduce costs and protect their market share at major airports, existing airlines reconfigured their point-to-point networks into a hub-and-spoke operation. This improved operational efficiency, increased flight frequencies, lowered costs and created 'fortress hubs' that effectively prevented the market entry of new carriers. They also enabled airlines to service a greater number of city pairs without the costs associated with flying point to point routes. The creation of hub-and-spoke systems improved consumer choice for the majority of travellers as it offered them a greater number of connection possibilities and lower fares (➤Chapter 9). However, scheduling flights in waves to optimise connection times at the major hub airports led to congestion and delays and increased the airline's vulnerability to disruptive events (➤Chapter 11).

The reorganisation and concentration of flights at key hubs combined with the new freedom of market entry and exit provided an opportunity for new carriers to enter the market and inaugurate flights on un(der)served routes. Many of the new operators adhered to the 'low cost' approach to doing business which had been pioneered by Texas-based Southwest Airlines (➤Chapter 8). The low cost airline philosophy provided only what was necessary for safe and efficient flight, and all unnecessary service elements were eliminated. This approach enabled low cost carriers to minimise their operating costs by filling their aircraft and passing the savings onto consumers in the form of lower fares which then stimulated further demand.

The introduction of new competition caused US airfares to fall by almost 20% and passenger numbers to rise by over 50% between 1978 and 1982. Business travellers, in particular, benefited from increased flight frequency, improved connectivity and the growth in the number of regional flights to major hubs, while leisure travellers benefited from lower fares and greater choice of providers.

However, despite these benefits, the contestable market was volatile, and many of the new entrant airlines quickly failed or were taken over by existing carriers who were keen to acquire the new competition. There was also concern that deregulation and the proliferation of new operators was leading to an erosion in safety standards and safety oversight as well as additional aircraft noise and air traffic congestion.

As well as transforming the spatial characteristics of airline networks, deregulation also impacted on airline labour relations (➤Chapter 19), airline business strategies (➤Chapters 8) and airline financial performance (➤Chapter 12). Encumbered by expensive legacy labour contracts and institutional inertia, some existing carriers found the transition to a highly competitive market challenging. Deregulation forced airline managers to actively control their inventory to maximise yields and revenues and price services according to consumer demand. Freed from government control (see Table 3.1), airlines were able to determine the fare levels, with associated services, amenities and restrictions, for a set of products in individual origin-destination markets, and this led to the emergence of a much wider range of fare and service combinations. A range of fare categories for travel on a particular route were developed, with fares varying according to time and class of travel (including peak/off-peak, First/Business/Economy), length of trip (including any minimum stay restrictions),

Table 3.1 Comparison of government controls in a regulated and a deregulated market

	Regulated market	Deregulated market
Safety standards	Yes	Yes
Price controls	Yes	No
Capacity controls	Yes	No
Domestic route controls	Yes	No
Remote/regional services subsidised	Yes	Yes*
Controls on service frequency	Yes	No
Freedom of market entry	No	Yes
Freedom of market exit	No	Yes
Product and service innovation permitted	To a limited extent	Yes

*To ensure the continued operation of air services to remote regions that were not commercially viable, subsidy schemes were introduced (➤Chapter 21).

the age of the travellers (children, infants, babies), and relevant demand factors (such as a major sporting, religious or cultural event). The new range of fare products enabled consumers to maximise their utility by comparing the price, schedules, ticket flexibility and service levels of different operators (➤Chapter 9). The advent and widespread use of the internet from the mid-1990s onwards made this process of price comparison even more convenient for consumers and opened up the market to more consumers.

Airline deregulation undoubtedly produced significant and long-term benefits to US travellers by increasing competition and lowering fares. However, for a time, these benefits were available only to domestic travellers as international services remained regulated by complex bilateral and multilateral air service agreements (➤Chapter 1).

3.5 Towards international open skies

The presence of bilateral air service agreements for international routes meant that international air service liberalisation could not be achieved as easily as US domestic deregulation. Moves towards liberalisation have usually taken the form of individual open skies agreements (OSAs) between specific countries. One of the first was signed between the Netherlands and the US in 1992 which led to the transatlantic alliance between KLM and Northwest Airlines. Further OSAs were agreed between the US and Canada (1995 and 2006) and the US and the European Union (2007, amended in 2010). These agreements liberalised the provisions of the bilateral agreements by permitting airlines to serve a greater number of destinations and by allowing airlines a greater degree of cooperation and coordination. This had the effect of improving connectivity between EU and US airports, increasing passenger numbers on transatlantic routes and reducing fares for consumers.

3.6 Airline deregulation in Europe

The idea of deregulating Europe's air transport market was first discussed in the late 1970s. As had been the case in the US prior to 1978, Europe's air transport system was oligopolistic in structure, and growing public dissatisfaction with high airfares combined with the rise of free-market philosophies and increased pressures on public spending raised the issue of airline deregulation and placed it on the political agenda. However, the European situation differed from that of the US in two important respects which made the formation of a unified European policy on air transport deregulation problematic:

- Unlike the US, Europe contains a large number of independent sovereign states, each with its own laws, currency, language and administrative and legal procedures.

- The European airline sector was dominated by state-owned national flag carriers who held bilateral traffic rights on key international routes.

The UK was an enthusiastic advocate of air service reform and signed an open-market bilateral air service agreement with the Netherlands in 1984 to stimulate competition on the route between London and Amsterdam. This agreement removed the existing capacity regulations, dissolved the duopoly operation of British Airways and KLM and led to the UK signing similar agreements with other countries. However, across Europe, there was concern that national economic self-interest would mean each country would pursue its own agenda and that this would result in the fragmentation of the European air transport market. It was decided that greater regional integration would be enhanced by new policies and legislation which could be coordinated by a single legal entity. In response, European governments, through the European Commission, sought to liberalise the continent's air transport market through three successive packages of measures.

- The first package, ratified in 1987, allowed European airlines to increase capacity and sell a greater range of discounted and promotional fares.

- The second package, passed in 1990, removed market constraints and granted fifth freedom flying rights to all European carriers.

- The third package (passed in 1997) created a single regulatory structure, introduced an open pricing regime and granted full freedom cabotage to European airlines which enabled EU-registered carriers to treat the whole of Europe as a single market.

3.7 The impact of European liberalisation

The liberalised operating environment was conducive to competition. New carriers formed to take advantage of emerging market opportunities, and existing airlines expanded their service offerings. Freedom of market entry and exit prompted new entrant low-cost airline operators to emerge and compete on price and routes with the existing operators. As in the US, many of the new airlines that emerged in the immediate aftermath of liberalisation pursued the low-cost business model pioneered by Southwest Airlines. These included using a single type of aircraft in an all-economy class configuration, operating from cheaper and less congested secondary airports and focusing on ancillary revenue generation (➤Chapter 8).

Liberalisation resulted in a more contestable market and consumers benefited from a greater range of services, fare products and flight frequencies. However, as well as creating new opportunities, liberalisation also posed new challenges to regulators. In order to protect consumer interests in the event of airline failure, severe delay or mishandled luggage, and to protect them from non-transparent pricing and misleading marketing, while safeguarding the environment and ensuring air transport is accessible to all, it was necessary to introduce new regulations of pan-European regulatory control and intervention. These regulations are published by the European Commission and include:

- *Regulation (EC) No 261/2004 of the European Parliament and of the Council of 11 February 2004 establishing common rules on compensation and assistance to passengers in the event of denied boarding and of cancellation or long delay of flights –* this regulation ensures passengers travelling to/from/within the European Union receive compensation for delays and denied boarding for which the airline is responsible. Compensation is dependent on the duration of the delay and distance flown. Compensation is not granted where the airline can prove the delay was due to an 'extraordinary circumstance' (such as severe weather or national security threat) that was outside its control.

- *Regulation (EC) No 1107/2006 of the European Parliament and of the Council of 5 July 2006 concerning the rights of disabled persons and persons with reduced mobility when travelling by air.* The European single market is designed to benefit all travellers, and passengers with mobility or other impairments have an equal right to access air travel under European law. This document stipulates the regulations governing the carriage of disabled passengers to ensure that they are not discriminated against and that any additional needs they may have to access air travel are free at the point of use.

- *Regulation (EC) No 1008/2008 of the European Parliament and of the Council of 24 September 2008 on common rules for the operation of air services in the Community (Recast).* This document establishes the regulations for the operation of the internal aviation market in Europe. It covers areas including air operator's certificates, route licensing, pricing and environmental protection.

3.8 Airline deregulation in other world regions

Starting in the early 1980s, countries in Australasia, Africa, Latin America, the Middle East and Asia have also initiated policies of airline deregulation and liberalisation. These interventions have sought to open up their respective air service markets to new competition, enable new carriers to emerge and have stimulated passenger demand through the availability of lower airfares. Australia and New Zealand both deregulated their markets in 1990 and saw increases in passenger numbers and a reduction in fares.

The liberalisation of individual air transport markets has also, over time, led to the emergence of regional aviation blocks who seek to achieve regulatory convergence and agreement in the field of international air services and who negotiate on behalf of their members (see ITF, 2019). The 1996 Mercosur Sub regional agreement on air transport services (involving Argentina, Bolivia, Brazil, Chile, Paraguay and Uruguay) and Members of the Andean Community Pact (Bolivia, Colombia, Ecuador and Peru) have both taken steps to deregulate air services within their combined territories by removing national ownership and control restrictions on Member State carriers.

The 1999 Yamoussoukro Decision created the conditions necessary for the liberalisation of intra-African air services within the African Union. In 2015, ten members of the Association of South East Asian Nations (ASEAN) established a voluntary open skies agreement. This policy removes restrictions on third, fourth, and fifth freedoms for airlines based in selected ASEAN Member States. The Association of Caribbean States is also exploring plans to deregulate air services in the region. However, despite such moves, barriers to entry still exist and the global economic regulatory regime for air transport services remains complex.

Stop and think

What factors might cause some countries to be more reluctant to deregulate or liberalise their air services than others?

3.9 Summary of deregulation's impacts

Any change in regulatory and economic policy has a range of consequences for producers, regulators and consumers, some of which may not always be anticipated.

The benefits of deregulation to consumers and national economies can be summarised as follows:

- Increased competition leading to:
  - lower average airfares;
  - a greater range of discount tickets being made available;
  - more passengers travelling by air;
  - enhanced service offering and greater levels of service innovation;
  - increased flight frequency;
  - greater choice of airports, routes and carriers;
  - inward investment in regions previously unserved or underserved;
  - continuing regulatory reform in other areas of air transport;
  - Increased opportunity for foreign direct investment in national airlines if ownership restrictions are relaxed.

The disbenefits of deregulation to consumers and national economies can be summarised as:

- market volatility and airline failure (with the associated impacts on airports, regional economies, airline employees, investors and passengers);

- declining airline profitability owing to increased competition, leading to lower returns on investment for owners and shareholders (➤Chapter 12);

- lower levels of state intervention and control, leading to possible reductions in safety standards and safety oversight (➤Chapter 13);

- environmental and social impacts of increased air traffic, particularly at regional airports (➤Chapter 18);

- potential deterioration of employment conditions for staff (➤Chapter 19);

- remote and regional services may require subsidies to ensure their continued operation (➤Chapter 21).

3.10 Barriers to further deregulation

At the international level, bilateral ASAs are still used as the legal basis for operating international routes, particularly where there is an uneven balance of demand between the two contracting nations. This mechanism still restricts competition. Other barriers take the form of airport slot restrictions and airport operating curfews, which limit the availability of desired departure and arrival times, political decisions regarding airport development, expansion and charging regimes and restrictions on the foreign ownership of airlines. The EU, for example, limits foreign ownership to 49.9% of total shares while the US imposes a cap of 25%. Furthermore, although many states have deregulated their airline markets and opened up airport operators and ground handling agents to commercialisation and privatisation (➤Chapter 6), the infrastructure on which air transport depends, namely the provision of air traffic services, remains wholly or majority state owned. Indeed, the majority of the World's Air Navigation Service Providers (ANSPs) remain state owned, and even for those which have been partially privatised (such as NATS in the UK), the government maintains majority ownership for reasons of defence and national security.

Furthermore, the widespread use of frequent flyer programmes, customer relationship marketing strategies, corporate discount schemes and travel agents (which direct customers and revenue to certain airlines) distort the market and disadvantage carriers who do not have access to them (➤Chapters 8, 9, 10 and 20).

3.11 The future of airline deregulation

Airline deregulation is widely considered to have improved consumer welfare by increasing efficiency and lowering airfares, yet the processes and implications are far from uniform or complete. Many world regions and countries, particularly in emerging economies, still maintain relatively high levels of economic regulation, although the process of deregulation leading to increased competition, new airlines, lower airfares, more passengers, greater price discrimination and enhanced production differentiation is now well established and its effects felt by almost 40 years of economic appraisal and analysis.

Deregulation and liberalisation of computer reservation systems and the increased use of digital and social media by consumers are creating new opportunities for airlines to develop and market new products to appeal to increasingly diverse customer segments (➤Chapter 10) while simultaneously helping airlines to lower their costs.

Key points

- The 1978 US Airline Deregulation Act was instrumental in changing the structure of the airline industry.

- Greater freedom of market entry led to the emergence of low cost carriers and an increasingly competitive market.

- This has resulted in lower airfares, increased connectivity, facilitated product innovation and enabled more people to travel to more places, more cheaply.

- Following the US's lead, other countries have deregulated and liberalised their airline markets.

- Barriers to further deregulation exist in the form of bilateral air service agreements, restrictions on the foreign ownership of airlines and Government desire to retain control of certain aviation functions.

References and further reading

Budd, L. (2018). Airline deregulation. In: J. Cowie and S. Ison, eds., *The Routledge handbook of transport economics*, Abingdon: Routledge, pp. 141–154.

ITF. (2019). *Liberalisation of air traffic*. ITF Research Reports Paris, OECD Publishing.

Rose, N. and Borenstein, S. (2014). How airline markets work, or do they: Regulatory reform in the airline industry. In: N. Rose, ed., *Economic regulation and its reform: What have we learned?* Chicago, IL: University of Chicago Press.

CHAPTER 4

Airfield design, configuration and management

Lucy Budd and Stephen Ison

LEARNING OBJECTIVES

- To identify the principal components of the airfield and understand the interrelationships between them.

- To understand the main factors that affect runway design and orientation.

- To describe the basic types of runway configuration and their effect on capacity.

- To demonstrate the management issues which arise from different airfield configurations.

- To recognise the environmental and social implications of developing, operating and managing an airfield.

4.0 Introduction

The design and configuration of an airfield directly affect its safety, usability, efficiency and environmental and social impacts. Given the importance of airfield design and configuration to safe and efficient air transport operations, this chapter will: identify the role of the principal components of an airfield and discuss the interrelationships that exist between them; examine the factors that affect the location,

Runway: a defined rectangular area of land on an aerodrome that is prepared for the take-off and landing of aircraft.

Taxiway: a defined path linking different parts of the airfield which is used by aircraft manoeuvring on the ground under their own power or by aircraft under tow.

Apron: a defined area of land on an aerodrome that accommodates the parking, loading/ unloading, refuelling and maintenance of aircraft.

Airside: areas beyond security control to which only authorised individuals have access.

Aerodrome: a defined area of land or water which is used for the arrival, departure and surface movement of aircraft.

Airport: an aerodrome with extended facilities such as terminal buildings and/or customs inspection posts that supports commercial flights.

Airfield configuration: the siting, number and orientation of runways, taxiways and apron areas.

orientation and physical dimensions of **runways, taxiways** and **apron** areas and how these affect capacity; demonstrate the principal management issues which arise from different airfield layouts; and consider the environmental and social implications of developing, operating and managing an airfield.

4.1 The airfield

An airfield includes all of the **airside** manoeuvring areas that are used by aircraft (including the runway(s), taxiways and aprons) and the spaces that are adjacent to them. The airfield can cover up to 80–90 per cent of an **aerodrome** or **airport**'s total land area, and it is usually protected by a secure perimeter fence to prevent unauthorised access.

Although every airport is unique in terms of its geographic location, physical layout, demand characteristics, built environment and air traffic mix, the design and configuration of every airfield needs to address three basic requirements. It needs to:

1 Facilitate routine safe and efficient air transport operations while complying with (inter)national safety and design standards.

2 Minimise and mitigate local environmental and social impacts as far as possible.

3 Be capable of future expansion, if required (➤Chapter 5).

The design of airfield infrastructure and the **airfield configuration** requires considerations of:

- the number, length, spacing and orientation of the runway(s);
- the number, location and design of exit taxiways and **rapid exit taxiways** (RET);
- the design and layout of taxiways;
- the design and layout of apron areas;
- the siting and interaction between these elements;
- aerodrome safeguarding.

The number, orientation, physical dimensions and configuration of individual airfield components are affected by: the physical space that is available for development; the capital cost of providing the infrastructure and likely return on investment; the type of air traffic that will use the facility both now and in the future; the nature of consumer demand; proximate local land uses and airspace constraints; and local environmental and social considerations. Airfield design is so important to the safety and efficiency of airport operations that international design standards have been devised to ensure that safety-critical infrastructure and associated assets including runway markings, airfield signs and aerodrome ground lighting standards are consistent around the world.

The regulations governing airfield design, configuration and management are strict. At the international level, ICAO's *Aerodrome Standards: Aerodrome Design and Operations* manual provides best practice design standards for airports worldwide. These Standards and Recommended Practices (SARPs) are used as a basis for supra-national and national regulations. In Europe, **EASA**'s *Certification Specifications and Guidance Material for Aerodromes Design CS-ADR-DSN* details the requirements in Europe, but the competent authorities in each EU Member State maintain certain discretionary powers for aerodrome licensing within their territory. In the UK, the CAA's *CAP 168: Licensing of Aerodromes* document specifies the standards that are required at UK airports, while the US's FAA *150/5300–13A-Airport Design* manual contains equivalent guidance on the requirements for US airports.

Collectively, airfield design standards stipulate, among many other characteristics, the length, strength, width, orientation, configuration, maximum permitted slopes, minimum sight lines, construction materials and pavement thicknesses of runways, taxiways, aircraft manoeuvring areas and aprons, as well as the provision of visual navigation aids, aeronautical ground lighting, airfield signs and markings, and asset installation and maintenance.

Although (inter)national design standards provide best practice guidance, many airports are constrained by their existing layout which, in some cases, may have evolved in an ad hoc way over time or been originally designed for military use, which may render (parts of) them incompatible with current regulations.

Five basic airfield design elements need to be considered when planning or expanding an airport:

1 The runway(s) have to be long enough and strong enough to accommodate the largest and heaviest aircraft that are likely to use the airport, both now and in the future.

2 The primary runway(s) should be aligned in the direction of the prevailing wind for maximum operational usability.

3 Taxiways and adjoining apron areas have to be wide enough and large enough to facilitate simultaneous aircraft movements, be positioned in such a way as to minimise **runway occupancy times** and taxiing durations and ensure the safe and efficient surface movement of aircraft.

4 The configuration of the aprons and the landside interface must maximise efficiency and minimise the potential for ground collision between aircraft, fixed infrastructure and mobile vehicles.

5 The runway(s), taxiways and aprons should be designed and operated in a way that minimises and mitigates any adverse local environmental and social impacts.

4.2 Runways

A runway is an airport's most important built asset. Runway design and configuration is determined by a range of physical, legislative and economic factors. An airport's geographic site and situation is of fundamental importance. The *site* refers to the surface area contained

Rapid exit taxiway: enables landing aircraft to vacate the runway at higher speeds to reduce runway occupancy times.

EASA: European Aviation Safety Agency – an organisation that develops common standards of safety and environmental protection for civil aviation in Europe.

Runway occupancy time: the amount of time (in minutes or seconds) individual aircraft spend on the runway rendering it unavailable for use by other flights. Airports try to decrease runway occupancy times to maximise the use of this facility.

Airfield elevation: The vertical distance the highest point of the landing area is Above Mean Sea Level (AMSL).

within the perimeter fence as well as the airfield's physical size, dimensions, **airfield elevation**, relief (whether the land is flat or undulating), sight lines and prevailing local weather conditions (particularly relating to wind direction and visibility). The *situation* describes an airfield's location relative to local topographical features (such as coastlines or mountains), urban areas and key markets, as well as to the configuration of surrounding airspace, the availability of land for future expansion, the existence of noise abatement or other environmental and social operating restrictions, the volume and type of air traffic that is handled and the performance characteristics of the aircraft that use it.

During the early years of powered flight, runways were not demarcated and pilots could take off and land in any direction. The relatively low mass of the early aircraft meant that there was no need to prepare a dedicated landing surface, and take-offs and landings could occur from any direction. Although this arrangement suited the operation of small single-engine aircraft, the introduction of larger airframes necessitated the preparation of dedicated runways that could support the increased weight of the aircraft and prevent them from getting stuck in wet ground. From the early 1920s onwards, cinders (crushed coal or embers), pulverised rock, gravel and ashes began to be added to the landing area to stop the runway flooding and to create a harder surface and greater surface friction for braking. As this process was expensive and time consuming, prepared runways were normally laid only along the line of the prevailing wind (see Example 4.1).

Example 4.1
The significance of lift and prevailing wind direction

In order to maximise airflow over an aircraft's wings and generate the maximum amount of lift, aircraft need to land and take off into the wind. In most locations, the wind blows from one or two directions year-round. The most frequent wind direction is called the prevailing wind, and runways are aligned into it to take advantage of this natural aid to flight. Some airports, which are subject to variable wind conditions, had to be developed with one or more crosswind runways which are aligned in different directions so that the airport can remain operational irrespective of wind direction. Examples of airports with crosswind runways include the domestic airport serving Reykjavik in Iceland as well as Halifax in Canada, Sydney Kingsford Smith, and New York Newark.

The introduction of progressively larger and heavier aircraft from the mid-1920s onwards required the preparation of more robust runways. The world's first hard-surfaced runway was reportedly constructed at Leipzig-Halle airport in Germany in 1926. A hard runway at New York Newark followed in 1928 and, from that date onwards, hard runways became an increasingly common sight at major airports. Although hard-surfaced runways offered a number of significant operational benefits, they were expensive to construct and maintain and their orientation required careful planning to ensure maximum operational usability. In the UK, many airfields were constructed for military use during the Second World War.

These facilities typically featured one main runway and two shorter crosswind runways on different alignments that intersected the main landing strip and formed a triangle-shaped runway layout. A good number of these old military airfields were subsequently transferred to commercial use and the legacy of the military runway layouts can still be seen at sites including Heathrow and East Midlands Airports.

As well as determining the prevailing wind direction and the optimum runway orientation (Example 4.2), other important design considerations include the number and physical dimensions of runways. The total number of runways should be sufficient to meet current demand during peak operations. Careful analysis needs to be conducted to ascertain demand during the peak period and to assess whether or not it is cost-effective to build runway infrastructure that may be underutilised at other times (➤Chapter 5). Runway length is determined both by the physical attributes of the site and the performance characteristics and weight of aircraft that use the airport.

Example 4.2
Runway orientation and identification

Every runway is identified by a two-digit number from 01 to 36 which is derived from the runway's bearing relative to magnetic north. These numbers are painted in highly visible white paint at the start of the runway to ensure that pilots are operating on the correct one. To determine a runway designator, the runway's compass heading (a three-digit number) is divided by ten and rounded to the nearest whole integer to give a two-digit number. A runway on a bearing of 092°, for example, would be identified as runway 09 (092° ÷ 10 = 9.2, which is rounded to the nearest whole number and prefixed by a zero to render it a two-digit number). A runway on a bearing of 147° would be runway 15. As runways are straight, the designation at the other end always differs by 18 (180°). In these examples, the runways are identified 09/27 and 15/33.

When an airport has two runways on the same alignment, the letters L (left) and R (right) are added after the two-digit number. When viewing a pair of parallel runways which are both on a compass heading of 272°, the runway on the left is designated 27L and the runway on the right 27R. Where there are three parallel runways, an additional classification of C (centre) is used (making 27L, 27C and 27R when viewed down their length). At airports with multiple parallel runways, runway designators have to be changed to avoid confusion. At Atlanta's Hartsfield Jackson International Airport, the five parallel runways, all of which are on a true heading of 090° and 270°, are designated 26L/08R, 26R/08L, 27L/09R, 27R/09L and 28/10.

A further complication is that the earth's magnetic pole is not static and runways may need to be redesignated. This normally occurs once every 40–50 years depending on latitude. In 2018, Geneva's main runway was redesignated 22/04 (it had been 23/05) and Cork Airport in Ireland changed its runway designation from 17/35 to 16/34.

Wake vortices: powerful columns of air that are generated when the low and high pressure systems that exist above and below an aircraft's wing during flight meet at and spin off from the wingtips. Wake vortices have the potential to dislodge roof tiles and endanger other aircraft.

Capacity: the maximum quantity or volume of passengers or aircraft that can be safely accommodated by an aspect of the airfield under agreed standards of service.

Runway capacity: the maximum number of aircraft movements that can be safely accommodated in a defined time period under specified operating conditions.

4.3 Runway configuration and capacity

The number and configuration of runways are usually determined by the volume and type of air traffic that an airport handles as well as the prevailing wind direction and any local airspace restrictions and interactions with other neighbouring airports. Aircraft size and weight are also important as aircraft generate powerful **wake vortices** when they fly, and distance- and time-based separation minima are used to sequence arriving and departing aircraft to allow the wake vortices to dissipate so they don't endanger other aircraft (see Section 4.4). Both the physical characteristics of an airfield and aircraft mix that use it affect **capacity**. **Runway capacity** varies according to the number and orientation of serviceable runways as well as local weather conditions, the availability and sophistication of air traffic control (ATC) facilities, aircraft mix, type of operations (fixed wing landplanes or rotary wing helicopters) and flightcrew behaviour.

While airports worldwide have many different **runway configurations**, four basic types can be identified (Figure 4.1). These are:

- single runway;
- parallel runways;
- intersecting runways;
- open-V runways.

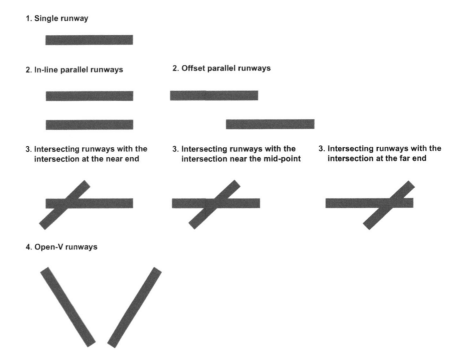

Figure 4.1 The four basic types of runway configuration

Single runway

A single runway is the simplest configuration as it consists of a single strip of land that is demarcated for the landing and take-off of aircraft (see Case Study 4.1). The **theoretical runway capacity** of a single runway airport, defined as **aircraft movements** per hour, varies according to whether flights are operating under **Visual Flight Rules (VFR)** or **Instrument Flight Rules (IFR)** as well as the nature of the traffic mix and demand. The **practical capacity** is often lower than the theoretical capacity. Owing to prevailing winds, one direction is usually used for the majority of movements.

At East Midlands Airport in the UK, for example, which has one runway (27/09), the prevailing westerly wind means the majority of aircraft use runway 27. Aircraft use runway 09 only when the wind is coming from the east. Some airports are so constrained by their physical location that aircraft can only ever operate from one direction. At Tenzing-Hillary Airport in Nepal, for example, aircraft can only land on runway 06 and take off from runway 24 owing to high terrain immediately behind the eastern perimeter of the airport.

Although single runway airports are common and generally easy to operate, there are a couple of notable exceptions. At Gibraltar Airport, the four-lane highway runs at 90° across the runway and road traffic has to be halted to allow aircraft to operate, while the runway at Gisborne Airport in New Zealand is bisected by a railway line and train drivers have to obtain permission from air traffic control before crossing it.

Runway configuration: the siting, number, orientation and layout of runways at an airport.

Theoretical runway capacity: the maximum number of aircraft movements that can be safely accommodated in a defined time period under specified operating conditions.

Aircraft movement: a single take-off or landing; 40 air traffic movements (ATMs) per hour means 40 aircraft land or take-off in 60 minutes.

CASE STUDY 4.1

SINGLE RUNWAY, LONDON CITY AIRPORT, UK

London City Airport, which opened in 1987, is situated in the Royal Docks, 11 km east of the City of London and just to the east of the financial district of Canary Wharf. The airport has a single 1,508 m (4,948 ft) long east-west runway (09/27) that was built on land between the Royal Albert and King George V Docks. In 2018, 4.8 million passengers were handled, making it the 14th busiest UK airport and the 5th busiest in the London area (after Heathrow, Gatwick, Stansted and Luton). Only aircraft and aircrew who are certified to fly the steep 5.5° approach are permitted. The largest aircraft that can operate from LCY is a specially modified A318.

Parallel runways

A parallel runway system consists of two or more operational runways aligned on the same magnetic heading. The **runway thresholds** of parallel runways may be located next to one another (so-called in-line runways) or offset to decrease taxi distance to/from the terminal building(s). See Case Study 4.2. Airports may also possess two or more pairs of parallel runways on different alignments. Parallel runways are further described as being close, intermediate or far depending on their distance apart.

Visual Flight Rules (VFR): under VFR, pilots navigate visually with reference to the ground. VFR is permitted only during daylight hours; pilots have to be able to see the ground and weather conditions (particularly visibility and cloud cover) have to be good.

Instrument Flight Rules (IFR): under IFR, pilots fly using flightdeck instruments. IFR permits operations during the hours of darkness and low visibility as navigation functions are automatically performed by computers.

Practical capacity: describes the number of aircraft movements that can be handled at a certain level or standard of service.

Runway threshold: Markings on a runway which denote the start and end of the safe area for landing and take-off under normal operating conditions.

- *Close parallels*: less than 2,500 ft (760 m) between centrelines. If the centreline spacing is too close, the runways cannot be operated independently and the airport effectively becomes a single runway system. Where space constraints dictate parallel runways are too close to one another to permit simultaneous operations, the runways might be staggered to enable independent operations. The parallel runways 23L/05R and 23R/05L at Manchester Airport, UK, for example, are only 1,280 ft (390 m) apart but are staggered by 6,070 ft (1,850 m) to permit landings and take-offs to be performed simultaneously from each runway. There are, however, occasions when the runways are operated in 'dependent' mode owing to weather conditions or an aircraft emergency. This means a departing aircraft has to be rolling when an aircraft is cleared to land within 2 miles.

- *Intermediate parallels*: 2,500–4,300 ft (760–1,300 m) between centrelines.

- *Far parallels*: over 4,300 ft (1,300 m) between centrelines.

The capacity of a parallel runway system depends on the total number and relative siting of the runways (ideally, they need to be positioned to minimise or eliminate runway crossings) and the centreline spacing between them. Whether or not parallel runways can support simultaneous arrivals and departures depends on their position, centreline spacing and local operating restrictions. In segregated mode, one of the parallel runways is used for arrivals while the other handles departures. In mixed mode operations, both runways can be used by arriving and departing aircraft. The capacity of parallel runway airports varies from 50 to 125 aircraft movements per hour under IFR, depending on local weather conditions, traffic mix, airspace constraints and local operating procedures.

CASE STUDY 4.2

PARALLEL RUNWAYS, ISTANBUL AIRPORT, TURKEY

The first phase of Istanbul's new airport opened in April 2019 on a 76.5 sq km greenfield site approximately 35 km north of the city centre. Initially, the airport opened with two pairs of parallel runways (34L/16R, 34R/16L and 35L/17R, 35R/17L). Once complete in 2025, the airport will have six runways, four terminals and an annual capacity of up to 200 million passengers.

Intersecting runways

An airfield with intersecting runways has two, or more, runways on different alignments that cross one another somewhere along their length. Intersecting runways are required where relatively strong winds frequently blow from more than one direction and would result in excessive **crosswinds** and a low usability factor if only one runway was provided.

Normally, the **primary runway** is longer than any secondary or crosswind runway(s), and it is used in preference to other runways when local weather conditions permit. If the wind strength, wind direction and local operating regulations allow, intersecting runways can be used simultaneously. While this offers a high degree of usability irrespective of wind direction, the interdependent nature of such operations requires careful control and coordination by ATC and acute situational awareness from flightcrew and does not always confer increased capacity (Case Study 4.3).

Crosswind: a wind which blows at an angle across the runway rather than along its length. Crosswinds can be dangerous, and aircraft are not allowed to operate if crosswinds exceed defined safety parameters.

CASE STUDY 4.3

WHERE TWO RUNWAYS ARE NOT NECESSARILY BETTER THAN ONE: CHHATRAPATI SHIVAJI MAHARAJ INTERNATIONAL AIRPORT, MUMBAI

Mumbai, India's second busiest passenger airport, handled 49.8 million passengers in 2018. It has two intersecting runways – 09/27 (the longer of the two) and 14/32. Between 2009 and 2013, both runways were used simultaneously to try to increase capacity. However, despite anticipating 45–48 ATMs per hour during dual runway operations, the airport rarely achieved more than 40 owing to the need to separate aircraft for safety reasons. In 2013, the airport switched to using 09/27 as the primary runway and used 14/32 only when 09/27 was unavailable. The airport now obtains an average of 42–46 ATMs in single runway mode.

The capacity of an intersecting runway system depends on the location of the intersection and whether it is near the take-off/landing threshold, near the midpoint or at the far end. The highest capacity is realised when the intersection is close to the take-off and landing threshold of the two runways. The further the intersection is from the threshold, the lower the capacity, as the risk of collision is increased and greater separation between aircraft is required. Depending on the location of the intersection, the capacity of an intersecting runway system can be as high as 90 aircraft movements per hour under VFR but some airports have found higher capacity is realised by operating as a single runway system.

Primary runway: the runway that is used in preference to all others.

Open-V runways

The fourth type of runway configuration is an open-V system. These consist of two runways which are aligned in different directions but which do not intersect at any point along their length. When winds are light, Open-V runways can be used simultaneously. The highest capacity (50–80 ATMs per hour) is realised when air traffic operations are divergent (aircraft operate away from the apex of the V). When operations are convergent (they are towards the apex of the V), capacity is reduced to 50–60 ATMs per hour for safety reasons owing to the increased risk of collision (see Figure 4.2). Open-V runways occupy a large land area and are typically found only where a lot of land is available for airport development (see Case Study 4.4).

CASE STUDY 4.4

OPEN-V RUNWAYS: JACKSONVILLE INTERNATIONAL AIRPORT, FLORIDA, US

Jacksonville International Airport is located in Duval County in northeast Florida. Served by a mix of low cost and full service airline operators, it handled 6.46 million passengers in 2018. The airport covers a site 7,911 acres (over 32 km²) in extent and has two runways – 08/26 and 14/32 – in an open-V configuration. In comparison, London Heathrow Airport, with two parallel runways and 80.1 million passengers in 2019, occupies a site of approximately 3,032 acres (12.3 km²).

Although there are four basic runway configurations, many airports exhibit a combination of these designs, and the configuration and operation of runways at individual sites are often a compromise between local topography, available land for development, local obstructions, airspace constraints, environmental and social considerations and optimal usability. As of 2019, the commercial airport with the highest number of operational runways in the world is Chicago O'Hare Airport, US, with eight (with a mix of parallel, open-V and intersecting runways).

Usability factor: the percentage of time that normal runway operations are not restricted by excessive crosswinds. A figure of 95 per cent or more is usually required.

Runway configuration and usability

While many physical, geographic and financial factors influence the provision, orientation, siting and configuration of runways, one of the most important considerations from a management perspective is the runway **usability factor**. An airport must ideally be capable of handling the aircraft it is intended to serve at least 95 per cent of the time, irrespective of local weather conditions. In the case of a single runway airport, if the usability factor falls below 95 per cent, it may be necessary to construct a crosswind runway to raise the usability factor.

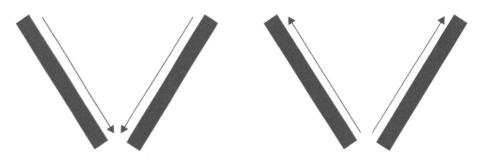

Figure 4.2   Convergent and divergent operations on open-V runways

Stop and think

With the aid of diagrams, identify the four basic types of runway configuration, and discuss their relative merits from a capacity and operational perspective.

4.4 Other factors that affect runway capacity

In addition to runway orientation and configuration, other factors can affect runway capacity. These include:

- the provision of ATC services and navigation aids ('navaids') such as instrument landing systems (ILSs) and high-intensity runway lights;

- aircraft performance characteristics;

- local meteorological (weather) conditions;

- environmental and social considerations and operating restrictions;

- flightcrew behaviour (pilots who are more experienced and familiar with an airfield may vacate a runway faster on landing, leading to reduced runway occupancy times);

- runway/taxiway/landside interface design factors.

The availability of qualified ATC personnel and the provision, type and sophistication of **navaids** and **ILS** (Instrument Landing System) for precision instrument approaches also affect runway capacity (Table 4.1). High-intensity runway lights and ILSs enable aircraft to continue to operate during hours of darkness and poor visibility and enhance the usability of an airport. Such systems are, however, expensive to install, operate and maintain, and airport managers must assess the relative benefits of providing them against the cost of their installation and operation.

Navaids: visual markers or electronic systems that provide navigation data.

ILS: A ground-based instrument approach system that provides lateral and vertical guidance to landing aircraft.

Table 4.1 ICAO/EASA ILS categories

ILS category	Decision height	Runway Visual Range (RVR)
I	Not lower than 200 ft	Not less than 550 m
II	Lower than 200 ft but above 100 ft	Not less than 300 m
IIIa	Lower than 100 ft	Not less than 200 m
IIIb	Lower than 50 ft or no decision height	Lower than 200 m but not less than 75 m
IIIc	No decision height	0 m

Note: IIIc is not used in Europe, as aircraft would need to be towed clear of the runway on arrival.

The performance characteristics of the aircraft that use an airport also affect runway capacity. The size, weight, speed, manoeuvrability, acoustic footprint, climb performance and braking capability of the aircraft, combined with the aeronautical skill and experience of the flightcrew and their familiarity with the airfield, all have the potential to enhance or degrade runway capacity. Individual aircraft types exhibit different approach and landing speeds and, on take-off, have different climb characteristics. Typically, larger and heavier aircraft require longer (both in terms of time and distance) **take-off rolls** and longer **landing rollouts** than smaller and lighter aircraft and, as a consequence, runway occupancy times for larger aircraft tend to be greater, leading to lower hourly capacities as fewer aircraft movements can be accommodated.

Larger and heavier aircraft also generate dangerous wake vortices that have to dissipate before the next aircraft can land or take off. All commercial aircraft are classified into one of five wake turbulence categories according to their weight and the wake vortices they generate. Strict regulations specify the length of time and/or distance that must be left between movements by different categories of aircraft. 'Heavy' aircraft include wide-body jets, such as the B747 (the A380 often has its own classification); the 'medium' category (which is subdivided into 'upper medium' and 'lower medium') includes A320s and B737s; 'small' aircraft include Dash 8 turboprops and commuter aircraft; while the 'light' category describes general aviation aircraft. The temporal and longitudinal (distance) based separation that must be observed in the UK between arriving and departing aircraft operating from the same runway is provided in Tables 4.2 and 4.3. The implications of aircraft mix for runway separation minima and runway capacity are significant.

An airport handling a mix of heavy, medium and light aircraft on a single runway may not be as efficient as one handling all the same category of aircraft. Runway occupancy time (and hence runway capacity) is also affected by the mix of departing and arriving traffic that is operating from the same runway and the proportion of fixed-wing and rotary traffic the airport handles.

Take-off roll: the period of time or the runway distance used between the aircraft's brakes being released and the aircraft becoming airborne.

Landing rollout: the period of time or the runway distance used between the aircraft's wheels touching the ground and the aircraft exiting the runway.

Table 4.2 Wake turbulence separation minima distances (in nautical miles) – aircraft arriving

		Following aircraft					
		A380	Heavy	Upper medium	Lower medium	Small	Light
Leading aircraft	A380	#	6	7	7	7	8
	Heavy	#	4	5	5	6	7
	Upper medium	#	#	3	4	4	6
	Lower medium	#	#	#	#	3	5
	Small	#	#	#	#	3	4
	Light	#	#	#	#	#	#

Source: Derived from CAA (2017) p. 13

= no wake turbulence separation minima required

Table 4.3 Wake turbulence separation minima (in minutes) – aircraft departing from the same runway or from a parallel runway that is less than 2,500 ft (760 m) away

		Following aircraft					
		A380	Heavy	Upper medium	Lower medium	Small	Light
Leading aircraft	A380	#	2	3	3	3	3
	Heavy	#	4 nautical miles or time equivalent	2	2	2	2
	Upper medium	#	#	#	#	#	2
	Lower medium	#	#	#	#	#	2

Source: Derived from CAA (2017, p. 14)

= no wake turbulence separation minima required

Stop and think

Why are wake vortices potentially dangerous, and why does aircraft size affect runway capacity?

Local meteorological conditions also have the potential to degrade runway capacity. Strong or gusting (cross)winds, poor visibility and low clouds can all reduce runway capacity by necessitating the implementation of **low-visibility procedures** (LVPs) and increased separation between aircraft. Rain, sleet and snow can affect the runway surface and its performance, leading to loss of friction and reduced braking action, which may increase landing rollouts and runway occupancy times. High surface temperatures lower the ability of an aircraft's wings to generate lift and require aircraft to have longer take-off rolls and/or use higher thrust settings or reduce payloads (which impacts on an airline's revenues). Strong headwinds can also reduce the landing rate and, in 2015, London Heathrow introduced an enhanced time-based separation system to improve runway capacity and reduce flight delays (Case Study 4.5).

Low visibility procedures: Enacted when visibility at an airfield falls below defined minima. It includes a range of procedural and operational changes to ensure safety such as reducing the landing and take-off rate, increasing separation between aircraft and increasing the intensity of aeronautical ground lighting.

Stop and think

Other than crosswinds and poor visibility, which other weather phenomena might affect runway capacity, and why?

ENHANCED TIME BASED SEPARATION AT LONDON HEATHROW AIRPORT (LHR)

In March 2015, LHR became the first airport in the world to replace distance based separation (DBS) with time based separation (TBS) for arriving flights. Under TBS, dynamic time based separation is used to sequence flights during periods of strong headwinds (over 20 knots on final approach). This decreases the distance between arriving aircraft but still ensures wake vortices can dissipate safely. Under TBS, LHR could recover a landing rate of 36–40 aircraft per hour. In Spring 2018 LHR introduced Enhanced Time Based Separation (ETBS) using the EU's RECAT aircraft wake turbulence categories. RECAT is more precise and more efficient at the heavier end of the traffic mix, which includes the majority of Heathrow traffic. ETBS enhances capacity and improves resilience during strong headwinds, and although it cannot eliminate the effects of wind, it enables LHR to increase its landing rate by 2.2 aircraft an hour in all wind conditions and by 4.2 aircraft an hour during strong headwinds. This is equivalent to extending Heathrow's operating period by almost 30 minutes a day.

Local environmental and social considerations may also affect an airport's runway capacity (➤Chapter 18). Environmental impacts of airport operations in terms of air quality, surface water drainage and water treatment and noise, together with the measures introduced to minimise or mitigate their impacts on local residents, are important. Noise abatement procedures and noise preferential routes (NPRs) that vector aircraft away from densely populated urban areas to reduce acoustic impact on the ground, night noise quotas that restrict the use of certain aircraft types or reduce the number of aircraft movements permitted at night, and other local operating restrictions may limit the use of one or more runways during particular hours of the day or certain days of the week. Wildlife activity on and around an airfield can also affect runway capacity, while growing incidents of unauthorised drone/UAV incursions into controlled proximate airspace have the potential to temporarily ground flights (➤Chapter 14).

4.5 Runway demand management

Many airports are operating close to their design capacity and have little flexibility to accommodate non-routine events or recover normal operations following disruption (➤Chapter 11). Where the demand for flights routinely exceeds the available supply (runway capacity), a system of **slot allocation** can be used to manage demand. Slot allocation is a complex topic which is detailed in Section 6.3 in Chapter 6.

4.6 Runway development

Growing consumer demand for air travel is driving the development of new and extended runways and airport infrastructure. This may require the acquisition of additional land beyond the existing boundary, involve substantial civil engineering works to overcome site limitations (Case Study 4.6), necessitate environmental and social mitigation measures and require substantial capital investment. Any runway development is expensive and the relative risk and rewards for the airport, its customers, surrounding communities, the local environment and the regional economy must be carefully considered.

Stop and think

Why do many airports want to extend or build new runways, and what are the risks associated with doing so?

!

CASE STUDY 4.6

CHANGI EAST DEVELOPMENT PROJECT, SINGAPORE

The Changi East Development project involves lengthening the airport's third runway (which used to be reserved solely for military operations) from 2.75 km to 4 km to permit commercial operations and constructing a fifth terminal. The development site is situated on reclaimed land made of soft marine clay – the runway extension works, which commenced in 2017, required a new drainage system and four new canals (the largest of which is 40 m wide) to drain water from the site.

4.7 Taxiway design

Irrespective of the siting, orientation and number of available runways, if adequate connecting taxiways are not provided, an airport cannot operate safely and/or efficiently. The principal function of a taxiway is to facilitate the safe and efficient surface movement of aircraft to or from the runway(s) and between terminals, maintenance areas and other parts of the airfield.

Taxiways need to connect all parts of the airfield that are used by aircraft and must be designed so as to:

- provide adequate separation between aircraft;
- not endanger, interfere with or delay arrivals, departures or other taxiway operations;
- provide the shortest practicable route between the terminal and the active runway(s) to minimise taxi distances;
- offer a sufficient number of runway entry/exit points, including high-speed turn-offs to enable landing aircraft to vacate the runway as quickly as possible;
- minimise or mitigate local environmental or social impacts.

Runways and taxiways should be located so as to provide safe separation between flying and taxiing aircraft and to minimise delays to landing, taxiing and take-offs. Taxiways need to be sufficiently wide and strong enough to accommodate the largest aircraft that are likely to use the airport, and their centrelines have to be sufficiently far apart to allow for simultaneous operation and ensure adequate wing-tip clearances. In addition, the radius for a taxiway needs to be such that it can accommodate the largest aircraft likely to use the airport and facilitate aircraft manoeuvring; otherwise aircraft may not be able to make a turn safely with the potential for conflict with other aircraft, airfield facilities or airfield equipment or for aircraft becoming stuck in the neighbouring soft ground. Commercial aircraft are assigned a classification code for taxiways and apron stands ranging from Code A (the smallest) to Code F (the largest) according to the span of the aircraft's wing and outer main gear wheels. Taxiways need to be designed to accommodate the largest aircraft that are likely to use the airport and, at some airports, Code F aircraft can use only a limited number of taxiways and stands.

Runway excursion: When an aircraft leaves the runway anywhere other than on a designated taxiway.

Wherever possible, it is desirable to build a taxiway that is parallel to the runway, but the danger of lateral (sideways) **runway excursion** needs to be considered. Where the airfield layout is severely constrained, it may be necessary to build a turning circle at one or both ends of the runway to allow aircraft to turn through 180° in preparation for take-off or to backtrack down the runway after landing. This practice, however, increases runway occupancy times and lowers runway capacity. Examples of runways with turning circles include London Luton and the airports serving the Greek islands of Kos and Skiathos.

!

Stop and think

Why are turning circles required at some airports, and what are their operational impacts?

The provision of appropriately positioned exit taxiways and RETs allows for higher runway utilisation and increased capacity. Right-angled 90° exit turn-offs require aircraft to decelerate to a low speed before vacating the runway, whereas shallower angles of 30° allow aircraft to exit at higher speeds, thereby decreasing runway occupancy time and improving runway capacity. The provision and location of exit taxiways depend on the mix of aircraft that use the airport, their relative approach speeds, the point of touchdown, their deceleration rate and the normal condition of the runway surface (i.e. whether it is usually wet or dry). Even when accounting for these factors, it is important to recognise that individual airlines have different standard operating procedures (SOPs) and that flightcrew will operate their aircraft according to local conditions. Consequently, the provision of a rapid exit taxiway does not oblige flightcrew to use it, and the configuration of some airfields may prevent their retrospective installation.

Environmental and social impacts also need to be considered in terms of taxiway operations, particularly where they are close to an airport boundary or local housing. A noise bund may be required to provide a physical barrier between the taxiway and local houses. A taxiway's operations can also be restricted during parts of the day/night, or single-engine taxi operations could be promoted as a way of reducing ground noise, improving air quality and lessening odour.

4.8 Aprons and the landside interface

So far, this chapter has concentrated on the design, configuration and management of runways and taxiways, but it is important that these components are not considered in isolation, as their interface with passenger and freight terminals and landside areas affect an airport's capacity. Wherever possible, apron areas and terminal buildings should be located adjacent to the principal runway and positioned in such a way that landing and departing aircraft do not pass directly overhead.

The apron is sometimes referred to as the 'ramp', and it is here that individual aircraft stands interface with terminal buildings and where aircraft are turned around between flights. Ramp areas should be designed so as to allow for the safe and efficient entry and exit of aircraft, and they should be free from obstacles or other restrictions but should allow easy access for service vehicles (including baggage handling equipment, fuel bowsers and in-flight catering vehicles). Some stands are configured to accommodate particular aircraft types (based on the aircraft codes referenced earlier), while others may be flexible and able to accommodate different-sized aircraft according to demand. Multi Aircraft Ramp Systems (MARS) allow airports to make their gates – and, by association, aircraft turnarounds – more flexible and efficient. MARS stands can be configured according to the type of aircraft that needs to use them, and this configuration can be changed throughout the day as the nature of the traffic demand changes. This system means that a MARS stand recently vacated by a single A380 could be subsequently filled by two smaller aircraft. The ability to adapt apron infrastructure in this way permits airports to maximise the efficiency of their stands while providing the flexibility to serve different types of aircraft.

> **Stop and think** !
>
> What are the advantages and disadvantages of MARS stands?

Again, there are environmental and societal impact issues to consider with an apron area and its aircraft stands, particularly in terms of noise and air quality. Fixed electrical ground power (FEGP) units are increasingly used as a quieter and cleaner alternative to aircraft auxiliary power units.

4.9 Airfield management

The airfield environment is dangerous, and health and safety is a critical issue. Passengers and airport employees can be injured or killed by jet blast and prop wash (the exhaust from aircraft engines), be crushed or harmed while using machinery and servicing aircraft, and slip, trip or fall. Airfields need to be regularly patrolled to ensure the perimeter is secure and to scare wildlife away from the active runways and aircraft manoeuvring areas. Airfield operations are also responsible for ensuring the structural and operational integrity of the airfield and must be equipped to respond to any eventuality, from removing foreign object debris (FOD) to escorting maintenance contractors out onto the airfield to repair damaged assets. Airfield operations teams must also compile detailed datasets to monitor defects and identify hazards.

It is imperative that all airside manoeuvring areas are monitored and essential maintenance is performed. Over time, airfield pavement surfaces will degrade and need to be resurfaced. Typically, this is done at night or during off-peak periods when the impacts are less severe. All routine maintenance work should be planned well in advance and communicated to the airlines who will be affected. Specialised contractors will need to be employed to undertake the work and any adverse impacts on airport operations or revenues need to be carefully managed.

4.10 Aerodrome safeguarding

Aerodrome safeguarding: active control of land use to protect surrounding land and airspace from developments which could affect the safe operation of an airport.

Aerodrome safeguarding seeks to protect the safety of aircraft and their occupants during take-off and landing and when manoeuvring within, and immediately around, an airfield. It involves the active control of local land use to protect proximate land and airspace from any development which could adversely affect the safety of aircraft operations.

Aerodrome safeguarding typically requires protection of:

- the airport's proximate airspace, which is achieved through establishing obstacle limitation surfaces (areas of sky that cannot be penetrated by tall buildings, trees or temporary construction equipment);

- radar installations and navigation aids by preventing the construction of buildings, wind turbines or other developments which could cause reflections, diffractions and distortions to radio signals;

- visual navigation aids and lighting by ensuring that residential street lights, advertising hoardings and floodlights do not obscure navaids or risk dazzling or confusing flightcrew;

- the airfield from wildlife hazards. All land uses that attract wildlife, including nature reserves, reservoirs, farmland and landfill sites, within a defined radius of an airport have to be actively monitored and managed to reduce the risk of wildlife strikes.

! Stop and think

What is aerodrome safeguarding, and why is it required?

Key points

- An airfield comprises all the airside aircraft manoeuvring areas and proximate land.

- The siting, orientation and physical characteristics of runways affect the capacity, operational efficiency and safety of an airport.

- There are four basic types of runway configuration, although some airports exhibit a combination of different types.

- Runways should be designed for maximum operational usability and be capable of handling the volume and type of air traffic that is using the facility now and in the future.

- Runway capacity is affected by the physical characteristics of the runway and the surrounding land and airspace as well as the nature of demand, the mix of air traffic, weather conditions and local operating restrictions.

- Taxiways and aprons must be designed to allow for the safe and efficient surface movement of aircraft around the airfield.

- Runway development is expensive and often contentious. Airport operators must ensure that local environmental and social impacts are minimised and mitigated as far as possible.

- Aerodrome safeguarding protects proximate air and land space from developments which could endanger aircraft.

References and further reading

CAA. (2017). *CAP 493 manual of air traffic services part 1*. 7th edn. London: Civil Aviation Authority.

CAA. (2019). *CAP 168 licensing of aerodromes*. 11th edn. London: Civil Aviation Authority.

EASA. (2017). *Certification specifications and guidance material for aerodromes design CS-ADR-DSN*. December 2017. Available at: www.easa.europa.eu

FAA. (2012). *150/5300–13A-airport design*. Available at: www.faa.gov/airports

Kazda, A. and Caves, R. (2015). *Airport design and operation*. 3rd edn. Bingley, UK: Emerald.

CHAPTER 5

Airport systems planning, design, and management

Richard de Neufville

LEARNING OBJECTIVES

- To understand that airport forecasts are 'always' wrong. Continuing volatility in the aviation industry means that what actually occurs differs, often substantially, from original predictions.

- To appreciate the need for flexibility in design, as a way of de-risking investments and future-proofing developments by enabling easy adaptation to future situations.

- To recognise the need for a comprehensive systems view of airports and the competition between them.

- To acknowledge that measures of capacity at airports are not absolute but depend on the nature of operations and on differing judgements concerning acceptable levels of service.

- To realise that the inherent instability of queues under peak loads can drastically reduce the acceptable maximum capacity of operations.

5.0 Introduction

Airport management: the range of activities needed to run airports effectively. These range from planning future facilities to implementing projects and supervising ongoing operations.

Airport management (➤Chapter 6) details the particular features that characterise airport behaviour. This chapter presents and explains the essential elements and consequences of these features.

All those concerned with planning, designing, and managing airports need to consider the following five points:

1 **Forecasts are 'always' wrong.** Experience shows that it is not possible to consistently anticipate future levels or types of traffic accurately. Air transport demand is highly volatile in the short run and continually in the midst of major changes.

2 **Flexibility in airport planning and design is essential.** Changes in the types of traffic and modes of operation constantly modify the requirements and performance of airport facilities. To be effective, airport managers need to be able to easily adjust the capacity and capability of their facilities to new conditions.

3 **Any airport is part of a competitive air transport system.** It faces other airports and airlines that compete vigorously. This means that airport managers need to be attentive to this competition and how it can affect the performance of their own facility.

4 **Measures of airport 'capacity' can be very misleading.** Measures of airport 'capacity' are conceptually challenging. They depend on specific assumptions about appropriate level of service and mode of operations. This means that the same physical facility can lead to different estimates of capacity. This can confuse managers and others trying to provide adequate capacity for airport facilities.

5 **Queues are at the centre of airport operations.** Passengers wait to check in, to pass security, to board aircraft; aircraft queue up to land, to move to a gate, to take off. Queues inherently reduce capacity. Managers need to carefully understand the ways in which queues behave to do their job effectively.

Experience worldwide indicates that these points are frequently overlooked and generally not understood. Failure to deal adequately with them is a continuing source of difficulty in airport systems planning, design, and management.

5.1 Forecasts are 'always' wrong

Forecast: a prediction, based upon past data and current best guesses as to future trends, of what might happen over the planning period. However calculated, forecasts are still guesses of future trends and are often disrupted by 'trend-breakers'.

The central fact is that **forecasts** of airport activities are almost certainly wrong. Again and again, the reality of what actually happens at an airport at some future time differs significantly from what forecasters anticipated. Study after study demonstrates the phenomenon by comparing forecasts with eventual results. ACRP (2012), and de Neufville and Odoni (2013) cite many such examples, such as those provided in this chapter. This observation can be verified by looking up airport forecasts and comparing these predictions with what actually happened in the forecast year.

Eventual actual outcomes deviate, often over a large range, from original predictions. The reality can be higher or lower than the forecast. It can even sometimes – rarely, given the possibilities – coincide closely with the forecast! In general, we must anticipate that future reality will differ greatly from original, inherently speculative, forecasts.

The extent to which actual future situations deviate from forecasts depends principally on three factors: the length of the forecast period, the level of detail, and the degree of stability of the issue. Thus:

- **The longer the forecast, the greater the possible deviations.** Patterns of activity are normally tied most closely to current habits; the farther we are from the present, the greater the chance that new patterns will have formed. This feature is particularly crucial since the most important airport investments – in runways and terminals – are long-term, capital-intensive projects designed to service traffic many decades into the future. Over-reliance on long-term forecasts has led to many poor investments when actual levels and types of traffic differed substantially from expectations.

- **The greater the level of detail, the greater the variability of the forecast.** As a general rule, overall forecasts of activity are normally less variable than their components. This is because gains in one component tend to balance out losses in other components. For example, overall leisure air travel from Europe is less likely to deviate substantially from a trend than travel to particular destinations. The demand for travel to the Canary Islands, Egypt, Greece, Spain, or Tunisia, for example, depends upon the vagaries of fashion, political unrest, terrorist attacks, and currency exchange rates. Holidaymakers can and do easily shift from one destination to another.

- **Some patterns have limited volatility.** For example, airlines do not change their aircraft fleets easily. They invest heavily in both physical assets and the training of pilots and other operational personnel. Likewise, the travel pattern of summer holidays in Europe is firmly fixed in labour contracts. Note, however, that airlines often ground aircraft or reduce route frequencies when customer demand decreases. Providers and customers of air transport can and do adjust their level of activity according to circumstances.

Overall, many factors influence the degree of uncertainty in airport forecasts. Importantly, the 10- to 20-year forecasts that guide major investments are likely to be highly unreliable.

A conservative estimate of forecast error is that actual traffic at an airport often deviates ±10 per cent from the forecast after only five years. This figure compounds over time, so that the deviation might be ±20 per cent after ten years, for example. This is the experience in the US, a well-established market that has little room for the rapid changes and greater uncertainties possible in developing markets. This figure derives from comparisons of the annual Federal Aviation Administration (FAA) five-year forecasts of airport traffic with the subsequent actual results.

Plus or minus 20 per cent deviations have significant implications for planning airport investments. This is because airports typically plan investments to cover prospective gaps between what they have and what they expect to need. The uncertainties in the forecasts of total traffic thus translate into much greater uncertainties in the estimates of the gaps (see Example 5.1).

Example 5.1
Forecast errors lead to greater planning errors

Consider an airport whose traffic might double over the planning period; for example, from 10 to 20 million passengers. If the passenger forecast is wrong by ±20 per cent, the total traffic might range from 16 million to 24 million passengers, and the planning gap – the difference between available and needed capacity – would range from 6 million to 14 million passengers. The point is that normal uncertainties in traffic forecasts can readily lead airport managers to face situations in which the range of their prospective needs differs by a factor of two.

Airport forecasts can be spectacularly wrong. See Case Study 5.1 and the following examples:

- Traffic forecasts for Dubai International Airport in 1984, when it had only one runway and Emirates airline did not exist, did not anticipate that 30 years later the airport would be in the top three worldwide in terms of passenger traffic and the busiest international airport.

- Conversely, 1985 forecasts anticipated that Boston Logan International Airport would soon double its passenger traffic, but 15 years later, it had barely risen above 25 million annual passengers. Meanwhile, against expectation, the number of annual aircraft operations decreased by 30 per cent, from over 500,000 to under 350,000. Now, 20 years later, Boston serves over 41 million, or 50 per cent more passengers – but the annual number of aircraft operations has just grown 10 per cent, to above 400,000 – still 25 per cent less than 35 years ago!

- St Louis (US) and Zürich airports are among several that saw their traffic collapse when their home airlines (TWA and Swissair, respectively) ceased operations. At St Louis, the total passenger traffic dropped 60 per cent, from over 30 million to around 12.6 million passengers – shortly after the airport had invested in a new runway designed to cope with traffic growth to over 40 million passengers.

INITIAL FORECAST FOR LONDON STANSTED AIRPORT

The forecasts for the development of Stansted Airport predicted its traffic would grow rapidly as it would absorb overflow traffic from Heathrow. (When Stansted opened in 1991, Heathrow served just over 40 million passengers a year.) The forecast for about 20 million annual passengers led to a significant initial investment featuring mid-field aircraft piers connected to the passenger terminal with a 'people-mover train' called the Track Transit System.

The forecast was doubly wrong, both as to the passenger growth and to the service needed. Stansted traffic did not reach 4 million passengers in its first five years. Neither passengers nor airlines wanted to move from Heathrow, whose traffic continued to grow steadily (it now serves over 80 million annual passengers).

Stansted traffic did eventually exceed 20 million passengers, but it was of a different type. Stansted grew because it became a base for Ryanair, a low cost carrier, which required completely different facilities from those initially built. Ryanair refused to use the expensive mid-field piers and their train, preferring an inexpensive ground-level pier accessed on foot.

In short, the forecast was wrong as to the timing and type of traffic. This led to a vastly premature and thus uneconomical investments in the wrong kind of facilities. Even when facilities were built specifically for the low cost carrier's requirements, they later became outdated when the airline changed its approach to passenger service level requirements.

Trend-breakers as a cause of bad forecasts

'Trend-breakers' cause most failures in airport forecasting. They disrupt established patterns of service, of demand, and of modes of operation. This phenomenon most immediately impacts airlines; changes in the airline industry then affect airport operations.

Many trend-breakers in the airline/airport industry stem from the economic deregulation of airlines (➤Chapters 2 and 3). Deregulation has led to the development of low cost carriers (LCCs), bankruptcies of many airlines, and the diffusion of airline connecting hubs. This continues to spread internationally and disrupt and transform patterns of use, of demand, and of requirements in the airport/airline industry.

Deregulation has changed air transport and airport systems planning and design. Before deregulation, government ownership and regulation encased the industry in rigid rules set by national laws and international conventions. Change required layers of regulatory approval and occurred slowly, if at all. Airlines were monopolies and enjoyed high fares. Although aircraft technology advanced, innovations in services, routes, pricing, and marketing were rare. Deregulation and liberalisation changed all that. Airlines now decide on fares and routes, and this generates new services. The resulting competition drives inefficient airlines into bankruptcy or consolidation with larger airlines, while more efficient airlines demand different airport services and different facilities (see Case Study 5.1). These waves of change resonate and reinforce each other as they spread across the world.

Technological development is also a trend-breaker (➤Chapter 16). Aircraft now routinely fly over 9000 miles (14,400 km). This has disrupted traditional routes, led to the creation of airport hubs in the Gulf, and downgraded the importance of traditional transoceanic gateways such as London.

Consequences of trend-breakers

Highlights of trend-breakers and their impacts include:

- **Rise of LCCs.** These airlines have set the standard of operations and are coming to dominate their markets. Southwest Airlines has become the leading carrier, by far, of domestic passengers in the US. Ryanair and easyJet have been establishing similar dominance in Europe. AirAsia provides an example of how this may occur in Asia. These new airlines have spurred remarkable growth in traffic worldwide (➤Chapter 8).

- **New airports and modes of operation.** LCCs demand low airport costs (➤Chapters 6 and 7). They turn their aircraft around quickly and so require fewer gates. They favour simpler terminals and cheaper secondary airports, such as Miami Fort Lauderdale, London Stansted and Bangkok Don Mueang. These pressures have disrupted traditional concepts of airport planning.

- **Consolidation of airlines.** Competition from efficient LCCs has driven some traditional airlines into bankruptcy and/or consolidation with others. Within 30 years, the traditional US airlines consolidated into three major operators: American, Delta and United. Continental, Northwest and US Airways have disappeared, along with Eastern, Pan American, and TWA. Independent national airlines are also disappearing in Europe: Air France merged with KLM; IAG consolidated British Airways, Iberia, and Aer Lingus; and Lufthansa took control of Austrian, SWISS and the remains of Sabena. These changes relocate flight patterns and airport needs. For example, after BA merged with Iberia, it rerouted much of its traffic from Madrid to London Heathrow.

- **Hub-and-spoke operations.** The development of transfer hubs is a result of the freedom of airlines to establish new routes easily. Hub operations provide more connections to passengers and facilitate the management of aircraft and crews. This development also reallocates traffic from some airports and airlines to others. Thus, Emirates is taking transfer traffic to and from India, Southeast Asia, and Australia away from Singapore in Asia, and away from London Heathrow and Frankfurt in Europe.

- **New competitors.** The liberalisation of air transport has resulted in the development of integrated carriers that collect and deliver parcels door-to-door. These have become the workhorses of online shopping and delivery. UPS and FedEx are among the largest airlines in the world and have the highest market value (see Table 5.1). These carriers are transforming the air transport industry as they require new types of airports and airport facilities. For a more detailed discussion of air cargo, see Chapter 17.

Table 5.1 Market value of selected airlines, July 2019

Innovative	US$ billion	Low cost	US$ billion	Traditional	US$ billion
UPS	71	Southwest	29	IAG (BA, Iberia, Aer Lingus)	12
FedEx	43	Ryanair	15	Singapore	9
Emirates	17	easyJet	5	Lufthansa	8

Source: airline corporate websites, 17 July 2019

Further trend-breakers supplement the continuing current of changes resulting from deregulation of the airline industry. These relate to the expansion of free-trade areas, such as the North American Free Trade Association (NAFTA), the European Union (EU), the Association of Southeast Asian Nations (ASEAN) and globalisation generally. The airport/airline industry is in the midst of continued market and operational volatility, and this is likely to continue for decades.

Stop and think

Why are demand forecasts invariably wrong, and what are the consequences of incorrect forecasts?

5.2 Flexibility is essential

The overall consequence of the forecasts 'always' being wrong is that it is impossible to know exactly what to plan for. This is very challenging: What do you do if you don't know what to expect?

The reality that it is impossible to know future conditions accurately also runs against standard practice. Indeed, Step One of standard guidance is: 'predict future conditions'. How do you proceed once you understand that the results from this task are unlikely to be accurate?

Standard practice thus does not develop consistently good plans. Indeed, the generic Step Two is: 'find the best plan to meet the forecast'. As reality differs, often substantially, from the forecast, the standard approach to airport planning often results in a 'wonderful plan for a fairy-tale future' (see Case Study 5.1). The results of conventional planning frequently range between having excess or inappropriate capacity to needing to expand facilities under difficult conditions. Such situations are wasteful and expensive.

Flexibility in design is the main way to deal with the problem of planning for an uncertain future. Flexible designs enable airports to adjust their capabilities or capacities to deal with unexpected levels of traffic, new clients, innovative operating procedures, and changing regulatory requirements. Case after case demonstrates that flexible designs can deliver greater expected value over time.

Flexibility in design: the capability of a design to add capacity or capability easily. Designers typically implement flexibility by providing unencumbered space for growth, or strength to deal with greater loads, or open spaces amenable to low-cost rearrangement.

De-risking: a means of doing things to lower the project risks. The primary focus is often on the financial implications of undesirable events that may degrade the performance of a project.

Future-proofing: The process of anticipating future scenarios and developing concepts for dealing with the range of possibilities. It complements de-risking that focuses on tactics for dealing with more immediate uncertainties.

De-risking uses design flexibility to avoid premature commitments that risk developing facilities that will eventually be inappropriate to the actual project's needs (see Case Study 5.1). It enables managers to provide for future requirements incrementally, in line with demonstrated experience. It involves choosing the earlier steps that enable developments that might be desirable in the future. Given the impossibility of predicting future issues, developers need to think of airport planning and design in the spirit of a game of chess: they need to think ahead many periods, recognise possible opportunities and vulnerabilities, and in each period develop good solutions while protecting themselves against future undesirable development (see Example 5.2).

Future-proofing uses design flexibility to maintain the possibility of alternative developments that might eventually be desirable. In practice, it prevents the creation of obstacles to future airport developments. For example, it discourages the practice of placing chiller and heating plants next to terminals: although such placements may be immediately convenient, they compromise the future efficient expansion of the airport's terminal facilities.

Example 5.2
Flexibility in airport planning and design

Land banking is an example of long-term, generation-ahead flexible planning. The idea is to set land aside for future airport expansion, acquiring it before metropolitan growth drives up prices and engulfs airport sites. Singapore, for example, reclaimed land next to Changi Airport so that it could double its capacity, as it began to do in 2014. Australia likewise reserved a site at Badgery Creek for its Western Sydney Airport to open in 2026, and India reserved the Navi Mumbai site for the Second Mumbai airport under construction in 2019. Such land-banking provides insurance that it would be easy to develop new airport capacity if needed, while in any case being a valuable long-term investment in land for other purposes as desirable.

Swing gates at airports provide short-term flexibility for dealing with variations in daily or seasonal traffic. These gates typically serve either domestic or international passengers depending on the flight. Designers implement them using doors and corridors that can be opened or closed so as to 'swing' a gate between domestic and international traffic as needed. Canadian airports have been noticeable users of swing gates. Vancouver International Airport uses them between domestic flights, transborder traffic to the US, and other international passengers.

In practice, managers can develop flexibility in design by following a three-step procedure:

1 Recognise that they and their successors need to deal with a range of possible futures, varying not only in levels of traffic but also in the operational requirements of their clients and regulators;

2 Consider what would be needed, in terms of capacity and capabilities, to meet the range of different possible conditions; and then

3 Develop plans that meet immediate needs and could be developed to deal with future requirements without foreclosing other developments that might eventually be desirable.

A flexible design will generally feature the following characteristics:

- It will be more modest and smaller than the design that would develop the full capabilities that projected forecasts indicate. This approach reduces the losses that would occur if some of that projected capacity *proved to* be unnecessary or obsolete.

- It will thus be less expensive than a design tailored to the most likely forecast. This is for two reasons. It will be smaller at the start and thus reduce initial capital expenditures ('capex'). It will also defer the cost of added capabilities for many years, and thereby save interest and lower the present cost of these expenses.

- It will allow room to expand each of the important facilities (runways, terminal buildings, passenger lounge, and baggage areas), being careful to locate supporting facilities so that these will not impede future expansions or capabilities that might be needed (such as larger buildings or automated passenger vehicles).

- The design of interior spaces will allow for easy reconfiguration so as to enable managers to adapt them for changing requirements due to new aircraft types, changes in government regulations and emerging economic opportunities.

Stop and think

Why wouldn't it be possible to adopt flexible design in all airports?

5.3 Airports are part of a competitive air transport system

Airport managers should recognise that they operate in a highly competitive system. The performance of any airport can depend considerably on competition elsewhere in the network. Managers of airport facilities need to be alert to the way changes elsewhere in the **airport system** – both physical and organisational – may impact their performance and thus require them to react effectively. Moreover, changes may occur rapidly and require prompt attention.

To illustrate how competitive decisions can have wide repercussions through the system, consider the following examples:

- **Airline mergers.** Corporate decisions to merge with other airlines generally lead to significant changes in routes and operations. When IAG, the parent company of

Airport system: the set of airports that affect each other in providing air transport. They may compete. They may also complement each other by providing connections in a network, much as Guangzhou Airport (China) is a hub on the FedEx network based in its main hub at Memphis (US).

British Airways, acquired Iberia and effectively merged their operations, this led to a 20 per cent drop in traffic at Madrid airport (from over 50 million to 40 million annual passengers). IAG decided to make their operations more efficient by routing transfer passengers through the UK instead of Spain. In general, hub airports compete with each other both for passengers and long-term bases for airlines.

- **Airline route planning.** As deregulation occurs, airlines can plan their destinations according to their economic interests. They can add or drop destination airports at short notice – and often do. An airline's main assets are the aircraft and staff, and these are highly mobile. So they may switch their routes to more profitable areas. Thus Delta abandoned its hub at Cincinnati, whose airport traffic dropped over 70 per cent from 23 million to 6 million annual passengers.

- **Passenger routing.** Passengers have considerable choice as to where they go and how they get there. Most obviously, the competition between leisure destinations affects airport traffic. The European holidaymaker might choose between the Red Sea, the Canary Islands, and the Caribbean, for example. Longer-distance travellers, who are likely to connect through transfer hubs, generally have many choices and make them not only on the basis of airline fares but on the airport itself. A passenger travelling from Vienna (Austria) to Boston (US) can, for example, connect through London, Frankfurt, Munich, Paris or Zürich. The route chosen may depend on the perceived performance of these facilities.

- **Airport shopping.** The most profitable duty free items – such as alcohol and perfumes – can be purchased from almost any duty free outlet. So the passenger has a choice: buy it at the departure airport, at some connection point or at the destination? Airport retail areas thus compete actively with each other. More attractive facilities and prices elsewhere affect local revenues and net income to the airport.

Airport managers need to take a comprehensive, systems view of airports and the competition between them. It is not enough to focus on one's own local facility (see Example 5.3). They need to recognise that they are actively competing with other airports for traffic and revenue. They should routinely monitor the competition and benchmark their performance against other facilities (➤Chapter 6).

Example 5.3
Airport competition

London Heathrow traditionally competes with Frankfurt Airport for passengers between North America and Asia. It now also competes for this traffic with Middle East airports in Abu Dhabi, Dubai, Istanbul, and Doha – as Etihad, Emirates, Turkish, and Qatar airlines offer easy and convenient connections between many North American and Asian destinations.

5.4 Measures of airport 'capacity' can be very misleading

Airports constantly have to deal with the problem of providing sufficient capacity. They regularly face complaints that part of the facility (for example, the runway or security inspection) is not providing enough capacity to their clients (aircraft or passengers, as the case may be). Moreover, as air traffic continues to increase, this problem may seem endless. Dealing with a bottleneck in one part of the airport soon reveals a new constraint elsewhere in the system.

Airport managers have to consider capacity issues carefully. There is a lot of confusion in this regard. This is because the concept of capacity for airports is problematic – as it is for transportation in general and a range of other industries. A central idea to focus on is that the capacity of a given airport facility is not a single number that anyone can determine absolutely. In general, with few exceptions, the capacity of an airport facility depends on how managers operate that facility.

Lack of **airport capacity** does not mean that the clients do not obtain a service or that the system turns them away. If the capacity of the security inspection system is insufficient, this does not imply that the passengers cannot get through security. What complaints about capacity signal is that the facility does not provide an acceptable **level of service (LOS)**, that it causes excessive delays, or that it is too crowded.

Airport capacity is not an absolute quantity that can be accurately measured. It is not comparable to the capacity of a physical container whose volume is definite and constant. Airports provide services, so airport capacity is contingent rather than absolute. Airport capacity depends both on customer expectations and on management practices which can improve – or degrade – capacity at short notice.

> **Airport capacity:** the capability of airport facilities to fulfil their intended function at an acceptable level of service.
>
> **Level of service (LOS):** the metric by which a particular aspect of airport service is measured.

Airport capacity depends on customer expectations

The prevalent practical concept of airport capacity depends on definitions of levels of service (LOS). These qualitatively characterise the service provided to passengers. A common scale for LOS runs from A (best) to F (unacceptable failure). This scale applies to both people in buildings and to highways. The LOS scale is three-dimensional. It considers crowding, ease of movement, and delays. For example, the rating for people moving along a corridor indicates how close people are to each other on average, the ease of flow (steady or stop-and-go), and the delays they may incur relative to unimpeded movement. The definition of any specific LOS is inherently subjective. (How else does one balance personal space and waiting time?) It also changes over time as airport experts discuss the issue and revise previous norms. Every recent edition of the IATA *Airport Development Reference Manual* has reflected these adjustments. The one constant is that a higher LOS provides more space and better service.

The logic of the connection between airport capacity and LOS means that any facility may simultaneously be considered to be at capacity or not. A waiting lounge may be over capacity from the perspective of providing a 'good' level of service (LOS = C) but may provide enough adequate capacity at LOS = D.

The capacity rating for runways also reflects subjective judgements about LOS. Historically, the definition of the 'practical annual capacity' for runways was a capacity such that aircraft would not have to wait more than four minutes for take-off. This corresponded to a very high LOS. Airlines and airports now agree that they can accept a lower LOS (average wait times for aircraft at busy airports greatly exceed four minutes!), and thus now rate the capacity of runways much higher than previously (➤Chapter 4).

Noticeably, as regards runways, the rated capacities of any given runway system differs between the US and the rest of the world. This is because the usual practice in the US is to schedule aircraft to Visual Flight Rules (VFR) appropriate for good weather rather than Instrument Flight Rules (IFR) suited for low visibility, as airports do in Europe and many other countries (➤Chapters 4 and 15). VFR allows for more landings and take-offs but results in greater delays when weather is bad. Thus, when a European airport declares its runway capacity, the figure is lower than would be acceptable in the US for an equivalent runway configuration.

Airport capacity depends on management practices

The capacity of airport facilities often depends on management practices (see Example 5.4). Specifically, capacity depends on the speed of the service, which managers can affect in many ways. This is obvious when we consider replicable units for providing a service, such as the number of available passport control booths or security check lanes. If more lanes are operating, the rate of service and capacity of the system increase. Management can thus influence the capacity of the service both by providing the possibility of more service (installing appropriate booths or scanners) and – more importantly – by appropriately scheduling staff to perform the service.

Example 5.4

How operational changes alter capacity

The capacity of many facilities depends on the way they operate:

- The capacity of a car park would seem simple to estimate: count the number of spaces! But this number does not tell you how many cars it can serve. This number depends on how long the cars are parked and thus the rate of turn-over. A short-term car park with 30 spaces might serve 500 cars a day if cars spend only 15 to 30 minutes in a space.
- The capacity of runways generally decreases when used by larger aircraft. This is because larger aircraft create bigger wake vortices and require greater separation between landings. Consequently, fewer aircraft can land per hour (➤Chapter 4).
- JetBlue almost doubled the capacity of the aircraft gates it inherited from United Airlines at Boston Logan International Airport. The reason was simple: JetBlue turned around aircraft in 30 minutes instead of an hour and thus

---

served twice as many passengers per gate. The gate areas were corre-
spondingly more crowded, but that was acceptable to passengers accus-
tomed to and expecting a low cost airline service.

The capacity of facilities also depends on what the clients consider an acceptable
LOS. This fact is a standard part of airport references. For example:

- A 'good' level of service (LOS = C) for passengers in a gate waiting area
  has been 1 m^2 per person. On that basis, a room of 200 m^2 would have a
  rated capacity of 200 persons. However, LCCs might accept an 'adequate'
  level of service (LOS = D) that calls for only 0.8 m^2 per person, and thus
  consider that the room has a capacity of 250 persons. Viewed as a VIP
  lounge (LOS = A), the capacity of the room would be half or less.

Management can also alter capacity by changing how staff operate. For example, some
airlines choose not to weigh checked-in bags. This speeds up the operation and increases the
capacity of the check-in counters. Note that since managers can also increase capacity by
scheduling more staff, they can generally choose between efficiency in operations, the
number of staff assigned, and the LOS they offer to customers in terms of their wait time and
length of the queues.

Management can alter the effective capacity of space by how they schedule customers into
and out of areas. For example, many airports, such as London Heathrow and Singapore,
open gate areas to passengers only a short time before the flight. The practical effect is that
some passengers are already boarding the aircraft while others are entering the gate area. The
result is that passengers move through the area quickly, reduce their **dwell time**, and lower
the number of travellers in the gate area at any time. This increases the capacity of the gate
area in terms of the LOS it provides.

Management practices in one part of the airport typically have knock-on consequences
for other parts. This is because airports provide their services through a sequence of processes.
Thus, changes to one process are likely to affect how passengers access and perceive the
capacity and LOS of other processes. As a case in point, the practice of limiting passenger
access to gate areas forces those who arrive early – from connecting flights or for other
reasons – to wait elsewhere, often in areas with inadequate facilities.

Airlines have an important role in defining the capacity of airport facilities. Indeed, one
of the important recent innovations regarding the capacity of airport terminals has been the
practice, pioneered by LCCs, of reducing the turnaround time of aircraft at their gates.
Airlines typically used to plan on taking at least an hour between the arrival of an aircraft at
a gate and its departure for the next flight. Southwest and other LCCs reduced this turnaround
time to around 25 minutes or less. They did this to increase the productivity of their aircraft –
less time on the ground means more time in the air carrying passengers and earning money.
They achieved this by introducing a range of practices such as having flight attendants collect
garbage during the flight. This innovation has had a significant knock-on impact on the
capacity of airport terminals (see Example 5.4). Lower dwell time of aircraft increases the

Dwell time: the typical
length of time
customers stay in an
area waiting for
service. When they
leave the space, other
customers can use it.
Dwell time therefore
translates into the
number of customers
that can use a space
per unit of time. Thus:
[Customers for a
space/unit time] = [1/
Dwell time].

capacity of each gate, and thus of the terminal building as a whole, regarding the number of aircraft it can serve.

Planning, designing and managing airport capacity

A proper estimation of the amount of capacity to provide for airport facilities and processes requires more than a technical analysis of the built environment. A complete analysis of capacity involves an understanding of the LOS that the range of airport customers will want and need. These clients are not all the same. Some airlines will want special lounges and facilities to promote their image and strengthen brand loyalty. Others specialise in providing good 'value for money' and prefer economical facilities that serve users at a lower LOS (see Case Study 5.1). Additionally, those who plan, design, and manage capacity need to recognise the importance of organisation and staffing in determining effective capacity.

Providing appropriate airport capacity is more than a matter of creating physical facilities. Construction and equipment can provide the potential for capacity, but it can be fully realised only through effective management, organisation, and service delivery. Airport managers should avoid the mistake of seeing construction as a solution to operational problems.

!

Stop and think

Why should airport managers avoid seeing construction as a solution to operational problems?

5.5 Queues are at the heart of airport operations

Queue: a number of clients waiting for service from some process. Often the persons or things in the queue actually line up in a conventional queue. More generally, they may simply have a ticket or an appointment for service and thus may be waiting in an area, ready to present when the service becomes available.

It is useful to visualise airports as a sequence of **queues** for service. Indeed, airports provide their services through a sequence of processes. Clients present sequentially at each process, wait until that process is ready to serve them, and then proceed. For example, aircraft proceed through a series of steps: they land, taxi in, and access a gate. Travellers likewise check in, pass through security, and go to their gate for boarding. Queues are central to airport operations. Good management of these queues is central to providing good airport service.

Good queue management is challenging. Queues can quickly become almost unmanageable. Whenever requests for service approach the nominal capacity of the process, queues lengthen, waiting times greatly increase, and the process becomes unpredictable. The following subsections explain how queues form, describe when and why they become chaotic, and provide advice on how to manage them.

How queues form

Queues arise because of a mismatch between the arrival of customers and the capability of the process to serve them. An obvious situation occurs when customers arrive continuously while the process provides service in batches – customers filter into the boarding lounge, but the bus to take them to the remotely parked aircraft does not proceed until full. The more general situation at airports is that both customers and the serving process are operating continuously – aircraft move into position for take-off at the end of the runway one by one; likewise passengers at the security check proceed one at a time. Continuous processes are subtler and require special attention.

Queues form in continuous situations because both the arrival of customers and the delivery of service are irregular. If each process were totally regular, like gears on a machine, there would be no delays (so long as the process had enough capacity). Everything would run smoothly – literally like clockwork. However, irregularities in the process mean that queues and wait times build up. Customers do not present at a constant rate: families and friends arrive together and may congregate in the check-in hall. The service may likewise be irregular: a passport control officer may wave some travellers through, yet take a long time with others. In short, queues and delays occur even when the facility has sufficient nominal capacity (see Example 5.5).

Example 5.5

How queues form

Consider a process that can serve exactly 1 customer per minute. Customers arrive irregularly, on average at the rate of 1 per minute. This might lead us to think that there is enough capacity to serve the traffic and that all should proceed without delay. Actually, queues build up, as this example demonstrates.

Suppose 5 customers arrive over a 5-minute period: 3 arrive at Minute 2, 2 more arrive at Minute 3. Now consider what happens:

Minute 1: no customers, no queue
Minute 2: 3 customers, 1 gets served, 2 are in queue
Minute 3: 2 more arrive, 1 gets served, 3 are in queue
Minute 4: 1 gets served, 2 are in queue
Minute 5: 1 gets served, 1 is in queue

Overall, although the system could process the customers without delay if they arrived like clockwork, 1 each minute, in this case the queue builds up to a maximum of 3 customers and the average wait time is 1.8 minutes.

How queues lengthen

Queues routinely lengthen when stressed and build up rapidly, leading to delays. Moreover, the overall performance becomes highly variable. Under apparently the same conditions, the system might perform reasonably well – or incur gridlock. Small, seemingly insignificant differences in the patterns of arrivals and service magnify into major changes in the length of queues and delays. This is a universal experience. You may have seen it yourself when driving: some days the traffic moves along steadily while at other times it is gridlocked – all it takes is a small traffic incident somewhere on the network. Managers need to understand the conditions that underpin these phenomena so that they can minimise the difficulties.

A few simple terms are needed to describe and understand the situation.

- The average arrival rate of customers (traditionally, λ or lambda).

- The average rate at which the process can serve customers (traditionally, μ or mu).

- The ratio between them, the utilisation ratio ($\lambda/\mu = \rho$ or rho).

- The delay multiplier $1/(1 - \rho)$.

The basic phenomenon is that queue length and wait time increase exponentially with the rate of arrivals, in proportion to the delay multiplier. This means that congestion and delays increase rapidly as the service nears full capacity. Small changes in the utilisation ratio lead to large increases in delays. An increase in utilisation from 0.75 to 0.80 increases the delay multiplier from 4 to 5, which translates to a 25 per cent increase in delay. Moreover, the effect gets far worse for higher utilisation ratios. If it moves from 0.90 to 0.95, the multiplier goes from 10 to 20, which implies that delays increase 100 per cent compared to the 25 per cent increase for the same change at the lower utilisation. Figure 5.1 shows the phenomenon graphically. Delays first increase gradually as the utilisation increases, and then very sharply as the utilisation ratio approaches full nominal capacity (at $\rho = 1$).

Most importantly, when utilisation is close to capacity, the queuing process becomes unstable and unreliable. This is because the variability of the process also increases with the delay multiplier (technically, the delays increase with the variance, which is the square of the standard deviation). As a practical matter, this means that neither customers nor managers

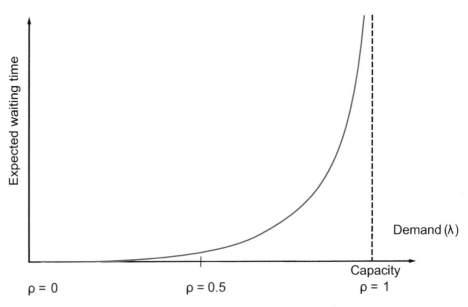

Figure 5.1 Relationship between utilisation ratio and expected waiting time

can count on the system performance. An average delay of 20 minutes at a security check might be 10 minutes on some days but 40 minutes on others. This variability makes it very difficult to schedule activities. In simple terms, the inevitable unreliability of queues under stress can cause chaos (see Case Study 5.2).

Queues manifest themselves too late

Queuing processes exhibit another feature that can confuse managers. Long lines and delays build up over time, so that the problem becomes most evident long after it really started. In this sense, the problem is rather like a disease that incubates. The factor that triggers the problem, such as an increase in arrivals that increases the utilisation ratio to a disruptive level, occurs quite some time before the peak length of queues.

This reality implies that managers must deal with queueing problems before they peak. If they wait until the problem is self-evident, they will have acted too late. For example, they should call in extra staff for security checkpoints well before the queues are excessive. Managers need to monitor their processes for signals that will alert them early on to take countermeasures against conditions that create long lines and excessive delays, severely degrade the LOS, and thus lower the effective capacity of the process and the entire airport through knock-on effects.

CASE STUDY 5.2

AUTOMATED BAG SYSTEM AT DENVER INTERNATIONAL AIRPORT

The original automated bag system at Denver International Airport was a management disaster. It did not deliver the capacity and reliability as intended, largely due to a failure to understand queuing systems.

The design consisted of carts running on a tracked network with multiple access and exit points serving many loading and unloading stations. Each of these processes created queues and consequent delays. Cumulatively, the sequences of queues guaranteed terrible performance.

The system never reliably delivered more than about 40 per cent of its design capacity. Beyond that amount, the average delays grew rapidly and the system became excessively unreliable – what was fine one day would lead to gridlock on another. This was completely unacceptable. At a transfer hub such as Denver, bags must make the connecting flight with their passengers. Managers could achieve this only by operating the system at a fraction of its nominal capacity.

The unreliable performance and the inability to operate near the rated capacity were quite predictable. All it took was an understanding of how queues behave, especially under stress.

Guidelines for managing queues

Simple guidelines for dealing with queuing processes are:

- Recognise that the rated capacity of a facility, as estimated by multiplying its average rate of service (such as bags per minute), is not a good indication of actual performance. The system will be capable only of delivering an acceptable, reliable service at a fraction of its rated service (see Case Study 5.2).

- Queuing processes are highly sensitive to the utilisation ratio. Managers face a trade-off between capacity utilisation and acceptable levels of service in terms of queue length, delays, and reliability. Greater utilisation degrades LOS.

- Increasing staff early is an important way to minimise the worst operational conditions.

!

Stop and think

Are queues an inevitable consequence of airport systems planning?

Key points

- Aviation and airport forecasts are 'always' wrong in that the continuing evolution of the airport/airline industry means that what actually occurs differs, often substantially, from original predictions.

- Due to the uncertainty of forecasts, there is a need for flexibility in design as a way of de-risking and future-proofing projects and enabling easy adaptation to actual future situations.

- Managers need a comprehensive systems view of airports and the competition between them.

- Measures of capacity at airports are not absolute but depend both on the detail of operations (especially for runways) and on differing judgements concerning acceptable levels of service.

- The inherent instability of queues under peak loads drastically reduces the acceptable maximum capacity of airport processes.

References and further reading

ACRP. (2012). *Addressing uncertainty about airport activity levels in airport decision making*, Report 76, Airport Cooperative Research Program, Transportation Research Board, Washington, DC.

de Neufville, R. and Odoni, A. (2013). *Airport systems planning, design, and management*. 2nd edn. New York, NY: McGraw-Hill.

de Neufville, R. and Scholtes, S. (2011). *Flexibility in engineering design*. Cambridge, MA: MIT Press.

FAA. (2013). *Airport master plans*, Federal Aviation Administration, Aviation Circular 150/5070–6C, Washington, DC.

FAA. (2015). *Airport system planning process*, Federal Aviation Administration, Aviation Circular 150/5070–7, including Change 1, Washington, DC.

IATA. (2019). *Airport development reference manual*. 11th edn. Montreal: International Air Transport Association (IATA) in association with Airports Council International (ACI).

CHAPTER 6

Airport management and performance

Anne Graham

LEARNING OBJECTIVES

- To identify the principles of airport management.

- To understand the importance and challenges associated with measuring airport performance.

- To appreciate the changing airport operating environment as it relates to privatisation, regulation and competition.

- To define performance metrics that can be used to measure key areas of airport operations and management.

- To evaluate mechanisms for managing scarce runway capacity.

- To understand the issues relating to surface access at airports.

- To assess future airport management and performance challenges.

6.0 Introduction

Airport operators provide the infrastructure, facilities and services that allow aircraft to take off and land safely and efficiently and that enable passengers and cargo to transfer between surface and air modes of transport. They increasingly also offer a broad variety of commercial facilities to satisfy the needs of passengers, employees and visitors and to generate additional revenue by, for example, selling products and services to other airports. Airports bring together a wide range of different organisations and companies, including airlines, air traffic control (ATC) providers, ground handling companies, government agencies and commercial concessionaires in order to perform their role as an essential component of the air transport system.

The aim of this chapter is to identify the principles and practices of airport management and to understand the different metrics or standards of measurement that can be used to assess airport performance. The chapter begins by considering the contemporary operating environment for airports. This is followed by an examination of the key performance areas of airfield operations, financial management and service quality. Two major issues for many airport operators, the management of scarce runway capacity and surface access, are then discussed in Sections 6.3 and 6.4. The chapter concludes by assessing future challenges for airport management and performance with a particular focus on technological opportunities.

6.1 Airport ownership and the changing operating environment

An airport operator has overall control and responsibility for the airport, even though the number of services and facilities that it provides directly can vary significantly. In the US, for example, the airport operator has more of a landlord role, with airlines, government agencies and others supplying most of the essential activities. In extreme cases, individual airlines actually operate their own terminals, as at New York's John F. Kennedy Airport. By contrast, in Europe, the airport operator is often more actively involved in the provision of services and may directly provide security, ground handling or ATC.

Historically, airports were government-owned, either at a national or local level. However, over the last couple of decades, there has been a trend towards shifting airport ownership and/or management into the private sector or to a private–public partnership. The reasons for this vary, although most often it is to improve efficiency and financial performance and/or to provide new funds for investment or access to capital markets. Table 6.1 provides a list of the most common reasons for airport privatisation.

Consortium: a group consisting of several different companies.

There are different types of privatisation models, ranging from the sale of part or all of the airport to investors on the stock exchange, as is the case with a number of European airports, including Frankfurt and Paris Charles de Gaulle, to handing over ownership (or long-term leases) of the airport to a **consortium** of investors, as with the major airports in Australia

---

Table 6.1 Reasons for airport privatisation

- To improve efficiency and/or financial performance.
- To provide new airport investment funds and/or access to capital markets.
- To bring financial gains to the government and/or remove the financial burden of operating the airport.
- To lessen government influence in airport operations.
- To improve airport service quality.
- To enhance airport management effectiveness.
- To allow diversification into new non-aeronautical areas.
- To encourage more competition.

(such as Sydney, Melbourne, Brisbane and Perth), Rome and Brussels, or allowing a consortium to operate the airport as a concession for a limited period, typically 20–30 years. This latter option is especially popular when the government wants to maintain ownership of the airport for strategic reasons but recognises the need to involve the private sector to provide investment and management expertise. Examples here include some of the major airports in India, Brazil (see Case Study 6.1) and other South American countries. North America is unusual in this respect as virtually all US airports remain in local public ownership, and in Canada the major airports are operated by local state-owned not-for-profit organisations.

CASE STUDY 6.1

PRIVATISATION OF BRAZILIAN AIRPORTS

Brazil has over 2,000 airfields, and until recently, all the main airports were managed by the state-owned Infraero organisation. However, the rapid expansion of the Brazilian economy, together with the selection to host the football World Cup in 2014 and the summer Olympic Games in 2016, necessitated urgent modernisation and expansion of airport infrastructure to cope with anticipated traffic growth and these special events. The three major international airports, São Paulo–Guarulhos (the main international gateway to Brazil and the largest airport in Latin America), São Paulo Viracopos/Campinas International Airport (São Paulo's third airport and a major cargo hub) and Brasília (serving the capital), were partially privatised in 2012 as concessions, with Infraero retaining a 49 per cent share in each. The time periods for the concessions range from 20 to 30 years. These three airports together handle around a third of all the passenger traffic in Brazil. A different consortium was chosen to run each airport, with one including the airport company Airports Company South Africa (ACSA), which had previous experience of handling World Cup traffic. Subsequently, Rio de Janeiro's Galeão International Airport and Belo Horizonte's Confins Airport were privatised in a similar manner in 2013 with concessions of 25 and 30 years respectively. Changi (Singapore) Airport has involvement with the Rio consortium, while Munich and Zürich airports belong to the winning consortium for the Belo Horizonte airport concession. Even though the worsening economic and challenging political situation has changed the environment for privatisation, more sales have taken place. This has included four middle ranking airports in 2017 (Porto Alegre and Fortaleza to Fraport – Frankfurt airport's operator; Salvador to Vinci – the global construction and concession group; Florianopolis to a consortium including Zurich and Munich airports). More privatisations took place in 2019, the first under President Jair Bolsonaro, when 12 regional airports, sold in three separate lots (Northeastern, Southeast and Central West), were bought by the Spanish airport operator Aena, Zurich Airport and the Brazilian consortium Aeroesete respectively. Over 20 more airports are planned to be privatised between 2020 and 2022.

By nature of their complex role in meeting the needs of airlines, passengers and other stakeholders, airports are subject to regulation globally (with regulation set by the International Civil Aviation Organization – ICAO), regionally (e.g. the EU) and nationally (e.g. the US Federal Aviation Administration – FAA). Many of these regulations relate to the operational, safety and security aspects of managing an airport. There are other areas such as the protection of employees and passengers (e.g. regulations relating to passengers with reduced mobility – PRMs), and airports are also increasingly subject to stringent environmental regulations, primarily due to the noise and air quality impacts that they produce (➤Chapter 18). This includes restrictions on noisy aircraft or night-time flying. Within the EU, examples include directives relating to the introduction of noise-related operating restrictions (Directive 598/2014/EC) and the assessment and management of environmental noise (Directive 2002/49/EC).

One major aspect of airport regulation is economic regulation, which is used to modify or control the behaviour of stakeholders involved in the delivery of the airport product. This primarily covers prices, the supply of services, and market entry and/or exit. The regulation of airport prices (through charges or fees) and slots are two key areas which will be discussed in this chapter. There are other areas, however, that can be considered. In the EU, for example, ground handling services have been regulated since 1996 in order to encourage a more competitive environment by preventing **monopoly suppliers** at all but the smallest airports.

Monopoly supplier: a single company that supplies a particular product.

With regards to airport charges, there are different levels of regulation. At the global level there are ICAO principles that state that the charges should be non-discriminatory, cost-related and transparent (ICAO, 2012). These must be incorporated into any national regulation. Within Europe there is an extra level of regulation with respect to charges which was introduced in 2011 and covers all airports serving more than 5 million passengers (EC, 2009). This builds on ICAO principles and includes additional features, such as the mandatory consultation and swapping of information between airports and airlines, and the requirement to have an independent supervisory or regulatory body.

At a national level, the amount of regulation concerning airport charges varies considerably. For some airports, typically those under public sector ownership, it may be a national government's responsibility to set charges or approve those set by the airport operator. In different cases, usually with smaller and often privately owned airports, there may be no specific state control. At the other extreme, there may be formal economic regulation, which typically occurs at relatively large airports, especially when they have been privatised. In this case, the types of regulation vary from so-called 'heavy-handed' approaches either when the regulator permits an airport operator to earn enough revenue to cover its costs and make a profit (rate of return or cost-based methods), which is comparatively rare but occurs, for example, at Amsterdam Airport Schiphol; or alternatively when there is incentive regulation, which is more focused on encouraging efficiency, with a **price cap** on the charges. The price cap is typically defined as **RPI +/− X**. This is a more popular practice and is used, for instance, at London Heathrow, Paris Orly and Dublin as well as the major airports of India, Mexico and South Africa.

Price cap: a form of airport economic regulation that sets a cap or limit on the price that airport operators can charge.

RPI +/− X: the price cap typically used with airport regulation, where RPI is the retail price index and X is the efficiency factor (sometimes the consumer price index (CPI) rather than RPI is used).

Regulatory till: the airport facilities and services, and the cost and revenue allocation processes, which are considered in the airport economic regulation process.

In setting the economic regulatory control, the **regulatory till** has to be established. Usually this involves deciding whether a 'single till' approach is used, when both the aeronautical and non-aeronautical/commercial airport revenues and costs are considered together as one entity in setting the charges, or whether a 'dual till' is adopted, in which the aeronautical and commercial aspects of airport operations are treated separately. There are also 'hybrid tills' which fall in between these two arrangements. In the US, there are 'residual' and 'compensatory' approaches which have some similarities to these two tills. The actual US regulatory system is significantly different from elsewhere, with charges depending predominantly on detailed conditions relating to revenues and funding, which are laid out in federal law and in **airport use and lease agreements** with airlines, rather than having the involvement of a specific regulatory body.

Some countries adopt a 'lighter touch' approach to the regulation of airport charges with price monitoring. Here, the threat of regulation rather than actual regulation is used. In this case, if airports and airlines cannot reach an agreement over charges, or if the airport operator is considered to be acting anti-competitively, the regulator will intervene; otherwise, no control will be exercised. A notable example of this type of regulatory regime can be seen at the major Australian airports, which shifted from a heavy-handed or strict regime in 2001, with some airlines consequently forming five-year agreements with the airports. More recently, a lighter touch or light-handed approach was introduced at London Gatwick (Case Study 6.2).

Airport use and lease agreement: a legally binding contract which details the fees and rental rates an airline has to pay, the method by which they are calculated and the conditions of use of airport facilities.

CASE STUDY 6.2

REGULATION OF LONDON GATWICK AIRPORT

London Gatwick Airport, as well as London Heathrow Airport, was traditionally regulated with a heavy-handed price cap. However when a new regulatory regime was introduced in 2014, the regulator (the Civil Aviation Authority – CAA) was given the powers to replace their control, if appropriate, with different forms of regulation, and to move away from the previous one-size-fits-all policy for all regulated airports. At the same time, Gatwick Airport had introduced its so-called 'contracts and commitments' initiative which involved agreeing a series of commitments with its airlines on price, service conditions and investment. With a few airlines (including Emirates, Norwegian and Thomson), the operator integrated these commitments into bespoke formal contracts. This had an influence on the CAA's decision to shift to a more light-handed price monitoring approach, with the CAA believing that such agreements could be better tailored to meet the needs of individual airlines and their passengers. However, price cap regulation has remained at Heathrow.

Economic regulation needs to be considered alongside airport competition because such regulation is primarily introduced when it is considered that there are insufficient competitive forces to ensure that airports do not abuse their market power (such as an ability to raise prices above what would prevail under competition). The amount and nature of competition varies from airport to airport. Table 6.2 identifies the ways in which an airport can commonly compete.

The weakest competition tends to exist when airports possess a unique catchment area, particularly if they are located on an island or in a remote location (➤Chapter 21). Regulatory or operational constraints may also limit the effective competition; for example, if an airport cannot offer long-haul services because of a lack of traffic rights or insufficient runway length. It has been argued that air service liberalisation combined with the development of commercially oriented and increasingly privately run airlines and airports, however, means that light-touch regulation, or even just normal competition law, is now the most effective way forward for airport economic regulation. This approach would eliminate the direct

Table 6.2 Types of airport competition

Nature of competition	Example
Shared local market in an urban situation	New York, Washington, London
Shared local market in a regional situation	Regional airports
Transfer traffic	Amsterdam, Singapore
Destination traffic	Tourism resorts, cruise embarkation points
Cargo traffic	Hong Kong, Dubai
Other transport modes	High-speed rail
Commercial facilities	Shopping malls, airline on-board sales

administrative and indirect market distortion costs associated with more heavy-handed approaches.

! Stop and think

What are the advantages and disadvantages of airport privatisation for airports, airlines and passengers?

6.2 Airport performance

Faced with this increasingly competitive environment, and combined with growing pressures on both physical and financial resources, it has become more important than ever for airports to effectively measure their performance. This task is challenging owing to the complex nature of airports with many different facilities and processes, and because of the existence of many different organisations that operate at an airport which collectively enable an airport to function. Traditionally these issues have meant that airports have monitored their own performance and have been cautious about benchmarking themselves against others, since each airport tends to be unique in the way it organises its operations. However, there does seem to be greater acceptance now that the benefits gained through the careful measuring of peer performance outweigh the shortcomings associated with less-than-perfect comparable data.

Performance can be measured against many aspects of airport management. Three key areas – airfield operations, financial management and the provision of service quality – are considered here. This is not an exhaustive list, and other areas of performance such as human resources, information technology, maintenance and planning/construction can be considered. One area which has received increasing attention is environmental performance, where airports have been measuring performance in areas such as their carbon footprint, energy use and waste recycling. (►Chapter 18).

As an airport's primary function is to support airlines and their passengers and cargo in departing to, and arriving at, their chosen destination, the broadest indication of an airport's performance can be assessed by looking at the number of aircraft movements, the passenger throughput, and the cargo which is loaded and unloaded. Table 6.3 presents the top ten airports in the world according to these three measures in 2018. While Atlanta has maintained its position as the world's largest airport in terms of passenger numbers, it is now being challenged by a growing number of airports outside of Europe and North America. These airports include Beijing, Dubai and Shanghai Pudong – the latter two being in the top ten for the first time only in 2012 and 2016 respectively. In terms of cargo, Memphis (the hub of FedEx) has lost its top position to Hong Kong, and other airports including Shanghai Pudong, Seoul Incheon and Dubai have seen their cargo volumes grow considerably in recent years (see ➤Chapter 17). With regards to movements, however, the US still dominates, since the average aircraft size is smaller due to competitive pressures, shorter sectors and the dependence on domestic traffic.

Airfield operations

When looking at airfield operational performance, a wide range of different features need to be considered. These include the number, length and configuration of runways; ATC services; instrument landing systems (ILSs); lighting and weather monitoring systems; ramp and apron space allocation; stand and gate provision; and fire, rescue and policing/security services (➤Chapter 4). Taking these into account, many factors will have an influence on the airport operator's ability to effectively handle inbound and outbound aircraft movements. Many of these are related to the critical functions of safety and security. These include the ability to cope with runway accidents and incursions (e.g. an occurrence on the ground that involves an aircraft, vehicle, person or foreign object) and to implement successful wildlife (usually airborne animals such as birds and bats) hazard management, emergency responses

Table 6.3 Traffic at the top ten airports, 2018

Passengers	(Millions)	Cargo	(000s tonnes)	Movements	(000s)
Atlanta	107,394	Hong Kong	5,121	Chicago ORD	904
Beijing	100,983	Memphis	4,470	Atlanta	896
Dubai	89,149	Shanghai PVG	3,769	Los Angeles	708
Los Angeles	87,534	Incheon	2,952	Dallas/FW	667
Tokyo HND	86,942	Anchorage	2,807	Beijing	614
Chicago	83,245	Dubai	2,641	Denver	595
London LHR	80,126	Louisville	2,623	Charlotte	550
Hong Kong	74,517	Taipei	2,323	Las Vegas	540
Shanghai PVG	74,006	Tokyo	2,261	Amsterdam	518
Paris CDG	72,230	Los Angeles	2,210	Frankfurt	512

Source: ACI (2019)

and bad weather operations (typically for snow and ice but also for hurricanes and thunderstorms).

Airport finance and productivity

Measuring the financial performance of an airport is also important, especially given the increased commercial focus of many airport operators. While there are standard generic financial measures – including profit measures such as earnings before interest and tax (EBIT) and earnings before interest, tax, depreciation and amortisation (EBITDA) – some of the unique characteristics of the airport industry necessitate specially defined measures. Typically this involves considering financial ratios (see ➤Chapter 12) that relate certain costs or revenues to aircraft movements, passenger or cargo volume – or a combination of these as with the **work load unit (WLU)**. With regards to cost, it is also essential to know the comparative importance of labour, capital and other inputs, especially given the relative capital intensity of the airport sector.

For revenues, the most useful distinction is made by dividing them into **aeronautical revenue** and **non-aeronautical revenue** (commercial) sources. Aeronautical sources include all the revenues generated from airline activities such as the landing charge (usually based on the weight of the aircraft), the passenger charge (based on passenger throughput) and the aircraft parking charge (usually based on aircraft weight or wingspan). Ground handling revenues will also be included if the airport operator directly provides this service. The commercial or non-aeronautical revenues are from commercial facilities such as retail, food and beverage outlets (which are usually outsourced, with a percentage of income being paid to the airport operator), car parking, car hire, advertising and rents (➤Chapter 7). Further disaggregation of these revenues, particularly in the commercial area (e.g. different types of retail revenues as well as different sales per square metre) is common practice when assessing performance.

Airport service quality

The quality of service that an airport provides is another key performance indicator that needs to be evaluated. The actual service delivery can be measured objectively and some operational performance metrics for the airfield area, including taxi time and gate delays, could be included here. Typically, within the terminal, the service delivery measures will relate to queue lengths and waiting time at check-in, security, border control and baggage delivery times, and check-in to gate time. Some of the more intangible aspects of service quality cannot be measured in this way, however, and these objective measures are unable to assess passenger expectations and perceptions. Hence many airports use performance measures gathered through passenger surveys related to passenger satisfaction as well. Such feedback can be divided into different areas, including the essential processes where passengers will be asked to comment on issues commonly relating to queuing, staff helpfulness, waiting and crowding. Commercial and other terminal facilities are also usually considered when questions are asked about seat availability, comfort, temperature and value for money. Other popular areas include wayfinding, cleanliness and availability of flight

Work load unit (WLU): an airport traffic output measure which is equivalent to one passenger or 100 kg of cargo.

Aeronautical revenue: revenue derived from aviation-related activity in the form of landing fees, passenger fees, aircraft parking fees and handling fees.

Non-aeronautical revenue: revenue derived from retail space, car parking, rents and leases. Non-aeronautical revenue is also sometimes termed commercial revenue.

information. Many airport operators undertake their own passenger surveys to gather this feedback and/or they participate in cross-airport studies, such as those undertaken by the Airports Council International (ACI) (see Example 6.1).

Example 6.1
ACI's airport service quality (ASQ) survey

ACI undertakes the largest global study of airport service quality. This covers more than 390 airports accounting for more than half the global annual passengers, with over 75 per cent of the world's top 100 airports in over 92 countries. It involves over 650,000 interviews. This departure survey covers 34 service quality areas including access, check-in, security, airport facilities, food and beverage, and retail. It involves a passenger self-completion questionnaire which is distributed at the departure gates, and a minimum of 1,400 passengers per year are selected to participate at each airport. In 2018 in the over 40 million passengers airport category, Singapore, Mumbai, Delhi and Shanghai Pudong airports achieved the highest scores in Asia-Pacific, Rome Fiumicino and Moscow Sheremetyevo in Europe, and Dallas Fort Worth and Toronto Pearson in North America (ACI, 2019). To complement this departure survey, ACI also introduced an arrivals survey in 2017, and in 2018 the best arrival experience airport was Kempegowda Airport in Bangalore, India.

Performance measures

Table 6.4 presents ten key performance measures for each of these three performance areas. These are illustrative of core indicators which can be included in an overall system of airport performance measurement. These will typically cover specific services and facilities and contain more detailed and disaggregated information.

Table 6.4 Examples of performance measures

Airport operations

- Runway accidents per thousand movements.
- Runway incursions per thousand movements.
- Public injuries per thousand passengers.
- Employee injuries per thousand hours worked.
- Average emergency response time.
- Wildlife/bird strikes per thousand movements.
- Average gate departure delay per flight in minutes.

Continued

Table 6.4 Continued

Airport operations

- Average time to taxi from gate to runway.

- Average time to clear runway of snow and ice.

- Number of airport closures due to adverse weather.

Airport finance

- Labour cost per passenger (or WLU).

- Capital cost per passenger (or WLU).

- Total cost per passenger (or WLU).

- Passengers per employee.

- Aeronautical revenue per passenger (or movement).

- Non-aeronautical revenue per passenger.

- Non-aeronautical revenue as percentage of total revenue.

- Profit per passenger (or WLU).

- Profit margin (profit as percentage of total revenue).

- Return on net assets (profit as percentage of net assets).

Airport service quality

- Queue lengths at check-in, security, border control.

- Waiting time at check-in, security, border control.

- Baggage delivery time (first and last bag).

- Average time from check-in to gate.

- Overall passenger satisfaction with the airport.

- Passenger satisfaction with processes.

- Passenger satisfaction with commercial and other facilities.

- Passenger satisfaction with wayfinding.

- Passenger satisfaction with flight information.

- Passenger satisfaction with cleanliness.

! Stop and think

How can airport performance be measured, and what are the relative merits of each approach?

6.3 Managing runway capacity

One of the most critical areas of performance concerns the airport operator's ability to effectively match demand for airport services with the available capacity. This role has become more and more difficult in recent years as air traffic has continued to grow faster than the available supply at many major airports, primarily due to environmental, physical or financial constraints involved with providing new or expanded capacity. While capacity constraints can occur at many places in the airport, including terminals and gates, arguably the most challenging area to manage concerns the runway.

In general the pricing mechanism is commonly used to ration demand when there is a shortage of supply. However, with airports, even though a few have adopted some form of peak pricing, the differential between the peak and off-peak prices, as well as the actual level of peak charges, is insufficient to significantly influence airline demand and 'clear the market' at peak times. This is particularly due to the fact that airline scheduling is complex, and that airport charges often make only a small contribution to total costs, which results in many airlines being fairly insensitive to changes in the charges.

In the absence of the use of an effective pricing mechanism to balance demand and supply, an administrative system for allocating slots has evolved. These slots are scheduling slots, which are different from the actual operational take-off and landing time slots that are assigned to the airline by the air traffic controllers. Three types of airports have to be considered. First there are level 1 airports, where there is plenty of runway capacity and gaining an airport slot at a certain time is not a problem. Then there are level 2 or schedule facilitated airports, where demand is approaching capacity but where slot allocation can be resolved through voluntary cooperation. Lastly, there are fully coordinated level 3 airports, where demand exceeds capacity and formal administrative procedures have to be used to allocate slots. In 2018, there were 204 slot coordinated airports worldwide that covered 43 per cent of global passengers.

The allocation of slots at level 2 and 3 airports is dealt with by the International Air Transport Association (IATA) scheduling committees and slot conferences. Within this slot allocation process, the most important feature is the principle of grandfather rights, which means that if an airline operated a slot in the previous season, it has the right to operate it again. This is as long as it meets the slot retention requirements, or so-called 'use it or lose it' rule, which states that the airline must operate at least 80 per cent of the flights associated with the slot. Lower-priority rules also apply which relate to giving preference to services for a longer time period, and to those which balance different types of services or markets at airports (IATA, 2019).

Within the EU, slot allocation comes under the 1993 regulation EU/95/93 (EC, 1993). These EU rules are a legal requirement, while the IATA coordination system is voluntary. The EU regulation retains the principle of grandfather rights but has introduced new concepts such as encouraging new entrants and financially independent coordinators. Over the last two decades, the regulation has been revised with the aim of making it more effective, but doubts still remain as to whether the processes it requires are the best to manage the scarcity in slots or to encourage competition. It certainly provides a stable environment for airlines and other stakeholders, but it is administratively burdensome and has encouraged inefficiencies, such as airlines flying uneconomic operations in order to preserve slots (so-called 'babysitting' slots) and not making full use of them. Hence, there has been considerable debate as to whether there are any alternative systems which would be more suitable. The options discussed broadly fit into two categories: first, maintaining an administrative system but changing the rules; and second, introducing some sort of market mechanism.

Within a reformed administrative system, priority could be given to different types of airlines rather than retaining the grandfather rights principle. For example, priority could be given to long-haul flights that normally have less flexibility in scheduling, or to those for which there is no surface access option. Alternatively, flights that have a smaller noise or emissions impact could be favoured, or airlines using larger aircraft could be chosen. Priority could be given to scheduled flights over charter flights, passenger flights over cargo flights, or new entrants could be given greater opportunities to gain slots. Another option could be to cap frequencies to a certain destination once a set threshold has been reached.

It has been argued, however, that even with a new set of administrative rules, the mechanism will still share the shortcomings of the traditional system in not ensuring that the slots will be used by the airlines who value them the most. This can be achieved only by using a pricing mechanism. Here consideration needs to be given to both primary allocation, when the slots are initially allocated, and secondary allocation, when the use of slots may be subsequently changed. For primary allocation, prices could be set for slots with very high 'market-clearing' values in the peak – much higher than any of the current peak charges at airports. Another option could be to use auctions to allocate the slots. Both these systems should theoretically provide for better use of the runway but may be detrimental to airline competition, as they will tend to favour the airlines with the greatest market power. In practice, they could also be disruptive to airline schedules, and a major issue of concern would be who should benefit from the revenue raised through the high charges or auctions.

With regards to secondary trading, in the 1993 regulation slot exchanges were permitted, but slot trading was not specifically allowed or banned. Proposed new legislation included provisions for the clear acceptance of secondary trading, but this stalled in 2012. At some airports, notably London Heathrow where capacity has become increasingly scarce, slot trading has become accepted practice. Indeed it was estimated that in 2010 the value of a slot pair was £30–40 million for pre-0900 arrivals and £10 million for 0900–1300 arrivals, with over 400 weekly slots being traded (Steer Davies Gleave, 2011). In 2016, Oman Air paid a record £58 million for a pair of slots it bought from KLM. The only other area outside of Europe which has experienced slot trading is the US, which has a different slot allocation process (see Case Study 6.3).

SLOT ALLOCATION IN THE US

In the US, there is generally no formal slot allocation process as this would be in conflict with US antitrust laws, which prevent predatory acts and anticompetitive behaviour. Instead, in most cases there is open access to the airports, with a first come, first served system. Airlines design their schedules independently by taking into account any expected delays, but when many flights are scheduled around the same time, there can be considerable congestion. The exception has been at the so-called 'high-density' airports of New York, Chicago O'Hare and Washington Reagan National.

Since 1969, the airlines at these airports have been given antitrust immunity to discuss their schedule coordination. In 1985, due to increasing traffic, slot trading was allowed at these airports. However, criticism of this system led to the allocation rules for high-density airports being withdrawn in 2002 at Chicago O'Hare (also due to the addition of new capacity) and in 2007 for the New York airports. In the latter case, both as a result of anticipating congestion (LaGuardia) and actual experience of severe delays (JFK and Newark), temporary slot control had to be introduced which involved a cap on the number of slots, a minimum 80 per cent use requirement and permission for secondary trading through leases but not on a permanent basis. To replace these temporary controls, auctioning of 10 per cent of slots was planned in 2009 but abandoned after fierce opposition, particularly from the airlines.

Slots at LaGuardia and Washington Reagan airports are now allocated on a continuing basis, primarily based on historic slots and a two-month usage requirement. JFK, which has a larger share of international flights, follows the IATA scheduling rules. These three airports have their number of slots controlled by the Federal Aviation Administration.

Stop and think

!

Why has it proved difficult to implement a pricing mechanism for allocating runway slots?

6.4 Airport surface access

A key airport management issue is surface or ground access. This has significant implications for the overall capacity of an airport and can affect a number of aspects of airport performance, such as the operational efficiency, the service quality/passenger experience and environmental protection. Historically, most passengers arrived at airports by private car or taxi, with only a small proportion using bus or coach transport. As a result ground transport made a major contribution to the overall airport noise levels and air pollution/emissions, and many airports have been developing public transport alternatives to the car. Whilst reducing car use is a key way in which airports can yield important environmental benefits, it can also reduce lucrative commercial revenues from car parking, and so difficult decisions about these trade-offs often have to be made.

The share of public transport use varies considerably between airports. The highest shares of public transport worldwide tend to be achieved at airports in Europe and Asia, where sometimes over half the passengers use public transport. In the US and Australia, higher dependence on the car generally and the smaller number of specific rail links mean that public transport use is generally less – even below 10 per cent for a number of airports

Consideration of surface access involves examining the main user groups. Firstly, there are the passengers who make up the majority of journeys. Generally passengers, especially those flying for leisure reasons, are fairly infrequent airport travellers and may be unfamiliar with, and hence anxious about, their surface access journey. Moreover, they may favour the car, viewing it as more comfortable, flexible, reliable and better equipped to transport large and often heavy amounts of luggage. Secondly, there are employees who travel to and from the airport on a daily basis. Often they may have to use the car, as public transport services may be inadequate given their shift, they may have unsociable working hours or their homes may be widely dispersed around the vicinity of the airport. The third main user group is visitors. This includes meeters and greeters, who undertake so-called 'kiss-and-fly' journeys, or others who may, for example, be just using the retail, food and beverage or event facilities at the airport. In terms of the environmental impacts, 'kiss-and-fly' visitors are particularly significant because for every flight there will be four vehicle trips (dropping off and collecting) rather than two if passengers drive themselves.

A number of options are available to encourage better use of public transport and smarter car use. In the longer term, if demand is dense enough, new infrastructure, for example associated with rail links, can be provided. More short-term strategies can include making public transport more accessible (e.g. with more routes/frequencies, concessionary fares, through ticketing), more reliable, more attractive (e.g. a better-quality service and a more secure waiting environment) or by making improvements in marketing, signage and the availability of information (especially real-time). Taxi-sharing (for passengers) and car-sharing (for employees) can also be encouraged. On the other hand, car use can be discouraged by limiting the car parking spaces, increasing the cost of parking or giving priority to public transport, for example by having dedicated bus lanes. Many airports set modal share targets when introducing such strategies. Notably Norwegian airport operator Avinor had a public transport share target of 70 per cent by 2020, but it reached this by 2015. Meanwhile, Heathrow Airport, with a planned new third runway, is aiming to have a public transport share of at least 50 per cent by 2030, compared to 40 per cent in 2018.

6.5 Future challenges

The changing airport operating environment has undoubtedly brought major challenges for airports as they strive to optimise their performance in all aspects of airport operations. One of the major reasons for these challenges has been the enhanced competitive pressures from airline deregulation (➤Chapter 3) and airport privatisation, together with an increased demand for a more sustainable, secure and quality-conscious industry. A key issue for airports is how to effectively serve increasingly diverse airlines and passengers. This has meant that, at many airports, the concept of 'one-size-fits-all' has been replaced with greater flexibility and adaptability, and a growing focus on providing facilities and services to suit the different needs of customers.

Technological developments, if used efficiently and appropriately, offer many opportunities for enhancing performance in the future and coping with the differentiation strategies that are needed. A good example for airfield operations is Airport Collaborative Decision Making (A-CDM) (see Example 6.2). In addition, within the terminal area there is considerable scope for technological improvements to be made to essential processes such as check-in, security and border control. These can potentially reduce airport costs and allow for the terminal space to be used more effectively, while at the same time providing a better-quality, simpler and quicker passenger experience. Particularly in the security area, there has been considerable discussion concerning the use of biometrics and more effective data sharing technology for risk-based security processes. A relevant example in the US is called TSA Pre✓®, where technology is used to allow certain members of airline frequent flyer programmes and US Customs and Border Protection (CBP) Trusted Traveler Programs, and any US citizen who applies to the programme and is evaluated to represent a low risk, to receive a known traveller number and expedited security screening benefits at over 200 US airports.

Example 6.2

Airport Collaborative Decision Making (A-CDM)

A-CDM was originally a European initiative that aimed to improve the overall efficiency of operations at an airport, with a particular focus on the aircraft turnaround and pre-departure sequencing process. This is achieved by the real-time sharing of operational data and information between the key stakeholders, such as airport operators, airlines, ATC and handling agents. Its goal is to optimise the interactions between these organisations and to lead to better punctuality, for example, by reducing taxiing time. Munich, Brussels and Paris Charles de Gaulle airports were the first to become A-CDM compliant in 2011. By 2019, 28 European airports were using A-CDM, and more plan to implement it in the future. In Asia there has also been considerable interest in A-CDM with 16 airports in countries such as India, China, Singapore and South Korea having implemented it. In the US there is a similar concept called Surface Collaborative Decision Making (Surface-CDM).

New technology also offers opportunities to improve service quality and the passenger experience in other areas, with product features such as remote and self-service check-in, electronic bag tags, electronic boarding gates and self-service transfers set to become the norm. Social media and smartphone developments have the potential to improve flight status and wayfinding information and to increase the attractiveness of the commercial offer for passengers, while at the same time providing opportunities for enhancing the airport's financial performance in non-aeronautical areas. Globally, IATA and ACI are collaborating in a so-called 'New Experience in Travel and Technologies (NEXTT)' initiative (see Case Study 6.4) to exploit the technological opportunities that exist.

THE NEW EXPERIENCE IN TRAVEL AND TECHNOLOGIES (NEXTT) INITIATIVE

In 2017, IATA and ACI agreed on a joint initiative called New Experience in Travel and Technologies (NEXTT). Its purpose is to examine some of the most probable key elements that will transform the complete end-to-end air journey over the next 20 years, and to develop a common vision to enhance the travel experience that will guide industry investment and help governments design regulatory frameworks. Overall it aims to foster a simplified journey by applying innovation and technology to eliminate hurdles, queues and service failures. It focuses on three emerging concepts, namely off-airport activities (flexibility in what can happen before and beyond the airport such as transferring on-site processes like security processing and baggage check-in/drop to earlier points in the journey at the home, hotel or on other transport services), advanced processing technology (increasing use of digital identity management, automation and robotics to improve efficiency, safety, security and the passenger experience) and interactive decision making (linking everything together with trusted, real-time data throughout the journey and better predictive modelling and artificial intelligence). Many airports (and airlines, handlers and others) have already trialled some of the new concepts, whilst some have yet to be fully developed. Numerous concepts exist including 'Smart Security' (a joint project of ACI/IATA), 'Smart Path' (a mobile app developed by the technology provider SITA that allows passengers to create a secure biometric credential on a mobile phone that can then be used at every touch point in the journey) and even 'Smart Stand' (a new generation ramp operation at Heathrow).

! Stop and think

What are the future challenges facing airport operators in terms of airport management and performance?

Key points

- Airports are arguably the most complex element of the air transport system since they bring together a range of different facilities and processes, many of which are provided by different organisations.

- Measuring airport performance is crucial in ensuring airports function efficiently, yet this can be a challenging task.

- Increased privatisation, economic regulation and competition have strengthened the emphasis on airport performance and management.

- A key area of performance relates to managing scarce runway capacity, but major uncertainties still exist regarding the optimal slot allocation mechanism.

- An effective surface access strategy is becoming an increasingly important aspect of airport management.

- Airports may be able to enhance their future performance by focusing more clearly on the specific needs of their customers and harnessing the potential provided by new technology.

References and further reading

ACI. (2012). *Guide to airport performance measures*, Montreal: ACI.

ACI. (2017). Airport ownership, economic regulation and financial performance, Montreal: ACI.

ACI. (2019). *World's top airports for customer experience revealed*, Press Report, 6 March, Available at: https://aci.aero/news/2019/03/06/worlds-top-airports-for-customer-experience-revealed/

ACI. (2019). *World's top five fastest growing airports for passengers and cargo revealed*, Press Report, 17 September, Available at: https://aci.aero/news/2019/09/17/worlds-top-five-fastest-growing-airports-for-passengers-and-cargo-revealed/

ACI Europe. (2018). *Guidelines for passenger services at European airports*, Brussels: ACI Europe.

EC. (1993). *Regulation (EEC) No 95/93 of the European Parliament and of the Council of 18 January 1993 on common rules for the allocation of slots at community airports*, Official Journal L14, 22 January, Brussels: EC.

EC. (2009). *Directive 2009/12/EC of the European Parliament and of the Council of 11 March 2009 on airport charges*, Official Journal L070, 14 March, Brussels: EC.

Graham, A. (2018). *Managing airports: An international perspective.* 5th edn. Abingdon: Routledge.

IATA. (2019). *Worldwide slot guidelines.* 10th edn. Montreal: International Air Transport Association.

ICAO. (2012). *ICAO's policies on charges for airports and air navigation services, Doc 9082.* 9th edn. Montreal: ICAO.

Oxera. (2017). *The continuing development of airport competition in Europe*, London: Oxera.

Steer Davies Gleave [SDG]. (2011). *Impact assessment of revisions to regulation 95/93*, London: SDG.

CHAPTER 7

The airport–airline relationship

Anne Graham

LEARNING OBJECTIVES

- To describe what is meant by the airport–airline relationship.
- To understand how and why the airport–airline relationship has evolved.
- To discuss the factors that affect the airport–airline relationship and the complexities that are involved.
- To appreciate the importance of the route development process for ensuring successful airport–airline relationships.

7.0 Introduction

The airport–airline relationship describes the formal agreements, business arrangements and daily interactions that exist between airports and the airlines that operate from them. These relationships are necessary from an operational perspective. Airports are dependent on airlines deciding to operate services from their facility, as without them they have no passengers or freight and no means through which to realise their market potential. Airlines, in turn, depend on airports to provide safe and secure facilities that can serve the needs of their passengers and aircraft, at a convenient location, at the required time and at the right price. Yet, despite the mutual dependency between airports and airlines and their shared aim to enhance customer satisfaction and the passenger experience, the airport–airline relationship is highly competitive. The different commercial objectives and priorities of airports and airlines may not always align. Airports need to optimise the use of their assets and maximise the revenues they can derive from airlines and their passengers, while highly mobile airline operators, particularly low cost carriers (LCCs), can and do relocate to other airports to take advantage of more favourable financial terms and lower airport charges.

Given the operational and financial importance of the relationships to both airports and airlines, this chapter examines the scope and changing nature of the interaction between airports and airlines. In particular, the growth of LCCs (➤Chapter 8) and the expansion of global airline alliances (➤Chapter 10) have changed the nature of the airport–airline relationship and necessitated new approaches to airport charges, airport terminal design and configuration, and the development of new services. The chapter explores the complexities of the airport–airline relationship, discusses how and why it has evolved and highlights some of the risks and opportunities involved from both an airport perspective and an airline perspective, with a particular focus on airport charges, airport services and facilities, airport–airline contracts and the route development process.

7.1 The changing nature of the airport–airline relationship

The nature of the relationship between individual airports and their airline partners differs according to geographic, political and commercial contexts. A major hub airport, for example, could have close to 100 full service network carriers (FSNCs) operating into it all year round, while a secondary or regional airport may be served by only a limited number of low-cost, charter or regional airlines (➤Chapter 8) at certain times of the year. Worldwide, airports and airlines are subject to different national regulatory regimes and degrees of government intervention, with some countries pursuing a more liberalised aviation policy while others are more protectionist (➤Chapter 3). Different airport and airline ownership patterns and levels of local competition also affect the nature of the relationship. The main factors which influence the nature of the airport–airline relationship can be identified as:

Privatisation: the full or part change of ownership or management from the public to the private sector.

Commercialisation: the imposition of commercial objectives on an organisation. Commercialisation can occur under public or private sector management.

- the extent of **privatisation** and **commercialisation** of individual airports and airlines;
- the regulation/deregulation of airline routes, passenger demand and forecasts for future traffic;
- the types of airline and airport business models and passenger expectations;
- the relative strengths, scale and market power of individual airports and airlines.

Historically, the common state ownership and operation of major airports and airlines meant that the airport–airline relationship was very close and developed in the joint interest. In the UK, for example, the British government owned both British Airways (BA) and the British Airports Authority (BAA) who operated a number of major UK airports, including BA's main base at London Heathrow, until they were both privatised in 1987. This meant that UK airport policy developed, at least in part, to meet the needs of the country's national airline. Moves towards deregulation and privatisation, particularly in Europe, from the late 1980s and 1990s onwards, rapidly dissolved the close relationship that had existed between airports and airlines and introduced new elements of competition and commercialisation (➤Chapter 6). This meant that both airports and airlines were suddenly competing for custom and seeking to develop new business opportunities. Even airports and airlines that still remain fully or partly state-owned have had to adapt to the changing commercial environment by investing in new infrastructure and new products to attract and retain airlines.

Under traditional regimes of state ownership, the commercial interaction between airports and airlines was a simple supplier/customer relationship. The airports supplied the infrastructure, in the form of runways, passenger terminals and departure gates and were primarily concerned with providing for airlines as opposed to passengers, since airport charges, mostly in the form of landing and passenger fees, represented their main source of revenue. The introduction of a more commercialised and competitive environment resulted in airports placing greater emphasis on non-aeronautical revenue streams that are derived from retail, food and beverage, car parking, property development and other commercial services. This has had the effect of complicating the traditional airport–airline relationship and, arguably, airports can now be considered as **two-sided businesses** (see Figures 7.1 and 7.2). This is because the positive interdependence between the airline and passengers means that airport operators are incentivised to compete for both airline traffic and passengers since these two groups influence both aeronautical and non-aeronautical revenue. A lack of passengers impacts the airline operators, who may then have to leave the airport. If airlines reduce or withdraw their services, this reduces passenger numbers and consequently the airport's non-aeronautical revenue.

Two-sided businesses: businesses that provide platforms for two distinct customers who both gain from being networked through the platform.

Another way of viewing the airport–airline relationship is by considering it as a vertical structure in which airports constitute a market which sells an essential input for the airline output. Within this context one might expect some formal vertical integration to occur. When there is state ownership of both the airline and airport when the government ultimately

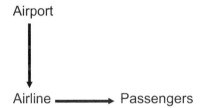

Figure 7.1   The traditional airport–airline relationship

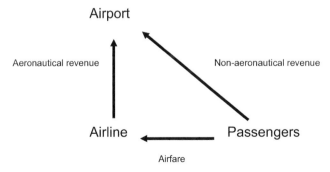

Figure 7.2   The contemporary airport–airline relationship

controls the airport–airline relationship, this integration can occur, as at Dubai, Abu Dhabi, Qatar and Singapore. Otherwise only a few rare examples exist, including the following:

- The US, where some airlines own/lease, manage and/or operate the terminals they fly from, as is the situation at New York's JFK Airport.

- Lufthansa's involvement with two German airports (see Case Study 7.1).

- Humberside Airport in the UK is owned by the Eastern Group, which also owns Eastern Airways which is based at the airport.

- Bangkok Airways owns three small Thai airports to enable it to effectively serve tourist destinations located near these airports.

In general, airport or terminal ownership by an airline can give the carrier exclusive use or greater control over the airport facilities and a stronger brand presence, but there may be a number of regulatory and competition issues that need to be considered to ensure this does not lead to discriminatory practices, especially if international traffic is involved. However, in the absence of a fully integrated industry, the distribution of risks and opportunities associated with the airport–airline relationship are always open to negotiation.

CASE STUDY 7.1

VERTICAL AIRPORT–AIRLINE INTEGRATION: LUFTHANSA AND GERMAN AIRPORTS

Lufthansa has invested in two German airports. This began with as a joint venture with the Munich airport operator to build and operate the second terminal which opened in 2003. The airport operator has a 60 per cent shareholding whilst Lufthansa has a 40 per cent shareholding. Lufthansa, its partner airlines, and airlines of the Star Alliance use this terminal. A new satellite facility, again developed and operated on a 60:40 basis, was opened in 2016. This expansion increased Terminal 2's capacity from 11 million to 36 million passengers per year. This airline involvement gives Lufthansa considerable control over its operations at Munich Airport. Lufthansa also acquired 4.95 per cent of Fraport (the operator of Frankfurt Airport) in 2005. It stated that this was to intensify its partnership with Fraport and to strengthen its position at Frankfurt (which is its major hub) at a time when competition in the industry was no longer between airlines and alliances, but between air traffic systems in their entirety, encompassing airlines, airports and air traffic control. It has subsequently increased its shareholding to 8.44 per cent.

The development of the airport–airline relationship can be challenging. Airports have high levels of sunk costs in infrastructure which, unlike aircraft, cannot be transferred to another location. This makes airports vulnerable to airline operators suddenly withdrawing services. Airports have to plan to be as flexible as they can owing to the degree of uncertainty in forecasting, the dynamic nature of the market (►Chapter 5) and the changing needs of airlines. Meanwhile, because of the long lead time for airport investment, airlines may have to start paying for new capacity a long time in advance. In essence, the planning time horizons for airports and airlines are frequently not aligned.

Greater competition between airports has also meant that neighbouring airports have increasingly attempted to attract airline operators away from their current airport by offering financial and other incentives. Airports may feel the need to meet the requirements of new airline operators and support the development and diversification of routes, but in doing so they will have to effectively manage possible tension between existing operators and new entrant airlines, particularly if the former perceive or can prove that the latter are receiving preferential terms. Moreover, at many major international airports where demand is strong and there is a high propensity to fly, the level of congestion can determine whether or not an airline is able to gain access to the market. Severe slot constraints (see ➤Chapter 6) can prevent market entry and protect incumbent carriers from the effects of new competition by acting as a barrier to entry unless an airline withdraws a service. The impact of this on the airport operator is that the airline may use the slots on the most lucrative routes rather than on those routes the airport operator would prefer they served.

> ## Stop and think
>
> How have privatisation and commercialisation affected the relationship between airports and airline operators?

7.2 Key aspects of the airport–airline relationship

The level and structure of airport charges

An important aspect of the airport-airline relationship is the level and structure of airport charges (see Table 7.1). These charges generate the airport's aeronautical revenue that represents a significant share of the airport's total revenue (➤Chapter 6). For the airlines, airport charges usually represent a relatively small part of their total operating costs (typically less than 10 per cent), but this varies considerably by type of airline.

Airport operators may not have total control over the airport charges that they levy, as these may be subject to some form of economic regulation (➤Chapter 6). Nevertheless many

Table 7.1 Examples of airport charges (note that not all charges will be levied by all airports on all airline operators)

Runway fee/landing fee	Airport development fee
Passenger fee/security fee	Environmental levies
Handling fee/infrastructure charge	
Handling fee/infrastructure fee	Terminal navigation fee
Aircraft parking fee	

airport operators have flexibility to make their charges as attractive as possible to the airlines, especially when new routes are being considered. Often this involves reducing the charges for a certain period. Such discounts will usually diminish as traffic grows and the service becomes sustainable. One of the most popular methods is to waive or reduce the landing fee in the first few years of operation so that the airline pays only for the passengers it carries. If demand at the start of a service is initially low, the airline will pay very little. This means the airport will share more of the risk when the airline is developing the route. There may also be discounts on passenger and parking charges. These discounts may be more generous for certain routes (e.g. long-haul, transfer) if the airport is particularly keen to attract this type of traffic (see Example 7.1). A key challenge, if an airport operator is offering incentives to attract new airlines, is that this can lead to pressure from established carriers for equivalent terms, and it may prove difficult for the airports to maintain good relationships with existing airlines. This may be addressed by offering incentives to existing airline operators to persuade them to offer more capacity, increase frequencies or use the airport as a base.

Example 7.1

International landing charge incentives at Kansai Airports 2017–2020

	Year 1	Year 2	Year 3
Network development (passenger flights)			
New route incentive			
Short haul	90%	60%	40%
Mid and long haul	100%	80%	40%
Growth incentive			
Short haul	80%	50%	30%
Mid and long haul	90%	60%	40%
Network development (cargo flights)			
Growth incentive	80%	50%	30%
Strategic route development (transfer flights)			
Route incentive			
Shorter route	100%	100%	100%
Longer route	As with passenger network development	As with passenger network development	As with passenger network development

Notes: Short haul = < 3,000 km from Kansai Airports; mid and long = > 3,000 km from Kansai Airports; the incentives represent the percentage of the normal landing charge that is not paid

Source: Kansai Airports

Airport services and facilities

Different airline business models (➤Chapter 8) influence the nature of the airport–airline relationship not only with regards to the airport charging strategies, but also with the airport services and facilities that are provided to meet the needs of these different carriers. Airports have to invest in passenger facilities and airside infrastructure in accordance with the specific traffic type of its airlines. Many FSNCs belong to global alliances (➤Chapter 10). As a consequence, they want to be able to share and achieve cost economies and brand benefits from operating joint facilities at airports, for example with common check-in areas and airline lounges. Moreover, where possible, they seek adjacent stand parking with alliance partners so as to allow for ramp transfers. This can be problematic if the relevant airport infrastructure was not initially designed in this manner. To compete effectively as a hub airport for transfer traffic, as is the situation with airports serving alliance airlines and others such as Dubai, the airports need to have an attractive **minimum connect time (MCT)** as well as passenger and baggage facilities that can effectively deal with this type of traffic.

By contrast LCCs have traditionally sought quick turnarounds, short taxi times and more basic facilities (with no airline lounges) to raise productivity and reduce costs. Their focus on point-to-point services means that more complex transfer passenger and baggage handling systems can be avoided. They choose to avoid facilities such as airbridges that they have to pay for and that can slow down their operations. A LCC that fills vacant capacity at a low marginal cost to the airport can be an asset, provided the airport is breaking even by assessing the costs and covering them. The moment new investment is needed to accommodate a low cost operator, an important decision is required by the airport management. This is particularly challenging when an airport serves both FSNCs and LCCs, as it may be a difficult task to meet the different and often conflicting needs of these two types of airline. One option that has been used is to develop a specialised dedicated low cost terminal (LCT) or pier facility designed to fulfil the needs of LCCs. These have a simple design, with lower service standards than expected in conventional terminals, with a lack of sophisticated equipment and facilities, namely airbridges, escalators, complex baggage systems and airline lounges. This usually results in the airlines that use the terminal being charged a lower passenger charge. Some of these facilities have been refurbished from existing facilities, including cargo or charter passenger terminals or maintenance buildings, but some have been constructed as dedicated new buildings.

Finavia, the governing body of the Finnish airports, was one of the first airport operators to open a LCT at the small airport of Tampere-Pirkkala airport in 2003. A number of low cost facilities were then built in Europe in the next ten years. Examples include Budapest, Amsterdam, Marseille, Lyon, Copenhagen and Bordeaux. One of the problems with dedicated low-cost terminals is that the simple design means than there are only limited opportunities to develop non-aeronautical revenues. This is one of the reasons why the low cost facilities at Copenhagen and Amsterdam are provided after the commercial facilities (at the pier level) in the passenger journey so that all travellers have access to the commercial facilities and the general departure lounge. Elsewhere in Asia, a notable development was the opening of two major LCTs in 2006, at Singapore and Kuala Lumpur. However, both of these original buildings have now been demolished to make way for new terminals which can handle more diverse traffic and have a wider range of commercial facilities.

Minimum connect time (MCT): the minimum time interval between a scheduled arrival and a scheduled departure for two services to be bookable as a connection.

Route churn: the frequency and volume of route entry and exit at an airport

Another common strategy used by LCCs to fully exploit their point-to-point low cost operations has been their use of secondary or regional airports. This typically reduces the costs and provides access to uncongested slots. Although this has led to significant growth at some formerly underutilised airports, it has also resulted in these airports being more vulnerable to the withdrawal of airline services. In particular, LCCs may swiftly withdraw from airports if the services prove to be unprofitable or new competition emerges, and this has resulted in **route churn** increasing at many of these airports.

In essence, while airline liberalisation introduced distinct airline business models (➤Chapter 8), it also led to the emergence of distinct airport models designed around primary and secondary airports. However, just as the airline models are converging to a certain extent, this is also the case with the airport models. In particular, some LCCs are now shifting into primary airports. For example, Ryanair moved from the secondary to the primary airports serving Barcelona, Brussels and Frankfurt in 2010, 2014 and 2017 respectively. This has been driven by a number of factors such as: greater airport convenience for passengers (especially those for business); a pricing premium to be gained; a desire to compete with network carriers or feed or code-share; the use of larger aircraft needing a greater volume of passengers; and the maturing of demand. This has had a major impact on the competition between the primary and secondary airports, and indeed has further challenged the viability of some secondary airports.

The partial convergence of airline models has led to the development of a variety of different approaches being adopted by airports to meet the changing needs of the airlines. Arguably, the popularity of dedicated LCTs seems to be declining, but there are many examples of airports that handle LCCs in different terminals, including Terminal 3 at Dubai, Terminal 1 at Mumbai, Terminal 1 at Tel Aviv, and Terminal 3 at Paris CDG. Others have shifted LCCs into older terminals, as has happened at Madrid, Milan Malpensa and Munich. Nevertheless, there have been some proposals for new LCTs, for example in Buenos Aires, Columbo and Nagoya. In general, each individual airport needs to decide how to weigh the costs involved with developing new dedicated facilities as opposed to having a standard offer; the flexibility that this affords the airport, given the increasingly volatile airline market; and the implications for both the passenger experience and commercial revenue generation.

Another consequence of the blurring of the FSNC and LCC model is the emergence of LCCs (such as Norwegian, Air Asia X, WestJet) that have long-haul routes (see ➤Chapter 8). Moreover some LCCs, which traditionally focused on just point-to-point services, are now offering connecting flights (some long-haul) or are entering into agreements with other airlines to allow this to take place. For example, Ryanair is now offering connecting flights at Rome Fiumicino, Milan Bergamo, Porto and Brussels Charleroi airports; and easyJet has its *Worldwide by easyJet* concept, through which connecting flights can be booked with certain easyJet flights or with partner airlines including Cathay Pacific and Emirates. Passengers are given missed connection protection provided by the travel search engine Dohop. A growing number of airports are also encouraging self-connection or so-called 'self-help' hubbing. This is when passengers buy two separate tickets and build their own connection, instead of it being arranged by the airlines. For this to happen, airport services and physical infrastructure may need to be adapted to make it easier for passengers to change flights. Ideally this should allow for the passengers to be kept airside and for their bags to be automatically transferred, as with conventional connections so that the transfer process is as streamlined as possible. The airports provide hosted transfers that offer self-connecting passengers a level of support which they would not get if they self-connected on their own. Gatwick, with its GatwickConnects service, and Milan Malpensa, with its ViaMilano self-connect service, are two examples (Case Study 7.2).

CASE STUDY 7.2

SELF-CONNECTION AT AIRPORTS: GATWICKCONNECTS

GatwickConnects is a self-connection service available for ten airlines (Aer Lingus, Air Transat, Aurigny, British Airways, Cathay Pacific, easyJet, Norwegian Airlines, TUI, Virgin Atlantic, and WestJet) and it supports the *Worldwide by easyJet* concept. It includes a free baggage service where baggage taken to the GatwickConnects desk in the baggage reclaim area can be checked in and loaded onto the connecting flight. There is also a possibility of booking the connecting flights via the GatwickConnects flight booking service (provided by the travel search engine Dohop) which, for a fee absorbed into the booking cost, will provide a protected connection (e.g. another flight, food, overnight accommodation) if the flight is missed. Premium security with a dedicated security lane and some discounts on commercial facilities are offered. Since its launch in 2015, the seats per annum sold through the GatwickConnects service has increased from 2,000 to 214,000 in 2018–19.

Airport–airline agreements and contracts

The stability, level of commitment and longevity of the airport–airline relationship will depend on the carrier, the competition and the context. On the one hand, an airline may decide to fly from an airport that it has not flown from before, maybe just to test the market, and so in this case there will be no initial long-term agreement made between the two bodies. The other extreme, however, is when an airport is an airline's home base (as is typically the case with the main airline and airport of a country) or a main hub, where there is a different permanent dynamic to the airport–airline relationship and the airline and airport can find themselves in a position of mutual dependency. Numerous examples include American Airlines at Dallas/Fort Worth, KLM at Amsterdam Schiphol, Emirates at Dubai, Lufthansa at Frankfurt and British Airways at London Heathrow.

Formal long-term contracts between airlines and airports are common in the US, where there are use and lease agreements that describe the nature of the operational and financial relationship between the airports and airlines (see Example 7.2). Elsewhere it has only been more recently that long-term contracts have been agreed at a number of airports. This sometimes may be the result of a light-handed regulatory system (➤Chapter 6), but not always. Long-term agreements are particularly popular with LCCs. For the airlines, such contracts can provide them with some certainty about the charges and facilities that they can expect in the future, which can be crucial for developing services. Meanwhile, for the airports, such airline commitment can provide a more stable environment for making future plans. Hence when these contracts work well, they can help alleviate the problem of airlines and airports having different planning horizons, with the former generally having a shorter-term outlook than the latter. Such contracts typically range from 5 to 20 years, where the airport operator will offer discounted charges, maybe with a volume discount, in return for long-term commitment from the airlines. There will be a number of other obligations on the airport operator, including the quality of service to be supplied (particularly regarding minimum turnaround times) and the requirement to undertake marketing on behalf of the airline. If the contract covers a long period, there might be commitments by the airport operator to undertake staged investment. In return, the airline will typically be obliged to guarantee that it will base a certain number of aircraft at the airport and provide a roll-out programme for adding additional aircraft. Sometimes, the airline will also have to guarantee a minimum number of passengers.

Example 7.2

Airport use and lease agreements in the US

Although not compulsory, a number of US airports enter into formal and legally binding use and lease agreements with their airlines. These use and lease agreements describe the nature of the operational and financial relationship between the airport and the airlines. They define the rights and responsibilities of each party, set charges, stipulate how the airport can be used, and define how the financial risks will be shared. These agreements typically consist of two elements:

- *Lease agreements* which concern an airline's occupation of airport buildings.
- *Use agreements* which detail an airline's use of airport facilities.

Owing to the capital intensive nature of airport infrastructure and operations, these agreements also define how the financial risk of current and future operations will be shared between the airport operator and the airlines, and in what proportion. Three basic models can be identified:

- *Residual agreements* in which airlines assume the financial risks of airport operations.
- *Compensatory agreements* in which the airport assumes the financial risk of operations.
- *Hybrid agreements* which combine elements of both models so that both airlines and the airport share the financial risk.

In recent years there has been a move towards compensatory agreements to reflect the more uncertain and volatile deregulated air transport environment.

!

Stop and think

What issues do LCC operations present to airport operators?

7.3 Route development

The development of new routes or the increase in frequency of existing services ultimately depends on the decisions of the airlines, but airports and airlines have a range of ways to establish new relationships to encourage this to happen. Airports and airlines can approach each other on an individual basis, but there are also route development forums that are organised to provide opportunities for airline route planners and airport operators to meet. The airports can sell the virtues of their traffic potential, facilities and services, while the airlines can explain their expansion strategies. There are now a significant number of these networking events with 'Routes' being the market leader (see Case Study 7.3). Other similar events include North America's 'JumpStart' and Europe's 'Connect'. In addition, there are route development opportunities online (e.g. the Route Shop, the Route Exchange) where airports and airlines can share relevant information electronically to assess opportunities for new relationships and services.

CASE STUDY 7.3

ROUTES: FACILITATING AIRPORT AND AIRLINE NETWORKING

The Routes company organises route development forums that provide networking opportunities for airline route planners and airport operators through one-to-one meetings. There is an annual global event (first held in Cannes in 1995) as well as region-specific meetings, namely Routes Americas, Routes Asia, Routes Europe and Routes Africa. In 2018, a total of 6,134 delegates attended Routes events, including 1,086 airline delegates, 2,239 airport delegates and 368 tourism association delegates. More than 16,800 face-to-face meetings took place over three days. Pre-scheduled, on-site and city pair meetings were held. These events are supported and complemented by the online platform for route development called Route Exchange. This site provides airports with the ability to promote their market opportunities to airlines with details about unserved/underserved routes and other information, including the marketing support, airport infrastructure and services. In September 2019, Route Exchange had 197 listed airlines, 155 listed airports, 11 listed destinations and 15 listed suppliers. In 2018, it had 28 million page views.

The extent to which an airport might negotiate with an airline will depend upon whether it is an existing or new operator, a major airline adding a new service, a low-cost operator promising a network of services or an independent airline offering a route for the first time. All these services pose a different level of risk and potential reward and will require different forms of support. When negotiating, airport managers will assess the risks and benefits on a case-by-case basis. They need to be flexible in their approach and understand and meet the needs of the different airline users (which may depend on the airline business model, airline nationality, route or time of the year). They will consider not only the aeronautical revenues but also any associated additional non-aeronautical revenues that might arise. They will also need to assess the impact on existing airline relationships and the infrastructure cost implications of accommodating new services.

Route development process: the process associated with attracting, growing and retaining air services at airports

Catchment area: the area or geographical reach from which most passengers at the airport are likely to be generated.

The **route development process** (also known as air service development process in some countries) is a very important activity that airport operators will commonly undertake to attract new airlines or encourage loyalty and grow services from current airlines. The overall aims are to identify potentially viable routes that are not currently being served or are underserved and, ultimately, to produce route-by-route forecasts and a feasibility assessment.

There are typically seven stages to the process. Firstly, the airport's **catchment area** needs to be defined. This is usually measured by isochrones of equal travel distance or time. This leads to a market assessment when the level of air travel demand within the catchment area is estimated. This will depend on factors such as the economic, business and tourist activity within the area, the demographic characteristics of the residents, and past immigration patterns. The assessment will also take into account neighbouring catchment areas of other airports, and estimates of how much traffic might 'leak' to and from them. For example, if the fare is low enough, passengers may be motivated to travel further (outside the natural catchment area) to access this lower fare with a different airline at another airport.

Once the airport operator has assessed the market within its catchment area and associated leakages, it can determine the adequacy of air services at the airport and identify routes that are not served satisfactorily. This is the third stage of the route development process. This is followed by the estimation of traffic forecasts for these routes and often an evaluation of the financial viability of the routes, or their revenue generating ability. This leads to the last stage of the process, when the airport will use this research to present the business case for the routes to the chosen airline or airlines (maybe at a route development forum) to demonstrate an understanding of the market, catchment area and other traffic characteristics. At the same time they will explain the financial incentives on offer and other marketing assistance as well as describe the merits of services and facilities available at the airport.

However, it is important to recognise that the route development process often goes beyond just the airport–airline relationship, as it may include other stakeholders who also have an interest in growing air services. Typically this may include tourist authorities or destination management organisations who are keen to grow tourism flows. Working together and offering incentives within such tripartite relationships and partnerships, often known as the 'golden triangle', can be more effective than focusing purely on the air service aspect. Local and regional governments may also want to get involved to complement the co-operation between airports and airlines, primarily to reap the economic benefits that growth in air transport may bring. This is often by offering additional support or incentives to airlines. Examples include minimum revenue guarantees in the US, where a mixture of airport incentives and local/government support guarantees protection of airlines from potential losses in the event that the flight does not produce the desired level of profitability. In Europe, there can be government route development funds that provide additional start-up aid for new services. Moreover, in remote locations, the national government may decide to subsidise airline services as part of a Public Service Obligation (PSO) route (in Europe) or Essential Air Service (EAS) in the US, for social and economic development reasons (➤Chapter 21).

Stop and think

How will the airport–airline relationship change in the future, and what factors will affect it?

!

Key points

- The airport–airline relationship is complex, dynamic, competitive and co-dependent. It is unique and varies according to local context.

- Factors which affect the airport–airline relationship include: the extent of the privatisation and commercialisation of individual airports and airlines; airline deregulation; and the relative strength and market power of the airport and airline.

- Two-sided businesses and vertical integration are aspects of the airport–airline relationship that are becoming increasingly important.

- Airports are fixed geographically, so they do not have the flexibility that airlines have. They need to consider the long-term implications of arrangements they enter into and capital investment decisions that they make. Airlines have typically shorter planning time horizons.

- Airport charges and the presence of charging incentives have a major impact on the airport–airline relationship.

- The development of airline business models has resulted in the evolution of different airport models.

- Long-term contracts between airports and airlines are becoming more common.

- The route development process is important for developing airport–airline relationships.

References and further reading

ACRP. (2010). *Airport/airline agreements: Practices and characteristics*, Washington, DC: Airports Cooperative Research Program, Transportation Research Board.

De Wit, J. G. and Zuidberg, J. (2016). Route churn: An analysis of low-cost carrier route continuity in Europe. *Journal of Transport Geography*, 50, pp. 57–67.

Dobruszkes, F., Givoni, M. and Vowles, T. (2017). Hello major airports, goodbye regional airports? Recent changes in European and US low-cost airline airport choice. *Journal of Air Transport Management*, 59, pp. 50–62.

Dziedzic, M. and Warnock-Smith, D. (2016). The role of secondary airports for today's low-cost carrier business models: The European case. *Research in Transportation Business and Management*, 21, pp. 19–32.

Graham, A. (2013). Understanding the low cost carrier and airport relationship: A critical analysis of the salient issues. *Tourism Management*, 36, pp. 66–76.

Graham, A. (2018). *Managing airports: An international perspective*. 5th edn. Abingdon: Routledge.

Halpern, N. and Graham, A. (2015). Airport route development: A survey of current practice. *Tourism Management*, 46, pp. 213–221.

CHAPTER 8

Airline business models

Randall Whyte and Gui Lohmann

LEARNING OBJECTIVES

- To identify the process of business modelling and its importance to airlines.

- To compare and contrast the business models of full service network carriers (FSNCs), low cost carriers (LCCs), charter operators, regional airlines, hybrid airlines and specialist operators.

- To appreciate the competitive threat LCCs pose and understand how FSNCs have responded.

- To explore the internationalisation strategies pursued by some FSNCs.

- To consider the motivations for, and impacts of, global airline alliances.

- To understand the reasons for airline failure.

8.0 Introduction

A business model describes a company's purpose, its approach to doing business, its brand proposition and its strategic corporate objectives. This chapter examines the critical operational and managerial characteristics that differentiate airlines as being either **full service network carriers (FSNC)**; **low cost carriers (LCC)** (or ultra-low cost); charter; regional; hybrid or specialist operators. The business model which is developed by each airline informs the target customers, the route network, the choice of aircraft, the revenue streams and forecasts, the marketing strategy and the operational cost structure of the company.

Full service network carrier (FSNC): an airline that offers high levels of in-flight service and connectivity, attracts a range of passenger segments to its network of short and long-haul routes, and operates a variety of aircraft types.

Low cost carrier (LCC): an airline that adopts a rigorous cost-minimisation strategy to keep its costs and fares relatively low.

Business model: a conceptual structure or plan that defines how a company conducts its business.

Brand proposition: a statement about what your brand has to offer.

Product differentiation: the process of distinguishing a product from those of its competitors by highlighting its unique attributes or qualities.

Seat pitch: the distance between rows of seats in an aircraft. The greater the seat pitch, the more leg room passengers have.

8.1 Airline business models

Airlines operate in a dynamic and highly competitive business environment. This demands flexible approaches to conducting business that can rapidly adapt to sudden periods of economic downturn as well as exploit new market opportunities during times of economic prosperity. Airline operations are both capital and labour intensive (➤Chapters 12 and 19) but also vulnerable to external factors such as fuel price rises, currency fluctuations, increased competition, weakening of consumer demand and political unrest, regulatory change, terrorism and threats of war. Consequently, airline profitability remains a major challenge. All airlines develop, operate and continually refine (and, in some cases, redefine) their **business model** in response to changing market conditions in order to remain operational and competitive.

Airlines are not unique in having business models. Indeed, business models are used by virtually all companies, from small to medium-sized enterprises and start-ups to major multinational corporations and from retail chains to financial service providers. In essence, a business model describes a company's purpose, its approach to doing business, its **brand proposition** and its strategic corporate objectives. The purpose of business models in the airline industry is to enable managers to identify target customers, develop marketing strategies, identify different revenue streams, establish a robust cost structure, specify margins and build flexibility into their organisation. Business models will often reflect the airline's country of origin, economic characteristics and its culture.

Airlines have become adept at segmenting their markets and using **product differentiation** to cater for the needs of different types of passengers. For example, British Airways on long-haul flights configure a four-class service compared to a low cost carrier with a single class (economy) service with a narrow **seat pitch**. As a consequence, the nature, distribution and cost of services that airlines provide differ substantially. Some airlines, for example, create a product and sell it directly to their consumers over the internet, whereas others also use commissionable intermediaries such as **wholesalers** and travel agents. For example, LCCs use the internet to sell their tickets, whereas charter airlines are more likely to use travel agents. Many airlines either belong to a strategic alliance comprising many airlines or engage in a strategic partnership with just one or two airline partners. An example would be Virgin Australia and Delta on routes between Australia and the US. Some carriers may position themselves in the market and decide to offer a **value proposition** based on high levels of service and multiple different travel classes or alternatively adopt a more streamlined no-frills approach to customer service. Some carriers fly a mix of aircraft types on long- and short-haul services to major airports, while others have identified a niche market operating particular types of aircraft into smaller regional airports or tourist destinations. This enables different types of business models to be identified.

Types of airline business model

Airlines can be categorised as adopting one of six basic types of business model. These are: full service network (or legacy) carriers (FSNCs), many of which are national flag carriers; and LCCs, hybrid carriers that are a crossover between LCCs and traditional carriers, charter, regional or specialist airlines, depending on the characteristics of the products and services they offer (see Figure 8.1). While this is a useful guide, these categories are not absolute, and some overlap exists between them (Moir and Lohmann, 2018); for example, when FSNCs operate in regional short-haul markets or where charter carriers have adopted attributes more commonly associated with LCCs.

One of the most significant developments has been the emergence and rapid expansion of a new type of airline business model that began with US-based Southwest Airlines in the 1970s. Southwest is now the largest airline in the US, and 163.6 million passengers flew with the airline in 2018 (BTS, 2019). As a consequence of the US Airline Deregulation Act of 1978 and policies of air service liberalisation in other world markets (►Chapter 3), a number of airlines and new entrant operators have applied the principles of Southwest Airlines with various adaptations. LCCs focus on minimising costs and offering a 'no frills' product and low fares in a streamlined operation especially in short-haul markets (one to three hours flight time); see Section 8.3.

FNSCs have more recently responded to the challenge from LCCs especially on short-haul routes, and in some respects both FSNCs and LCCs have converged their offering with a standard economy class product. As later sections in this chapter will show, LCCs have been adept at growing market share by stimulating demand for services through the provision of lower fares, exploiting point-to-point routes rather than a hub-and-spoke network, entering markets that have either been vacated or ignored by FSNCs and offering low fares. However, when it comes to long-haul markets, LCCs have had only a minimal impact as (a) different types of aircraft are required and (b) the different operating characteristics of long-haul markets mean that the same cost savings attainable on short haul operations cannot be achieved on long-haul operations (Whyte and Lohmann, 2015).

Wholesaler: a company which buys large quantities of goods from multiple producers (in this case, airline seats) and sells them to retailers (for example, tour operators), typically at a low price.

Value proposition: an innovation or service feature which aims to make a company's product attractive to potential customers.

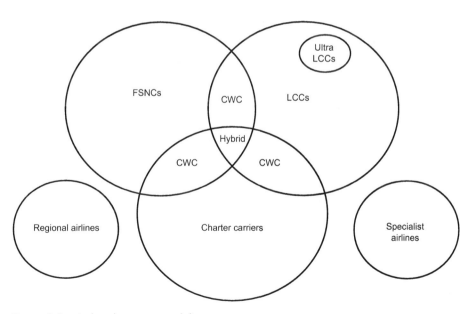

Figure 8.1 Airline business models

The need for airlines to focus on their business models has been driven by:

- the rapid growth of LCCs in most aviation markets since the late 1990s and the resulting erosion of FSNC market share;

- the need to address and reduce costs across the business and focus on yield management (➤Chapter 9);

- the reticence of national governments to provide bailouts and subsidies to national airline operators (as well as regulations governing State Aid in Europe);

- a reassessment of route network, capacity and flight frequency, including building partnerships and codeshares with alliance partners;

- the need to fulfil fleet requirements through financing and leveraging debt (➤Chapter 12);

- airport capacity and congestion and the availability of slots at major airports;

- the need to adopt a product differentiation strategy suited to the destination, length of haul and market served;

- the cost of marketing and advertising and using new technologies in an increasingly competitive market (➤Chapter 20).

Every airline has to assess both its current position in the market and where it would like to be. Certain markets are nearing maturity; others are emerging, for instance Asia; while some (particularly in Africa) are restricted by regulations that effectively prevent the adoption of more innovative and financially sustainable business models. A complete business analysis should include actual and predicted route performance, such as the number of passengers, classes of travel, load factor; yield (revenue per seat kilometre); seasonal fluctuations; route profitability; product and service offering; competitors' strengths and weaknesses; and the different operating characteristics of short-, medium- and long-haul routes, where applicable. In addition, labour costs (➤Chapter 19), the second highest expenditure after fuel, is an important area to review, as is how and where maintenance of the aircraft is conducted.

8.2 Full service network carriers (FSNCs)

Many of the world's major FSNCs, including British Airways, Lufthansa, and KLM, were originally established as state-owned 'flag carriers' with extensive global networks. Over time, they have developed partnerships with other operators and have adapted their business model in response to a changing competitive environment. Table 8.1 illustrates the key characteristics of FSNCs in terms of their historic and current operations.

Financial pressures on governments combined with growing moves towards deregulation, privatisation and maturity of the industry led many governments to sell airlines and airports to private companies from the 1980s onwards. The mid-1990s and the first decade of the 21st century saw the emergence of a new competitive threat in the form of LCCs, especially in short-haul regional markets, and intense competition in almost every market. FSNCs have

Table 8.1 A comparison of the historical and current characteristics of FSNCs

Characteristic	Historical situation	Current situation
History	Most were state-owned enterprises and supported by governments.	Governments have fully or partly privatised national airlines and shares are traded on the stock exchange.
Route network	Operated extensive worldwide networks to 'fly the flag' but were often loss-making. Bilateral Air Service Agreements and reciprocal landing rights were commonplace.	Free of direct government intervention; have a clear focus on markets; are more efficient and customer focused with attention to costs. The move towards 'open skies', deregulation and liberalisation has enabled airlines greater freedom to choose where they fly to; however, bilateral agreements still apply in some markets. The hub-and-spoke operation is a key component of their operation.
Aircraft types	Governments dictated aircraft choice and aircraft purchases were often influenced by national interests such as to 'buy British'.	Most airlines determine fleet choices according to the merits of each type; route network and ideal capacity such as short, medium, or long haul; and financing requirements. Leasing is now more widespread.
Airports	Major airports used but governments invariably directed airlines to specific airports.	Major airports still predominate driven by traffic volumes and frequencies plus securing valuable (and expensive) slots.
Alliances	Before the advent of the modern airline strategic alliance, many airlines operated 'pool' agreements.	The majority of FSNCs are members of one of the three main alliances – Star, oneworld, or SkyTeam. Alliances enable code-sharing; an extensive route network; a seamless service; loyalty programmes; joint marketing; and aircraft purchasing and engineering agreements.
Industrial relations	Highly unionised workforce.	Enterprise Agreements now prevail between the different work groups, and subcontracting and casualisation are common. Long-serving employees are often still protected from changed conditions by legacy agreements.

met the challenge from LCCs and have adapted their product along with cost cutting so as to be competitive in order to retain market share along with reducing cost and inefficiencies. New technologies such as check-in kiosks at airports and self-online check-in have facilitated efficiencies and reduced the demand for labour. Contracting out and the casualisation of the workforce has been a further cost-reduction strategy, and greater fuel efficiency with new generation aircraft has greatly benefited airlines.

FSNCs have adopted a more flexible pricing strategy, such as a more flexible suite of pricing options according to the day of the week and the time of the flight and one-way prices along with stringent yield management practices (➤Chapter 9). Some FSNCs offer a limited number of 'economy lite' seats which are cheaper than regular economy fares but have more stringent terms and conditions attached to them.

Some elements of the FSNC model include: using major airports; developing one or more hubs that fundamentally differentiates FSNCs from LCCs; seeking route expansion through alliance agreements and/or code-sharing, operating a frequent flyer programme, contracting out ground handling, catering and engineering, and using travel agents and other commissionable sources as a distribution channel. One of the strengths of FSNCs is being a member of a strategic alliance. This enables a seamless service and gives airline consumers a wide choice of options (Case Studies 8.1 and 8.2). However, certain airlines choose to remain non-aligned or merely have a limited alliance. Examples include Dubai-based Emirates and Abu Dhabi-based Etihad Airways.

Customer relationship marketing has become a significant marketing strategy for FSNCs. Unlike LCCs, who continuously seek new customers and whose relationship with their customers may be merely a single transaction bought on the basis of price and/or schedule, FSNCs use customer relationship marketing to retain existing customers and obtain repeat business. This group includes mainly the corporate and government sectors, who may have to fly on peak days at peak times and are therefore high yielding passengers who are likely to belong to an airline's frequent flyer programme. The notion is to develop loyalty and reward repeat purchase, with incentives such as points earned for travel, and other benefits such as tickets to sports events, stage shows and exclusive retail offers (➤Chapter 10).

CASE STUDY 8.1

QANTAS

Founded in 1920 as the Queensland and Northern Territories Aerial Services Ltd, Qantas has grown to become Australia's largest domestic and international airline. The Group has two complementary airline brands – the premium full service brand Qantas and low cost Jetstar. Qantas operates a mixed fleet of narrow and wide-body Airbus and Boeing aircraft, including the A380. As a member of oneworld and its alliances with Emirates, China Eastern and American Airlines, Qantas links over 1,200 destinations in more than 150 countries. In 2018, Qantas carried 22 million domestic and 8 million international passengers on almost 5,000 weekly flights. In 2018, Qantas inaugurated a new service from Perth to London with a B787–9 aircraft. This was the first non-stop direct flight between Australia and Europe.

CASE STUDY 8.2

FINNAIR

Driven by Finnair's Asian-focused growth strategy, Finland (population 5.5 million) has been able to establish Vantaa-Helsinki Airport as a northern gateway to/from northern and central Europe. Finnair operates to ten Asian destinations. Unlike Singapore and the Middle East that are at crossroads between Australasia/Asia and Europe, Helsinki is different in terms of its geographic location. Almost half of the 24 million passengers flying to/from Asia in 2017 were transferring through Helsinki (Finnair, 2017). Finnair offers competitive flight times and fares, convenient scheduling and ease of connections to facilitate passengers through Helsinki. A key part of Finnair's strategy is fleet modernisation, reliability, product innovation, product upgrades and price competitiveness.

8.3 Low cost carriers (LCCs)

LCCs, or no-frills budget airlines, have become an established model of operation in domestic and regional markets worldwide as a result of deregulation and liberalisation of the airline sector (►Chapter 3) owing to ease of market entry and selecting routes with market potential. The pioneering LCCs, Southwest, Ryanair and easyJet, exploited first mover advantage to grow their business. In 2018, LCCs carried an estimated 1.3 billion passengers (31 per cent of all passengers on scheduled flights worldwide). Market penetration of LCCs is 36 per cent in Europe, 35 per cent in Latin America/Caribbean, 30 per cent in North America and 29 per cent in Asia/Pacific (ICAO, 2018).

LCCs apply a simple, streamlined business model where the control of costs is all important. While there are many common features of a LCC, the sector is not homogeneous, and differences occur between carriers according to location and the state of the market. For example, in Asia, LCCs have attracted many first-time air travellers, whereas in North America and Europe, the majority of passengers have already flown. However, although LCCs have had only limited success in long-haul markets, they have been successful in stimulating new demand for air services in emerging countries (Bowen, 2019). One such example is Colombia (Case Study 8.3).

CASE STUDY 8.3

VIVA AIR, COLOMBIA

Viva Air commenced flight operations in 2012 and, in 2019, flew around 40,000 flights, carrying over 6.3 million passengers, on 35 domestic and international routes using a fleet of 20 A320 aircraft. The carrier has adopted a low cost approach to its business and has been responsible for reducing fares and stimulating demand on domestic services within Colombia as well as on its international flights to Lima and Miami.

LCCs generally opt for a single-aisle aircraft such as the Boeing 737 or Airbus A319/320 variants. Fleet commonality enables these airlines to negotiate bulk purchase discounts with manufacturers as well as minimise staff training, maintenance and spares costs. The aircraft are operated in single-class configuration with a narrow seat pitch and limited galley space to add extra rows/seats. They tend not to use expensive terminal facilities such as airbridges and generally avoid major airports because of congestion and time delays. Instead, they may use cheaper and less congested secondary airports in order to reduce costs by increasing the utilisation of their aircraft as well as being able to access incentives that may be offered by the airport operator (►Chapter 7). Turnaround times are kept to a minimum, with 25–30 minutes being common. Other cost-reduction strategies include outsourcing and contracting out services to reduce fixed overhead costs; keeping advertising and promotional messages short and simple; and bypassing travel agents and encouraging customers to book directly with the airline. Technology has had a major impact, increasing the information available to consumers and giving them more market power. The unbundling of the product has been a marketing strategy deployed by LCCs in which 'add-ons' are charged separately. It is now common amongst LCCs to offer passengers preferential seats for an additional fee, limited carry-on baggage and extra charges for checked baggage. Unlike FSNCs, LCCs charge for food and refreshments and in-flight entertainment. The principal differences between FSNC and LCC business models are detailed in Table 8.2.

Another way in which LCCs seek to minimise costs is through their use of airport facilities. Table 8.3 shows the facility expectations of LCCs and FSNCs.

Table 8.2 FSNC and LCC business models compared

Business model element	FSNCs	LCCs
Aircraft fleet	Operate different types of both narrow (single aisle) and wide-body (two aisles) aircraft according to the route flown and required capacity.	Generally operate one type of narrow-body aircraft, such as the Boeing 737 or Airbus A319/320/321/320neo.
Aircraft utilisation	Lower than LCCs, due to delays, connecting flights, consolidating passengers at hub airports, time zones on long-haul operations leading to dwell time, including crew rest periods.	Aim for 11+ hours utilisation per aircraft per day through short-haul operations (1–3 hours flight time) and fast and efficient 30-minute turnarounds.
Route network	Operate a hub-and-spoke network to and from (more expensive and often congested) major airports. Transfers and interlining are common.	Operate point-to-point services between cheaper and less congested secondary regional airports. No transfers or interlining.
Product offering and in-flight service	Offer two-, three- or four-class cabin configurations according to route and demand. Seat pitch in economy is generally 32–34 inches (81–86 cm). Meals and in-flight refreshments are usually included in the ticket price (especially in premium cabins), as are headsets for in-flight entertainment.	One-class all-economy configuration, narrow seat pitch (28–30 inches; 71–76 cm); some LCCs apply no pre-allocated seating. Meals/refreshments and in-flight entertainment are offered as pay-for extras.
Target market	Corporate accounts, business and government travel. Medium- and long-haul markets (all classes of travel). Leisure travellers prepared to pay a fare above LCCs for perceived service enhancements.	Predominantly leisure travellers and less time-sensitive travellers. Increasing number of business travellers.
Pricing strategy (►Chapter 9)	Flexible fares set high, catering for corporate accounts and business and government travel. Other fares set according to day of week/time of day/holiday periods. Have evolved from round-trip fares to offering one-way fares in response to LCCs.	Simple fare structures. Offer one-way fares and use promotional offers to stimulate demand.

Business model element	FSNCs	LCCs
Strategic alliances (➤Chapter 10)	Most major airlines – with some exceptions – belong to an alliance to enhance operational and marketing position and to gain traffic.	Usually remain independent, although some subsidiaries of larger airlines do belong to an alliance.
Checked baggage allowance	Varies according to cabin class and ticket. Premium cabins typically include some hold baggage allowance.	Hold baggage typically incurs an added charge.
Advertising and promotion (➤Chapter 20)	Brand image and reputation is important. May use desirable aspects of national stereotyping (such as punctuality, reliability, good food and hospitality) in their advertisements.	Message is simple – a destination, a price and a brand.
Distribution	Online bookings are encouraged, but travel agencies and corporate bookings (which charge commission) remain important.	Use digital technologies to handle online reservations and avoid commissionable sources.
Frequent flyer programmes (➤Chapter 10)	Most offer a loyalty programme to reward frequent travel and incentivise repeat purchases.	Not always offered.
Labour and industrial relations (➤Chapter 19)	Often have legacy agreements that specify job functions and govern the rate of pay and overtime. Often highly unionised.	Pay lower wages and try to avoid collective union bargaining. Greater use of outsourcing and contracting to lower costs.

Ultra-low cost

Some LCCs, notably Hungary's Wizz Air, US Spirit Airlines and Singapore's Tigerair, are termed ultra-low cost carriers (ULCCs), which means they have unbundled the product to its fundamentals (a seat that does not recline and one small carry-on bag) and make passengers pay for any extra product or service they require such as faster boarding, stowing a hold bag, checking in using an airport counter and any in-flight food or refreshments. These add-ons are excluded from the advertised fares but generate important ancillary revenue for the airlines (➤Chapter 9). They may also be prepared to outsource IT and accounting functions if it saves them money.

ULCCs seek to reduce airport charges and costs and do not want expensive facilities. Local and regional councils have encouraged LCCs to use secondary airports often for no or little airport landing charge, as they consider airlines bring new businesses to the airport as well as visitor numbers.

Table 8.3 Facility expectations of LCCs and FSNCs

Expectations	LCCs	FSNCs
Ground access (►Chapter 6)	Location of secondary importance. Good road and rail links not essential but preferable.	Convenient city location essential for efficient ground access.
Terminal	Automated check-in processes. Encourage online check-in and printing of boarding pass.	High profile ticket desk and check-in areas reflecting corporate image and presence.
	Quick and efficient check-in required.	Check-in convenience and profile is of great importance.
	Terminal services such as food and retail of little importance.	Important that passengers feel purchasing needs are met.
Gate	Low tech gate facilities (air stairs).	High tech gate facilities (air bridges).
	Passenger seating areas at gate not necessarily required.	Business and first-class lounges required in addition to economy space (separation of different classes essential to the product's image).
	Need to separate incoming and outgoing passengers.	Need to facilitate connecting passengers.
General	Minimal catering facilities required.	Facilities for preparation of in-flight food essential since it forms part of the in-flight service offering.
	Cleaning staff required less frequently – cabin crew collect on board waste.	Aircraft cleanliness and presentation essential before boarding commences.
	No standby aircraft.	Standby aircraft required.
	Efficient removal and loading of aircraft baggage. No baggage transfers.	Efficient delivery of arriving baggage including transfers to connecting flights.

Stop and think

What are the potential benefits to airlines of adopting a ULCC business model?

8.4 Charter airlines

Charter airline: an airline that provides point-to-point services to popular holiday and leisure destinations, often as part of an inclusive tour (also known as a package tour).

Charter airlines originated in Europe and offer a distinctive type of service. Charter airlines operate on a demand-driven basis, specialising in special interest groups and tours. Seats on charter flights may be available only as part of an inclusive tour, and the charterer may purchase an entire flight's seat inventory and undertake to fill it for an agreed price. They often operate from cheaper and less-congested secondary or regional airports that are not otherwise served by scheduled airlines. See Case Study 8.4.

TUI

TUI Airways is the world's largest charter airline, offering scheduled and charter flights from the United Kingdom and the Republic of Ireland to destinations in Europe, Africa, Asia and North America. The airline carried 11.2 million passengers in 2018, making it among the largest UK airlines by total passengers carried.

Charter airlines' business is subject to seasonal demand. Charter airlines may lease their aircraft (➤Chapter 12) to foreign operators during leaner months of the year when demand is low. Many have also unbundled some of their products to compete with the low cost operators.

8.5 Regional airlines

A **regional airline** operates short-haul low-density routes often to where the major carrier chooses not to fly. Examples include Bangkok Airways within Asia, Dragonair (Hong Kong) and Hong Kong Airlines. Regional airlines can be independent (such as Eastern Airways in the UK), government owned (such as Aurigny of Guernsey in the Channel Islands) or **franchises** (such as Air Nostrum, which operates short-haul routes with smaller capacity aircraft as a franchisee of Iberia and Binter Canarias). See Case Study 8.5.

Regional airline: an airline that operates frequent short-haul routes within particular geographic regions, usually with a fleet of small regional jets or turboprops.

Franchise: an independent airline that uses another airline's branding and operates services on its behalf.

WIDERØE

Norwegian carrier Widerøe is Scandinavia's largest regional airline. Founded in 1934, it currently flies 2.8 million passengers a year on around 400 daily services to 50 domestic and international destinations. It operates all four versions of the Dash-8 turboprop which seat from 39 to 78 passengers. Importantly, the -100, -200 and -300 series aircraft can operate from the short runways that are found at some of the more remote airports in Norway. The airline was also the global launch customer for the E190-E2 jet aircraft which seats 114 passengers. The airline operates important lifeline services that link remote communities to larger cities by air.

Regional airlines form an important link to remote destinations which are difficult, expensive and/or time-consuming to access by surface transport modes (➤Chapter 21). In many respects, they are niche operators, flying secondary routes and providing the 'spokes' to/from a main hub. Regional carriers may have higher costs than FSNCs because they cannot take advantage of the same economies of scale. Regional airlines may choose to tender for routes and be awarded a government contract and a subsidy to provide an air service to a more remote location (➤Chapter 21). In Europe, regional airlines collectively serve over 1,700 short-haul point-to-point routes and operate 1.1 million flights a year, with each flight lasting an average of 75 minutes and seating 79 passengers (ERA, 2019). In order to access larger computer reservation systems (CRSs), broaden their distribution, raise their profile in foreign markets and grow their traffic, a number of regional carriers have become affiliate members of major alliances and/or entered into codeshare agreements.

8.6 Hybrid airlines

Hybrid airline: an airline that does not adhere to one single strategy but which adopts attributes from different airline business models.

The term **hybrid airline** evolved from the recognition that the 'one-size-fits-all' business model descriptors do not adequately explain airlines that exhibit characteristics from several different business models. Hybrid airlines are a crossover between an LCC and a network airline that, in addition to offering lower fares, blends low-cost traits with those of traditional FSNCs. In the US, New York-based JetBlue has broadened its route portfolio by operating to Central and South America. Its premium intercontinental product, 'Mint', offers lie-flat beds and arguably has more in common with a traditional FSNC than an LCC. Some LCCs that are subsidiaries of their parent airline company fall into the category of hybrid airlines as they operate both domestic and international routes including different aircraft types and a premium economy class with a limited number of seats.

8.7 Specialist operators

Specialist operators undertake low density but vital services, such as flights flown as part of Public Service Obligation (PSO) routes to/from remote airfields. Specialist operators often use particular aircraft (including helicopters) that can operate from short and/or unprepared runways (➤Chapter 21). Another form of specialist operator exists in Australia that takes engineers, technicians and other mine workers from the nearest airport to the mining site. A special term applies to this segment called 'fly-in, fly-out' (FIFO). These are lucrative contracts for specialised aviation companies that tailor services according to their client's requirements.

Stop and think

Detail the operational characteristics of the different airline business models and assess the relative merits of each.

8.8 The carrier-within-a-carrier (CWC) model

The 'carrier-within-a-carrier' or 'airline within an airline' model had an original objective and strategy to create a subsidiary airline akin to a low cost "no frills" budget airline to both defend market share from rival LCCs and to counter new LCC start-ups. Airlines in both the US and Europe that established new offshoots, however, invariably failed to gain market traction and failed. Examples include Delta subsidiary Song, United's Ted and SAS's Snowflake. A major contributory factor was that staff who transferred from a full service airline could not adapt to the different culture and operation of a low cost airline. Some full service airlines have made a success of their low cost subsidiary airlines. Qantas launched Jetstar in May 2004, in response to Virgin commencing operations in Australia in August 2001. Jetstar has grown to account for 9 per cent of all departing passengers from Australia and is a major profit contributor to Qantas.

The second example of a successful CWC strategy is the Singapore Airlines Group. who have established two LCCs: Tigerair, which operates short to medium hauls from/to Singapore and falls into the category of ultra-low cost; and Scoot for long-haul routes, including routes from Singapore to Australia, China, South Korea, Japan and into Europe. South African Airlines launched their low cost subsidiary, Mango, in 2006. These carriers are seen as complementary to the parent airline, as they are designed to cater to different market segments.

Stop and think

What are the challenges faced by FSNCs when launching a low cost CWC, and why have few CWCs succeeded?

8.9 Strategic alliances

Strategic alliance partners have become an integral marketing strategy and part of most major international airlines. One only has to look at an airline's airport departures or arrivals board to see a range of codeshare flight numbers related to the actual airline operating the service to highlight the extent of alliance agreements. Code-sharing has given airlines the ability to market an extensive network and offer airline consumers a 'seamless' service from origin to destination.

Strategic alliances are considered a form of competitive strategy offering more destinations and frequencies, with each group seeking a competitive advantage. Most importantly, membership of an alliance allows access to markets that would otherwise be difficult and costly to access owing to airline ownership regulations and air service agreements. Smaller airlines have the most to gain because they can receive feeder traffic at a hub point which both airlines serve for transfer into their spoke network. For larger airlines, feeder traffic coming into a major hub enables the carrier to build its capacity. Alliance agreements overcome, in part, the lack of access or traffic rights to a particular country or where demand is such that the carrier's own service is not justified. For example, apart from British Airways, no EU airline operates to Australia but through their alliance partners can transfer traffic at hub points such as Bangkok, Hong Kong, or Singapore for onward carriage, thus creating a 'seamless service' from origin to destination that will include check-in and boarding passes at the origin of the journey via any en-route transfer point when there is no stopover involved.

Strategic alliance: a business agreement in which airlines combine resources and efforts to jointly achieve common objectives while remaining separate entities.

Alliances are aimed primarily at business travellers and to those belonging to a frequent flyer programme (FFP). This accords passengers a certain status, such as priority boarding, priority airport check-in, lounge access, priority baggage handling and additional baggage allowance. Points earned with a partner airline can usually be transferred to a member's FFP.

Several factors have driven the establishment of airline alliances. Through their alliances, the airlines seek to:

- create a 'seamless network service' for the travelling passenger from origin to destination, especially for business travellers;

- be competitive through economies of scale and scope amid the intense competition which characterises most international airline markets;

- reduce costs and obtain operating efficiencies;

- eliminate or minimise existing barriers to accessing international markets.

Airline alliances also offer the opportunity for cross-selling each other's services and expanding access to a broader market, with the ability to mobilise network resources through the various partners.

! Stop and think

What are the relative merits for airlines joining a strategic alliance?

8.10 Airline failure

So far this chapter has identified the factors that characterise different airline business models. It is important to note, however, that merely copying an approach that has proved effective elsewhere is no guarantee of long-term success. Business models are not static and must adapt to changing market conditions.

The main reasons for failure include under-capitalisation, failure to achieve a competitive cost base, rapid and/or unsustainable expansion, operating the wrong type of aircraft, a lack of airline management experience and difficulty in penetrating existing markets. No airline is immune from failure. The impacts of failure are felt widely, such as the immediate impact on passengers who have bought tickets, employment, ground handling contracts, suppliers and other unsecured creditors. Airline start-up and failure must be seen as part of a natural cycle of deregulation and open markets that permit an airline to make money or to lose money and go out of business. Failure can occur to any type of airline, and many full service (and often well-known) airline brands, including SABENA (Belgium), Malev (Hungary), flybmi (UK) and Pan Am (US), have left the market as have numerous low cost, charter, regional and hybrid operators (see Case Study 8.6).

THOMAS COOK AIRLINES

Thomas Cook Airlines was a charter and scheduled airline based in Manchester (UK). It operated services to popular leisure destinations from eight UK airports using a fleet of A321 and A330 aircraft and carried over 8 million passengers in the calendar year 2018. In September 2019 the airline and its parent company ceased trading. The failure was caused by a number of financial, social, and even meteorological factors, not least the company's debt (in May 2019 it reported a £1.5 billion loss for the first half of the financial year), creditor banks demanding increased finance, customers increasingly buying their holidays online as opposed to high street travel agents, terrorist attacks in key markets, and the hot UK summer in 2019 which reduced demand for foreign holidays. Over 140,000 passengers were repatriated using the Air Travel Organiser's Licence (ATOL) scheme which protects customers in the event of an airline failure. This meant that Thomas Cook customers who were on vacation overseas (and who were ATOL protected) were flown back to the UK using aircraft that the UK government had chartered from other operators around the world.

If it is possible to define ingredients for success, the following offer a good start: creating a strong brand and product; being an early entrant into a market and basing operations in northwest Europe; adhering to the Southwest Airlines model; securing the backing of an existing airline; operating point-to-point routes of one to three hours' flight duration with either Boeing 737 or Airbus A320 family aircraft; and avoiding direct competition with rivals.

Stop and think

!

Detail the reasons for airline failure, and discuss the management strategies that can be employed to avoid it.

Key points

- A business model seeks to explain an airline's purpose, goals and the way it conducts its business.

- Six main types of airline business model can be identified, but there is a degree of overlap between them.

- Hybrid airlines adopt elements of different business models and often use product differentiation to attract both price-conscious passengers and business travellers.

- Airline management has focused on cost reduction and embraced technology to further lower costs and to become more efficient.

- The 2000s have been characterised by consolidation and cost-cutting for FSNCs.

- An airline business model needs to be adaptable, flexible and innovative. All airlines need to be vigilant, cost conscious, adaptable and customer service oriented if they are to succeed in a competitive and demanding market.

References and further reading

Bowen, J. (2019). *Low cost carriers in emerging countries*, Amsterdam: Elsevier.

Budd, L., Francis, G., Humphreys, I. and Ison, S. (2014). Grounded: Characterising the market exit of European low cost airlines. *Journal of Air Transport Management*, 34, pp. 78–85.

Bureau of Transportation Statistics (BTS). (2019). *Airline traffic statistics*. Available at: www.bts.gov

Dobruszkes, F. (2016). The geography of European low-cost airline networks: A contemporary analysis. *Journal of Transport Geography*, 28, pp. 75–88.

ERA. (2019). *2018 Handbook European Regions Airline Association*. Surrey, UK: ERA Communications, Lightwater.

Fageda, X., Suau-Sanchez, P. and Mason, K. (2015). The evolving low-cost business model: Network implications of fare unbundling and connecting flights in Europe. *Journal of Air Transport Management*, 42, pp. 289–296.

ICAO. (2018). *Solid passenger traffic growth and moderate air cargo demand in 2018*. Available at: www.icao.int/Newsroom/Pages/Solid-passenger-traffic-growth-and-moderate-air-cargo-demand-in-2018.aspx

Moir, L. and Lohmann, G. (2018). A quantitative means of comparing competitive advantage among airlines with heterogeneous business models: Analysis of US airlines. *Journal of Air Transport Management*, 69, pp. 72–82.

Whyte, R. and Lohmann, G. (2015). The carrier-within-a-carrier strategy: An analysis of Jetstar. *Journal of Air Transport Management*, 42, pp. 141–148.

CHAPTER 9

$ Airline pricing strategies

Peter Hind and Gareth Kitching

LEARNING OBJECTIVES

- To understand the development of airline pricing strategies and the difference between traditional and one-way pricing.

- To identify the difference between long-haul and short-haul pricing.

- To appreciate passenger profiles and their implications for revenue management.

- To recognise the difference between point-to-point and connecting passengers and understand the revenue management implications of these journey types.

- To understand the pricing strategies of low cost carriers (LCCs) and how they utilise low headline prices to stimulate demand.

- To examine emerging trends in revenue management and the implications these may have on passenger bookings and airport infrastructure.

9.0 Introduction

The disciplines of pricing and revenue management are central to airline business models (►Chapter 8). They are intrinsically linked functions that, when working in harmony, enable an airline to maximise the revenue from each flight and give it the best chance of maximising profitability.

Recent years have seen significant changes in how airlines sell their product, both in terms of the underlying pricing strategies that are used and the way in which it is distributed. The evolution of airline pricing has been an essential response to the dual impacts of deregulation and the advent of the internet

Tariff: a list of pre-determined prices that seats can be sold at.

Revenue management: the process of predicting customer behaviour to optimise product availability and price to maximise revenue growth.

(➤Chapter 3), both of which have fundamentally changed the industry in a way that could not have been foreseen at the turn of the millennium.

Pricing relates to the fares that are charged for seats and which are sold on a seat-by-seat basis for each flight. Typically, each route will have a price range that determines the minimum and maximum fare levels that are charged and, within that price range, a series of pre-determined prices (or fares). This is sometimes known as the **tariff**. On a short-haul route, the price range may be £19.99 to £299.99 and, within that, there will be a number of fares (see Table 9.1). This price range changes infrequently – sometimes once a year, sometimes less.

The principles of **revenue management** apply to any market segment where there are time limitations to the supply of a product or service and where pre-booking of that product or service is possible. Examples include hotel rooms, railway seats, hire cars, car parking spaces or airline seats. Each of these has a value up to a particular point in time, after which it cannot be sold and has no value.

In the airline industry, revenue management is the method through which an airline determines how many seats to sell on each flight at each tariff. For full service network carriers (FSNCs) that operate a frequent flyer programme (FFP) (➤Chapter 10), there is the additional consideration of FFP redemption tickets (how many tickets are to be made available for FFP redemption in each cabin class). There are a number of considerations that influence the price passengers are prepared to pay, and using a revenue management system, an airline can understand these factors and manage the sale of seats accordingly. Airline websites may also track the number of times a potential customer has clicked through the website to view available fares. Some of the key factors to consider are the time of the flight, the day of the week, the time of the year and whether any one-off events are occurring. Such events may include a major business convention or sporting occasion which will increase demand and therefore the price that passengers may be willing to pay (Case Study 9.1). Successful airlines pay close attention to these since they can be the difference between profit and loss.

CASE STUDY 9.1

2019 UEFA CHAMPIONS LEAGUE FINAL

On 1 June 2019, Tottenham Hotspur FC played Liverpool FC in the final of the UEFA Champions League at the Wanda Metropolitano Stadium in Madrid. Supporters of the two clubs complained that fares for flights from the UK to Madrid around the date of the match had risen to more than £1,000 instead of the usual £250 and accused the airlines of profiteering. The airlines explained that their pricing regimes are demand led and that the higher fares were a natural consequence of huge customer demand for tickets.

The data in Table 9.2 shows how, from the same tariff, the airline can offer very different fares to passengers. In the off-peak winter example, it sells over 50 per cent of its seats at less than £39.99 because demand is low and it needs to keep prices low in order to sell seats. In the school holiday example, on a flight operating to a typical holiday destination, demand will be strong, and the airline may not sell any seats below £59.99.

Table 9.1 Example of a price range (tariff) for a short-haul flight

Fare level	
Lead-in fare	**£19.99**
Fare 1	£24.99
Fare 2	£29.99
Fare 3	£39.99
Fare 4	£59.99
Fare 5	£79.99
Fare 6	£99.99
Fare 7	£129.99
Highest fare	**£299.99**

Table 9.2 Applying revenue management profiles to a tariff

	Tariff	Percentage of seats sold at each fare level		
		Off-peak winter flight	Mid-season business destination	School holiday leisure flight
Lead-in fare	**£19.99**	25	0	0
Fare 1	£24.99	15	0	0
Fare 2	£29.99	15	10	0
Fare 3	£39.99	15	15	0
Fare 4	£59.99	10	20	10
Fare 5	£79.99	10	20	20
Fare 6	£99.99	5	15	20
Fare 7	£129.99	3	10	25
Highest fare	**£299.99**	2	10	25

9.1 Pricing

Traditionally, airlines tried to differentiate the price of airfares by identifying passenger journey characteristics and pricing them accordingly (➤Chapter 10). Generally, return fares were offered at a lower total price than the sum of two single fares, on the basis that the airline

Round-trip pricing: pricing based on a passenger making a return journey purchase with the same airline/same booking.

Minimum stay rule: a booking criterion related to the length of stay (or days of stay) which must be met before a certain tariff is shown to the customer.

wanted to fill seats in both directions and most outbound passengers are likely to need to return. This is called **round-trip pricing**. More return (or round-trip) fares were available than one-way fares.

Using this approach, which is referred to as the traditional pricing model, the airline will offer a range of different fares, each of which have different booking or travel conditions, in order to maximise revenue from each journey. The more flexibility the passenger needs, the higher the price they must pay. Sometimes pricing differences are obvious; for example, when the class of travel is different – business class is more expensive than economy class as the in-flight services are more expensive to provide. Other times they are more subtle; for example, if the passenger books a long time in advance of departure, known as 'advanced purchase', it may be cheaper than booking on the day of travel. If they include a Saturday night stay in their itinerary (**minimum stay rule**), it probably indicates the passenger is travelling for leisure purposes and may have less to spend than a passenger travelling for business purposes.

Another way of differentiating the product is to offer a lower price in exchange for more restrictive travel conditions – the passenger may have to stick to the booked flights and not be able to change to an alternative, or the ticket price may not be refundable in the event of cancellation. Fares in the traditional pricing model are usually grouped into normal and special. Special fares are those with restrictive conditions, whereas normal fares have no restrictive conditions but are significantly more expensive.

One-way pricing: pricing each seat on a one-way basis.

The more modern approach, and one which is now almost exclusively observed in short-haul travel, is **one-way pricing**. When this approach began in the mid-1990s (introduced by low cost carriers (LCCs) as a core element of their pricing strategy and subsequently adopted by FSNCs in response), it marked a significant departure in pricing strategy for the airline industry, and one which has ultimately led to a more competitive marketplace which offers better value for customers and, importantly for the airlines, fuller (and more profitable) flights.

In the one-way pricing model, the airline makes little attempt to use conditions to determine the fares that are available to passengers; it simply prices each journey on a one-way basis with the same price for adults and children. If the passenger is making a return journey, the fare will be the sum of the prices for A to B and B to A. This is a much simpler approach, but one that relies more heavily on revenue management to ensure the airline maximises revenue. Some airlines may choose to make one-way tickets more expensive than return fares.

In both traditional and one-way pricing models, the airline will use a pricing system to maintain its tariff, and this will, in conjunction with the reservations and revenue management systems, calculate and display the applicable price for a flight when a passenger searches online or through a travel agent. It also explains why prices for the same flight will often change, sometimes quite considerably, over time.

!

Stop and think

What are the main differences between round-trip and one-way pricing, and what are the implications for passengers?

9.2 Revenue management

Revenue management, sometimes referred to as **yield management**, is the practice of maximising the revenue from each flight by controlling the number of seats that are sold at a particular price or in a particular market (see Table 9.2). It is central to the industry and is an approach that has evolved beyond recognition in recent years. It stimulates competition, enables aggressive marketing and attractive lead-in prices and, when done correctly, ensures the airline sells every seat on every flight at the highest possible price.

In the early years of computerisation, a simpler approach – called **seat inventory** (or space) control – existed. This was a way of making sure that the number of passenger tickets sold was no more than the inventory (the number of seats available on each flight). In order to control the number of seats available for sale, each flight on each day it operated would be set up in the airline's computer reservation system (CRS) with the seating capacity of the aircraft. Each time a reservation was made on the flight, the number of available seats would be reduced accordingly, ensuring that the airline could not 'overbook' the flight by selling more tickets than it had available seats.

This system has subsequently evolved into a centrepiece of commercial operations, liaising between the marketing and pricing departments to manage and control how the airline sells its seat capacity. While the principle is straightforward, in practice the more an airline understands about passenger behaviour, the better it can optimise its price availability for each flight, which, in turn, maximises revenue.

Typically an airline revenue management department will comprise a group of route analysts (who have responsibility for maximising revenue on the routes that they manage) who use a revenue management system. The revenue management system uses complex algorithms that seek to forecast demand for each flight at each price-point by analysing past purchasing behaviour. The system proposes how many seats should be sold on each flight and at what price in order to fill the flight at the optimal revenue. Over time, the revenue management system will re-calibrate its algorithms based on the booking profile of previous flights to constantly refine the booking profiles.

One of the key considerations of the revenue management function is to make sure that each flight operates to a high level of occupancy but not with more passengers booked to fly than there are seats on the aircraft. Some airlines will deliberately sell more seats than the aircraft holds, a practice known as **overbooking**, because experience has shown that a certain proportion of passengers will not turn up. These are known as 'no shows'. If the average no show rate on a route is 10 per cent, the airline may sell 5 per cent more seats than it actually has, to ensure the flight will be almost full. Occasionally things will go wrong and too many passengers may turn up for the flight. In Europe, this carries financial penalties for the airline (through EU Regulation 261/2004), which awards compensation to passengers who are unable to fly as planned. EU 261 legally requires airlines to financially compensate passengers if they are denied boarding (for example, due to overbooking), if their flight is delayed by at least two hours or if their flight is cancelled at short notice. Compensation amounts vary according to the sector length of the affected flight and may not be paid if the flight is delayed or cancelled due to 'extraordinary circumstances' beyond the airline's control. Airlines often cite the imbalance of EU 261 compensation compared to other

Yield management: a variable pricing strategy aimed at selling (perishable) products to achieve revenue management principles.

Seat inventory: the number of seats available on any given flight.

Overbooking: the practice of selling more seats than the aircraft holds on the assumption that a certain percentage of passengers will not turn up for their flight.

transport modes to/from/within Europe (e.g. ferries, EuroTunnel) which are not subject to the same financial compensation criteria.

Other revenue management complexities include: using point-of-sale-pricing to maximise revenue, whereby an airline seeks to charge more in a particular market than another because it knows customers from that country or city have greater disposable income than others; and special event pricing, where one-off events – conferences, sports events or even a large societal event such as a major wedding – create a spike in demand. For example, providing it has the required traffic rights on a route (➤Chapter 1), an airline based in Country A may charge less for a service to a city in Country B that originates in Country C (but transits through Country A) than it does for a direct flight from A to B.

! Stop and think

What are the benefits to airlines of operating a revenue management system?

9.3 Pricing strategies of FSNCs and LCCs

At the outset of the low cost revolution, there were clear differences between the pricing strategies of full service network carriers (FSNCs) and LCCs. FSNCs used the traditional pricing model, whereas LCCs developed the one-way pricing model. However, FSNCs have been forced to adopt one-way pricing in the markets where they compete with LCCs. This means that, for short-haul flights, there is little difference in the pricing models of most airlines. Conversely, long-haul routes are more commonly operated by FSNCs and retain the traditional pricing model.

The pricing strategies of airlines are based on the type of passengers being carried and the nature of the passenger's flight itinerary (➤Chapter 10). Airlines who target the ultra-price-sensitive leisure market will operate on a volume-driven model, where the objective is to fill every flight to as near 100 per cent load factor as possible; whereas airlines serving largely business-driven markets may be content to operate nearer to 80 per cent occupancy but at a higher average fare.

It is easier to differentiate passengers by looking at the purpose of travel. Generally, there are three main reasons for travelling – business, where there is no real choice in destination or the date of travel; leisure, where the passenger often has a level of discretion in when they travel, and sometimes where to; and visiting friends and relatives (VFR), where travel might be discretionary but the destination is fixed. Thus, these three groups show different characteristics in their choice, as shown in Table 9.3.

The pricing and revenue management strategies of FSNCs and LCCs are also influenced by differences in their business models:

Connecting: (a passenger) flying to a hub airport and connecting onto another service.

- FSNCs generally fly both short- and long-haul services;

- FSNCs carry point-to-point traffic as well as passengers **connecting** between short- and long-haul flights at hub airports; and

- FSNC passengers may interline between two different airlines, paying one through fare that covers all flight segments/airlines; whereas

- LCCs generally handle only point-to-point passengers (although a small number do offer through fares via their main airports).

Table 9.3 Pricing considerations of different passengers

Passenger type	Booking profile	Journey considerations	Other considerations
Business	• Tend to book closer to date of travel. • Do not necessarily make the booking themselves. • Destination usually driven by need rather than desire.	• Less price sensitive. • Frequency of service important (enabling a day return). • Convenient airports preferred (shorter surface access journey times).	• Frequent flyer programmes and premium airport products (such as lounge access and fast-track security).
Leisure	• Tend to book further in advance of the travel date. • Range of motivations for travel (destination, tourist offering, price). • Potential flexibility with regards to destination.	• More price sensitive. • More willing to accept disadvantages, including secondary airports and poorer flight schedules.	• Some demand for premium (but rarely first class) travel in long-haul markets, particularly in luxury market segments such as flights serving cruise ships.
VFR (visiting friends and relatives)	• Tend to book long in advance of travel. • Flexible in date and time. • Destination driven by location of friend/relative.	• More price sensitive. • Seek most convenient destination airport.	• May travel frequently or use premium cabins for trips of longer duration on long-haul.

Stop and think

Consider the extent to which passenger segmentation is important as part of a revenue management system.

9.4 Point-to-point revenue management

This relates to generating revenue from a passenger who is being transported between two points and no further. This is the operational model for the majority of LCCs (while some LCCs such as Norwegian do offer through ticketing, this is only a small aspect of their operations), whereas it forms only part of the operational model of FSNCs. Essentially, the only way to fill a flight in the point-to-point model is with passengers who want to fly between the two airports served by the flight. Thus, in periods of low demand, where there is the possibility of the seat flying empty, the carrier will use the pricing and revenue management process to stimulate traffic to attract people to fly when they may not have considered it through offering very low fares. The skill is to manage revenue to ensure that enough people are stimulated/attracted to travel with that airline through low fares while ensuring that revenue is maximised for that sector. Example 9.1 details the impact of booking profiles on LCC fares.

Example 9.1
Impact of booking profile on LCC fares

In 2019, the largest route (by seat volume) operated by LCC Ryanair from London (including Luton, Gatwick, Southend and Stansted airports) was to Dublin Airport. Both Dublin and London are major tourist destinations as well as being large business centres. Consequently, both business and leisure passengers use the route. Fare data, collected from Ryanair's website, shows that at three months and one month prior to the date of travel, fares are around 50 per cent lower than the fare advertised one week prior to travel.

Origin	Destination	Three months (€)	One month (€)	One week (€)
Luton	Dublin	26.26	33.09	66.33
Gatwick	Dublin	27.80	39.79	76.39
Southend	Dublin	18.88	21.42	36.06
Stansted	Dublin	26.63	32.04	56.01

Average three months/one month (€)	Difference from one week
29.68	−55%
33.80	−56%
20.15	−44%
29.34	−48%

Source: www.rdcapex.com, July 2019. Data indicates average for April to June 2019, one way including government tax, weighted according to the number of tariffs on each route.

The nature of passenger trips and booking profiles means that, on point-to-point services, airlines need to find a balance between:

Price sensitivity: how sensitive customers are to changes in the price of a product.

- attracting (primarily leisure) passengers to book early (through lower fares);

- increasing fares closer to the date of travel to take advantage of business passengers' need to travel, often at short notice, and lower **price sensitivity**.

Through careful management of this process, both load factors and revenue are maximised. Higher load factors are important, especially for LCCs, as having more people on board not only results in higher ticket revenue but also increases the opportunity to generate extra income through the sale of ancillary items such as seat assignment, on-board purchases, hold luggage fees and priority boarding. This important additional revenue stream enables airlines to bolster their ticket revenues by between 20 and 30 per cent (see Section 9.7), and in the case of very cheap headline fares, may be the difference between a loss-making and a **breakeven** passenger.

> **Breakeven:** when the revenue covers the costs of providing the seat.

For LCCs, each individual flight may be considered as a stand-alone 'profit centre', with route analysts being challenged to manage routes and even flights to profitability (i.e. where ticket plus ancillary revenues are greater than the sector operating costs). As a result, the pricing policy of LCCs is almost always on the one-way basis, as this helps to ensure that the strategy of maximised revenue/load factors is achieved on each flight. It is also a much less labour intensive way of managing pricing.

FSNCs have a more complex series of considerations in their pricing and revenue management. Each flight will contain a mix of point-to-point and connecting passengers, and potentially a mix of pricing strategies – traditional and one-way. Typically, passengers flying non-stop between two points are more cost-effective to carry than connecting passengers, in that they are less likely to miss flights or lose bags (which would result in costs for the airline in terms of compensation and additional administration), and they incur lower administration costs in terms of pricing complexity. The network effect, however, gives carriers the scope to operate in considerably more markets if they carry connecting passengers than a point-to-point airline, especially if they offer interlining or codesharing opportunities (►Chapter 8).

As Figure 9.1 shows, an airline flying six routes from airport G using a LCC point-to-point model offers six routes (one each from A, B, C, D, E, and F to G). Under a network model, whereby the airline offers connections at airport G, it is able to offer 21 city-to-city combinations – six point-to-point (outlined previously) and 15 connecting – A to B, C, D, E, and F; B to C, D, E, and F; C to D, E, and F; D to E and F; and E to F (see Table 9.4).

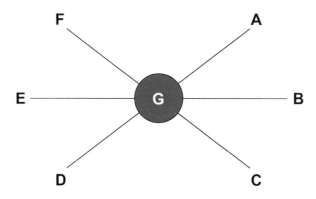

Figure 9.1 Example network and city combinations

With the requirement to price in both directions, the LCC will be managing 12 flights and associated tariffs, whereas the network carrier will be managing 42, although it serves the same number of airports.

Table 9.4 Possible network connections via Airport G

Origin airport	A	B	C	D	E
Transit airport	G	G	G	G	G
Destination airport	B, C, D, E, F	C, D, E, F	D, E, F	E, F	F

Although this means more management time and complexity for the network carrier, it also gives the airline greater scope to be tactical in how it fills its aircraft, as it does not have to rely on the traffic flows on six city-pairs because it can carry connecting passengers on a further 15 city-pairs. For the network carrier, the situation is more complex – not only will point-to-point demand and revenue have to be considered as part of the route performance analysis, but also how to allocate the revenue from connecting passengers who pay one fare but fly on two (or more) sectors.

9.5 Connecting passenger revenue management

In general, long-haul services are operated by a network carrier to or from their main hub airport(s). Demand for long-haul services is also lower than for short-haul services for a number of reasons – long-haul flights are more expensive to operate and therefore more costly to sell; people tend to take holidays and mini-breaks within a relatively close proximity to their home country; and there are often additional requirements with long-haul travel in the way of visas and the need for vaccinations that further reduce the frequency with which passengers consider long-haul flying. In turn, airlines tend to focus their resources on a small number of origin or hub cities around which they develop a network of air services. In fact, many airports cannot be effectively connected with direct non-stop services (even with modern aircraft types with extended range) because market sizes do not warrant frequent, year-round flights, and so it is seen to be more cost-effective to transfer passengers between flights via the hub airport. In addition, due to operational considerations, long-haul services (and the larger aircraft that are required to operate them) are likely to be consolidated at larger airports that have the necessary infrastructure.

As network airlines will be carrying transfer passengers via a hub airport (and carrying passengers who are using that service only due to the ability to connect), the allocation of revenue for a booking is generally split using a prorate agreement, which uses a weighted-mileage-based formula to ensure that the OD fare is appropriately allocated to the various individual segments of the trip itinerary (one or more of which may be operated by a different airline due to a codeshare arrangement). This may mean that revenue allocated to that passenger on a short-haul feed sector (which is connecting on to a long-haul service, such as a domestic feed on to a long-haul international service) will be significantly lower than what a passenger flying point to point on the same short-haul sector may pay. Example 9.2 provides an example of how revenue from connecting tickets may be allocated.

Revenue management for connecting passengers is therefore different from that of point-to-point passengers. The revenue management team within an FSNC must review many more combinations of itinerary to work out how best to fill each flight. These combinations of itinerary are known as origin and destination (OD), and the airline pricing and revenue management analysis will evaluate the pricing on each OD before deciding which markets to prioritise over others. Route analysts will ensure that fares are competitively priced to attract passengers to their airline (over their hub) whilst remaining competitive against other airlines flying the same OD (maybe via a different hub). The analyst also needs to consider the opportunity for filling a seat with a passenger from another connecting service which might pay more, and thus a wider network view on profit and loss needs to be considered, rather than looking at single routes in isolation.

Example 9.2

Example of prorate revenue allocation for connecting tickets

A passenger flies from Glasgow to New York Newark via London Heathrow, paying US$368 one-way for their fare.[1] The revenue, which is generated from one ticket for the entire journey, needs to be allocated between the two sectors being flown, and this is achieved according to the relative share of the distance of each sector to the overall journey using weighted mileages established by IATA. These mileages are weighted towards shorter sectors, where the operating costs are proportionately higher per mile than longer flight sectors.

Glasgow to Heathrow is 345 miles, and the sector length between Heathrow and Newark is 3,465 miles, meaning the Glasgow to London sector accounts for 9% of the total mileage. However, IATA's Prorate Manual (which sets the appropriate weighting to be applied to various segments of an itinerary) states that the Glasgow to London sector should be weighted to represent 734 miles and the London to New York sector weighted to 4,444 miles.[2] This has the effect of increasing the proportion of revenue allocated on the Glasgow to London route to 14% of the total (i.e. 734/[734 + 4,444]). Therefore, applying this 14% to the total fare of US$368 results in US$52.16 of the fare being allocated to the Glasgow–Heathrow sector based on the sector weighted miles divided by the total weighted miles of Glasgow–Heathrow–Newark. The remaining US$315.84 of the fare is allocated to the Heathrow–Newark sector.

However, this revenue allocation is much lower than the average point-to-point fare between Glasgow and Heathrow (US$118).[3] The purpose of carrying the passenger from Glasgow to Heathrow is not primarily to contribute revenue to the first short-haul sector, but to the long-haul sector from Heathrow to Newark.

Sources: [1]ba.com June 2019, half-return fare minus UK APD
[2]IATA Prorate Manual, June 2019
[3]rdcapex.com, average fare for travel in June 2019
Distance based on great-circle distance.

9.6 Other revenue management considerations

How an airline approaches its pricing and revenue management will be defined by the market segment it is trying to occupy. Ryanair is an example of an airline that for many years has used price as its main selling point, using slogans such as 'Fly Cheaper' and 'The Low Fares Airline' as centrepieces of its marketing activity. By creating this brand image, the airline has to offer the low fares it promotes. Other airlines adopt more generic straplines. 'Keep Discovering' (Emirates), 'To Fly. To Serve' (British Airways) and 'Journeys of Inspiration' (KLM) are just some that have been used to create a brand proposition rather than a pricing statement (➤Chapter 20).

These marketing positions are important because they determine how a carrier approaches its price-point. Some airlines will want to be market leaders or 'price makers', meaning they have the brand power to be the first to reduce or advertise low prices or fare sales – and are likely to be the first to increase prices as demand grows, while maintaining the perception of lowest cost. This is particularly true of ultra-low cost carriers (ULCCs) such as Wizz Air. ULCCs use their position as being the cheapest airline to fly between two cities as a key promotional tool to capture initial demand for a service. ULCCs are able to be market leaders in this regard due to their exceptionally low (and industry leading) cost base. ULCCs benefit from the market's perception of them being the lowest cost airline to operate on a particular route in that customers will often go to these airlines' websites before any other airline on the assumption that they will be the cheapest (among other factors).

On the other hand, other carriers are left being 'price takers', meaning they generally react to moves made by the market leaders or price makers – and they need to constantly monitor the prices these carriers offer in order to increase or decrease their own fares to stay in line with the market.

The other segments that influence revenue management are block bookings and off-tariff agreements. The first of these is where an airline sells a block of seats to a third party to sell on, usually a tour operator looking to offer inclusive package holidays with flight and accommodation included. This type of arrangement falls into two categories – hard and soft block space agreements. Under a hard block space agreement, the third party pays for the seat allocation at a pre-agreed volume and price, meaning the airline has some certainty of revenue for part of its flight inventory. Under a soft block space agreement, the third party is allocated a variable number of seats on any given flight and pays only for seats it sells (thus a greater level of risk remains with the airline). In both instances, however, the airline revenue management analyst has fewer seats to sell to the general public, and therefore more skill is required to optimise revenue for a flight.

Off-tariff is where an airline makes a contractual arrangement with a potential provider of traffic in exchange for offering discounts on the standard fares. As with block bookings, this is often a tour operator or travel agent, but one who doesn't want to guarantee a specific number of passengers on a flight. The other type of off-tariff agreement, one that is essential to the large FSNCs, is the corporate agreement. These are typically negotiated deals with the high-volume corporations that have large, regular amounts of business traffic on single routes or across a range of destinations (for example, oil companies who regularly need to access oil-rich cities, regions and countries such as Aberdeen, Texas, and Nigeria respectively).

The airlines seek to lock in as much spend as possible from these businesses, which can provide significant volumes of traffic and revenue. Again, the pricing and revenue management analysts need to work closely to ensure that the right balance of competitive price and seat availability can be offered to such companies.

A further challenge for airline revenue managers is whether particular airline brands are able to command a price premium over their competitors in future years as travellers develop a greater awareness of price comparison sites and **online travel agents (OTAs)**. Particularly in the price-sensitive leisure segment, passengers are often tempted by the lowest headline price rather than a specific airline. This places a greater emphasis on the price-makers to monitor competitor pricing to make sure they are always close to the top of the price comparison site search results.

The ways in which an airline sells its services (i.e. aircraft seat capacity) to customers are called **distribution channels**. Distribution channels can be owned and operated by the airline (such as the airline's own website or high street travel agency) or owned and operated by another distributor (such as an online travel agent (OTA) or Global Distribution System (GDS)) which will sell the seat capacity on behalf of the airline and charge commission or a fee respectively for the sale. Price comparison sites also offer consumers an additional way to find and purchase travel. Each method has advantages and disadvantages for an airline (Table 9.5).

Online travel agent (OTA): travel companies that sell airline tickets over the internet.

Distribution channel: The method(s) by which goods or services are transferred from the manufacturer/operator to the end consumer.

Table 9.5 The advantages and disadvantages of different distribution channels for airlines

Ownership of distribution channel	Distribution channel	Advantages	Disadvantages
Own channel	High street travel agency	• Highly visible branding/marketing. • Opportunity for giving customers a 'personal touch'. • Opportunity for upselling (persuading a customers to buy additional and/or more expensive products or services). • Opportunity for influencing a customer towards a certain product. • Enables airline/tour operator to retain customer and their data.	• Highest cost of sale in comparison to other methods. • Declining high street footfall. • Not the preferred method of shopping for the younger demographic (and therefore future customers).
	Website	• Cheapest cost of sale for the airline/tour operator. • Often a consumer's first point of search for a product. • Opportunity for upselling. • Opportunity for influencing a customer towards a certain product. • Enables airline/tour operator to retain customer and their data.	• Increased target for cyberattacks (resulting in loss of sales if sales channels are disrupted, or large fines from regulatory bodies if data protection laws are breached). • Not every potential customer will be aware of the website. • Needs to appear in multiple languages for international customers. • Not all customers can access or use the internet.

Continued

Table 9.5 Continued

Ownership of distribution channel	Distribution channel	Advantages	Disadvantages
	Online travel agent (OTAs)	• Strong brand presence and can be the first point of search for many consumers.	• Customers can easily compare the price and service offering of a range of providers. • Not all customers can access or use the internet.
Third party channels	Global Distribution System	• GDS enables a wide and large range of smaller companies to sell inventory in multiple geographic locations, maximising market exposure. • Used by companies which manage corporate/business travel (thus gives airlines access to customers who will purchase higher-priced tickets).	• Higher distribution costs than own-channel web sales. • Loss of control over the interaction with consumer.
	Price comparison site	• Can help a low-priced fare stand out against other airlines. • Strong brand presence and can be the first point of search for many consumers.	• Targets price-sensitive consumers, thus driving the sale of low-value fares. • Drives other airlines to price match – good for consumers, but lowers revenue generation opportunities for all carriers. • Higher distribution costs than own-channel web sales.

9.7 Contribution of ancillary revenue to total revenue

Ancillary revenue: the revenue generated from passenger spend on secondary services not associated with the ticket price, such as hold baggage and on-board food and drink purchases.

Ancillary revenue describes additional purchases that may be made by passengers on top of their ticket price. This is a common practice in the LCC one-way pricing model. Examples of primary ancillary services include the purchase of hold baggage, allocated or extra-legroom seating and on-board food and drink.

In the traditional pricing model, these items and services were usually included within the ticket price, particularly on long-haul routes where it might be considered unreasonable to charge passengers for checking in baggage or in-flight meals. These services were first unbundled from headline fares by LCCs as they sought to reduce fares to the lowest possible level in order to stimulate demand, gain customer interest and compete with traditional FSNCs. Lower ticket prices are often justified by charging extra for services which the passenger may not necessarily want and therefore shouldn't be obliged to pay for.

The revenue generated from these services is a major source of income for airlines, especially for LCCs (Table 9.6). Different passenger types will have different needs/desires for ancillary services. For example, a leisure passenger is more likely to be staying away for a longer period than a business traveller and therefore may require additional hold baggage.

Table 9.6 Top 10 airlines, 2018 – ancillary revenue as % of total revenue

Airline	Country of origin	Business model	Ancillary revenue as % of total revenue
Viva Aerobus	Mexico	LCC	47.6
Spirit	US	ULCC	44.9
Frontier	US	ULCC	42.8
Allegiant	US	ULCC	41.2
Wizz Air	Hungary	ULCC	41.1
Volotea	Spain	LCC	34.8
Volaris	Mexico	ULCC	32.3
Ryanair	Ireland	ULCC	31.7
Jet2.com	UK	LCC	31.1
Air Asia	Malaysia	LCC	29.0

Source: Adapted from Sorensen (2019, p. 8)

Families are more likely to pay for pre-assigned seats to guarantee they sit together, while a business passenger may value a seat at the front of the aircraft to ensure quicker disembarkation on arrival. If each LCC sector is to be considered on a profit and loss basis, revenue management analysts need to consider the potential ancillary revenues from passengers if the overall level of revenue is to be maximised.

FSNCs have started to increase the number of ancillary services they offer (which were traditionally included in ticket prices), especially on short-haul sectors. This is partly in recognition of the level of competition they face from LCCs but also in recognition of the change in travel habits of passengers – as LCCs introduced more and more ancillary services, passengers became more used to them, and now the practice of charging for certain ancillary services, especially assigned seating and hold baggage, is now considered the norm. The most recent trends in pricing have seen the re-bundling of fares to include some of the most popular ancillary products. This varies according to the type of ticket being purchased, but whereas LCCs used to offer only seat tickets and some/all ancillaries were extra, some now offer fares/products which do offer bundled products, such as easyJet Plus (which offers allocated seating, dedicated bag drop, security fast track, speedy boarding and some ticket flexibility for an annual subscription) or Ryanair Flexi Plus tickets (which include additional cabin bags, security fast track, reserved seating and free airport check-in amongst other benefits).

Typically, long-haul travellers still receive certain ancillary products for free as part of their booking, though short-haul passengers purchasing standard tickets on both LCCs and FSNCs pay for most additional services (see Table 9.7).

Table 9.7 Free or fee? Selected airline's economy class fare products, 2019

Carrier Type	Airline	Free airport check-in	Free priority boarding	Free advanced seat allocation	Free hold baggage allowance		Meals/refreshments included	
					Short haul	Long haul	Short haul	Long haul
FSNC	Aer Lingus	✓	N/A	✗	✗	✓	✗	✓
	British Airways	✓	N/A	✗	Varies	✓	✗	✓
	KLM	✓	N/A	✓	✗	✓	✓	✓
	Lufthansa	✓	N/A	✗	Varies	✓	✓	✓
	American Airlines	✓	N/A	✗	✗	✓	✗	✓
	United Airlines	✓	N/A	✗	✗	✓	✗	✓
	Emirates	✓	N/A	✗	N/A	✓	N/A	✓
LCC/ ULCC	easyJet	✓	✗	✗	✗	N/A	✗	N/A
	Jet2.com	✗	N/A	✗	✗	✗	✗	✗
	Norwegian	✓	✗	✗	✗	✗	✗	✗
	Ryanair	✗	✗	✗	✗	N/A	✗	N/A
Charter	Thomson	✓	N/A	✗	✗	✗	✗	✓

Note: N/A = not applicable

! Stop and think

What factors have led to ancillary revenues being an important source of revenue for airlines, and how might this phenomenon develop in the future?

9.8 The impact of the internet

LCCs were market leaders in terms of developing online booking systems in the late 1990s/ early 2000s, moving away from traditional bookings made through Global Distribution Systems (GDSs) used by travel agents and other booking agencies.

Online booking systems gave LCCs a way to directly interact with their customers, cutting out additional costs associated with third-party sales platforms and therefore helping to offer lower fares. This development also supported the addition of ancillary service sales to customers, as well as creating new ways of generating revenue not directly associated with the passenger's flight (e.g. through advertising and through receiving commission payments from other companies when the airline's customers book hotels or car hire with third-party providers).

The development of the internet dramatically changed the way in which airlines could interact with customers. Not only were online bookings cheaper to manage, but for the first time airlines had complete control over what the customer was shown in terms of sales options, destination choice and direct marketing opportunities and had more flexibility to react quickly to changes in competition and market pressures. Customers also had the ability to check multiple flights to find the cheapest option for their journey.

The ability for airlines to highlight cheaper fares for flights of interest and new destinations helped them to maximise load factors across their networks, especially for leisure passengers who have more flexibility in terms of flight timing and destination. The promotion of new routes (without passengers having to search for them) opened up new destinations and tourist markets (both inbound and outbound), giving customers a wide choice of options on their expenditure.

Loss leader: a product marketed and sold at below operational cost to generate interest in the product among consumers.

LCCs promoted the lowest possible headline fare to generate interest and drive potential customers towards their website. In many instances, this has led to LCCs selling fares below operational cost and at prices that are **loss leaders** (see Example 9.3).

Example 9.3

Loss-leading fares

In August 2019, Ryanair was advertising one-way fares between London Southend and Dublin Airport for £17.13 (inclusive of taxes and airport charges, but excluding optional ancillary services and paying by debit card). From this fare, the airline would need to pay the UK government the standard rate Air Passenger Duty (APD) for a Band A flight of £13 per passenger, which would bring the ticket revenue down to £4.13. From this remaining amount, the airline would need to pay airport charges, fuel costs, crew salaries, lease and insurance costs, and aircraft handling fees. These costs would outweigh the remaining £4.13 left after APD was paid, resulting in that fare being a loss leader.

The airline's commercial model also includes selling extra products such as assigned seating, checked bags and on-board refreshments. These ancillary revenues are usually cheap to deliver and represent a high margin, meaning that although only a proportion of passengers will buy them, they increase the marginal revenue for each passenger with relatively little cost. This additional revenue supports the loss leaders.

In directing customers towards online booking systems, airlines have been able to collect key information about their journeys and travel preferences, which in turn has aided online marketing strategies. Not only is direct online marketing cheaper than traditional methods of engaging with customers, but the information gathered as part of previous booking processes could be tailored to give more intelligence behind marketing messages and stimulate more customers to book with the airline.

Booking systems continue to evolve as technology advances, resulting in changes in the operation of flights. Airlines such as Ryanair and easyJet pursued online check-in as a way of reducing their core operating costs by reducing the number of staff required to process passengers checking in at the airport. This has subsequently evolved into mobile apps, with boarding passes now being stored electronically rather than needing to be printed out, enabling passengers to manage their travel online. Continued customer engagement with apps gives further information to the airline for marketing purposes (▶Chapter 20).

The resulting impact of these changes means that airport operations have changed too. For example, with more passengers checking in online and travelling with less hold baggage, fewer check-in desks are required. This frees up terminal space and potentially reduces the number of check-in staff that are required but requires Wi-Fi and/or QR readers to operate.

! Stop and think

How has the internet changed practices of airline pricing?

9.9 Emerging trends in airline pricing

When LCCs first entered the marketplace, they offered a genuinely different product from network airlines, particularly in the creation of a completely new pricing strategy, followed by the unbundling of services, so that all ancillary services were an additional cost. FSNCs, after losing market share and revenue to their LCC competitors, started to adopt these practices too, albeit where it was not detrimental to the overall service level offered to their main customers, while LCCs have started to loosen their approach towards completely unbundled products. Some LCCs have started to offer fare products that now include hold baggage, allocated seating, priority boarding and flexible tickets, which are more akin to the old traditional pricing approach originally offered by the FSNCs. While it is unlikely that the two pricing models will converge into one universal approach, it is likely that both types will continue to evolve.

LCCs have also moved away from being the sole sales channel for their fares, with both the main European LCCs (easyJet and Ryanair) completing deals with GDS providers to make their products available across third-party sales channels and evaluating participation in the interline and off-tariff markets. This not only gives LCCs visibility to corporate booking agencies (which rely upon GDS feed for their systems) but could also open up the opportunity for LCCs to accept transfer passengers or indeed for LCCs to start their own hub operations in a more formal way.

As the internet is now the primary booking channel in many major markets, there is likely to be a continued shift towards providing internet-enabled services that facilitate easier journeys for the customer while simultaneously maximising the type and volume of information airlines are able to gain from them. However, the initial advantage of direct booking is now being eroded by the price comparison sites. Initially focused on simply comparing airfare prices, these online platforms now offer a range of travel solutions from flights to hotels, car hire and travel insurance. The 'one-stop shop' appeals to the new generation of traveller, who are just as likely to look at Skyscanner, Opodo, Kayak or Expedia as they are to shop around on individual airline and hotel websites. Airline pricing managers now need to ensure their product is included in this online platform – and priced competitively. As such, the customer benefits from highly competitive pricing as airlines and price comparison sites compete for customer loyalty.

Advances in internet technology may offer new opportunities for airlines to maximise their revenues; for example, through IP tracking which is the practice of using a computer's internet protocol (IP) address to obtain the geographical location of that computer. This information can then be used to tailor products, services and future marketing copy to that consumer.

While not precise, IP tracking could give airlines the chance to differentiate prices by point of sale, thus enabling them to better match prices by customer resource (e.g. a ticket could be priced lower for a passenger who lives or works in a particular country). Equally, IP addresses can be used to automatically direct passengers to a specific website to enhance the customer experience by ensuring that the correct local language and any region-specific branding and promotions are shown to customers.

Personalisation is a key development which all travel companies (not just airlines) are pursuing in order to maximise revenue opportunities from each sale and to retain loyalty by ensuring customers' expectations are met during the booking process. This may include, for example, the use of web cookies to target specific fare and ancillary products in adverts and during the booking process, the bundling of ancillary products based on a customer's previous bookings and suggesting specific fares or travel times based on the profile of a customer. To enable this personalisation to take place, the collection and retention of customer data are key. However, new legislation such as the General Data Protection Regulation (GDPR) in the EU can affect the ways in which companies collect, store and use customer data.

A future challenge airlines will face is the ability to further reduce operating costs, particularly with increasing fuel costs and the arrival of carbon offsetting through ICAO's Carbon Offsetting and Reduction Scheme for International Aviation (CORSIA) (➤Chapter 18). The 'quick wins' have already been adopted, and finding further cost efficiencies is becoming increasing challenging. This in part is the reason behind the move towards more homogenised fare products. If LCCs are unable to further drive cost efficiencies and more pressure comes from the network side of the industry on base fare costs, LCCs will need to be even more focused on revenue management to ensure each flight is profitable. FSNCs, in contrast, need to be aware of their cost base and the historical legacy of their (traditionally more expensive) IT and HR systems (➤Chapter 19).

Key points

- Pricing and revenue management are fundamental to airlines, and carriers adopt different approaches depending on their business models.

- Long-haul and short-haul markets are priced differently due to the different cost of producing them and the different customer segments they attract.

- Airlines have moved from traditional and more restrictive pricing models towards more flexible one-way pricing regimes.

- Network operations provide greater potential for connectivity and revenue generation than point-to-point routes, but they are more complex to manage.

- LCCs, in particular, may advertise loss leaders to stimulate demand.

References and further reading

Belobaba, P., Odoni, A. and Barnhart, C. (2009). *The global airline industry: Chapter 4 – Fundamentals of pricing and revenue management.* 1st edn. Chichester: Wiley.

Doganis, R. (2019). *Flying off course: Airline economics and marketing.* 5th edn. London: Routledge.

IATA. (2019). *Ticketing handbook.* 51st edn. Montreal: IATA.

Ozlap, O. and Phillips, R. (2012). *The Oxford handbook of pricing management.* 1st edn. Oxford: Oxford University Press.

Sorensen, J. (2019). *2018 Top 10 airline ancillary revenue rankings*, Shorewood, WI: IdeaWorksCompany.com.

Vasigh, B., Fleming, K. and Tacker, T. (2018). *Introduction to air transport economics: From theory to applications.* 3rd edn. Abingdon: Routledge.

CHAPTER 10

🚶 Airline passengers

Andreas Wittmer, Gieri Hinnen, and Erik Linden

LEARNING OBJECTIVES

- To identify airline passenger segments based on situational, socio-economic, demographic, and psychographic characteristics.

- To understand the concept of customer value.

- To recognise different interpretations of passenger utility.

- To identify differences between airline loyalty programmes.

- To be aware of the relative merits to airline passengers of airline loyalty programmes.

- To develop an understanding of the impact of an ageing population on the air transport sector.

10.0 Introduction

Airlines create value for employees, investors, governments, customers, and consumers. This chapter focuses on the **customers** and **consumers** of the airline product. Airline customers, who include passengers and **corporate travel providers**, exert considerable market power in the air transport sector through their purchasing decisions and travel behaviour. As a purchaser of a commercial aviation product, customers are responsible for stimulating product and service innovation, and their purchasing decisions ultimately decide which airlines succeed and which fail. As all airlines offer the same basic product – safe carriage by air from A to B – they seek to differentiate themselves from their competitors in terms of price and service. Airlines will be successful only if they can create sustainable value for their customers. Whereas customers make purchasing decisions, consumers experience (consume) the product or service that has been bought. Satisfying the needs of passengers, who may be both customers and consumers, is inherently challenging and complex. This chapter focuses on three management tools that airlines use to achieve sustainable competitive advantage: passenger segmentation, passenger value, and passenger retention.

10.1 Passenger segmentation

To effectively tailor their products and services to the needs of particular customers, airlines segment their passengers into different groups. The process of **passenger segmentation** seeks to identify groups of customers who share common characteristics. The resulting market segments contain customers with similar preferences and/or buying behaviour. Customer preferences differ between market segments. Criteria including nationality, age, and trip motivation (business or leisure) are typically used to segment passengers. Individuals travelling on business, for example, may have different needs from if they were travelling for leisure purposes. Market segments can be categorised based on situational, socio-economic, demographic, or psychographic criteria.

Situational criteria relate to the context in which the customer travels. Typical **situational segmentation** variables include:

- sales channels (such as travel agents, online or telephone);
- time/date of flight;
- time of booking;
- location and access to origin and destination (OD) airports;
- seat and ticket availability;
- ticket flexibility;
- loyalty/frequent flyer benefits;
- airport services;
- in-flight services.

Socio-economic and demographic segmentation considers the personal characteristics of individual travellers. Typical criteria are:

- gender (passengers may exhibit different needs and priorities on account of their gender; Korean Air offers a female-only area of their airport lounge at Incheon Airport, while All Nippon Airways, Japan Airlines, and GOL have women-only lavatories on selected aircraft);
- nationality;
- religion;
- age;
- physical (dis)abilities (which may require special assistance and the use of wheelchairs or other assistive or adaptative technologies);
- relationship status;
- income;

Customer: a person who purchases a good or service for personal use (in the case of a passenger) or on behalf of another person (in the case of a corporate travel provider).

Consumer: a person who consumes the product or service that has been purchased. In the case of air travel, passengers can be both customers and consumers.

Corporate travel provider: a specialist travel company that arranges business travel on behalf of other organisations.

Passenger segmentation: the grouping of passengers according to their stated or revealed preferences and buying behaviour.

Situational segmentation: the grouping of passengers according to booking preferences and travel requirements.

Socio-economic and demographic segmentation: the grouping of passengers based on personal and social characteristics.

- first language;

- occupation;

- education/qualifications;

- whether passengers are travelling alone, in a group, in a family group, or with babies or young children.

Psychographic segmentation focuses on trip motivation, engagements, values, attitudes, interests, opinions, personality, behaviour, and lifestyle characteristics. These characteristics might indicate why a specific product category is preferred but not why a specific product was chosen. The biggest challenge with psychographic segmentation is that these criteria are often more challenging to measure than demographic segmentation criteria. Psychographic variables include:

Psychographic segmentation: the grouping of passengers based on travel behaviour, motivation, values, attitudes, interests, opinions, personality, and lifestyle criteria.

- trip motivation: the reason for traveling, such as business or leisure or visiting friends and relatives (VFR);

- destination;

- length of flight: short- or long-haul;

- length of total time away from home;

- travel class: economy, economy plus, business, or first class;

- travel experience: frequency of flying;

- the cultural background of the passenger;

- airline preference, which may be based on the business model of the airline (see ➤Chapter 8), its perceived safety and service standards, its brand value and reputation (high quality or low cost), its cultural resonance and familiarity to the passenger, and whether or not it belongs to a global airline alliance;

- membership of airline or airline alliance loyalty programme, account balance, and status level;

- seat preference (whether for a particular seat, an extra-legroom seat, an aisle seat, or a seat in a child-free quiet zone);

- environmental considerations: age of the aircraft and the airline's environmental credentials.

The aviation sector uses different variables to segment its passengers. The European aircraft manufacturer Airbus, for example, segments passengers into six groups according to socio-economic, demographic, and psychographic variables (see Table 10.1).

Once segments are identified, airlines need to identify the requirements of these groups and explore the most effective way to target and engage them. Conventional market research techniques, such as questionnaires, opinion polls, and focus groups, have long investigated

Revealed preference: how customers behave.

Stated preference: what customers say they prefer.

Big data: describes large data sets detailing human behaviour or interactions which are too big, complex, or dynamic to be handled by traditional data processing methods. Big data also refers to the sophisticated and fast-changing technologies that are used to analyse these data sets.

the relative importance of different product dimensions (including comfort, convenience, and price). However, such methods may not accurately reflect actual decision behaviour, where financial resource restrictions and trade-offs play an essential role. Contemporary market research applies sophisticated methods which enable the identification of **revealed preferences** (as opposed to **stated preferences**) or motives which sometimes customers are unable or unwilling to disclose. Innovations in data analysis techniques and the exploitation of **big data** will enable more sophisticated customer segmentation. Airline operators are increasingly using advanced statistical methods to analyse large amounts of customer data in the search for patterns and trends that identify new customer segments and so enable them to tailor their products to their needs.

Table 10.1 Airbus's passenger categorisation

Bargain travellers	Mainstream travellers
• Functional/cheap chic	• High street shopper/travel experiences
• Age: 20–45	• Age: 25–45
• Marginal business flyer	• Occasional business flyer
• Occasional leisure flyer	• Occasional leisure flyer
• Segment size: major	• Segment size: major
Traditional travellers	Trendsetting travellers
• Conservatives/habituals	• Cosmopolitans/discoverers/globetrotters
• Age: 55+	• Age: 18–99
• Marginal business flyer	• Frequent business flyer
• Occasional leisure flyer	• Frequent leisure flyer
• Segment size: major	• Segment size: medium
Senior travellers	High-society travellers
• Corporate seniors	• Company leaders/VIPs/celebrities
• Age: 45–70	• Age: 20–99
• Frequent business flyer	• Very frequent business flyer
• Frequent leisure flyer	• Frequent leisure flyer
• Segment size: medium	• Segment size: niche

! Stop and think

Why do airlines seek to segment their passengers, and what is the value of the resulting information to their business?

The segmentation of passengers represents a critical business component. Segmentation is often hard to achieve, despite initially appearing to be obvious and straightforward. Extraneous variables, such as price and market conditions, as well as individualisation and rapidly changing passenger behaviour (including demands, values, motives, and beliefs) make it a challenge to define specific and distinct customer groups and then identify patterns of similarity within these groups. One recent example is the development of 'Flightshaming', which seeks to reduce passenger demand for air travel for environmental reasons (➤Chapter 18) and which might affect passengers' behaviour and motivation on an individual level, making it nearly impossible to find patterns of similarity within groups of people.

The solution might be new approaches to segmenting passengers, such as creating personas that are representative of a large number of customers. The internet and database technologies generate large volumes of market and customer data, enabling researchers and managers to deploy metasearch and cognitive capabilities in addition to traditional linear segmentation approaches. Further, airlines require new and complementary approaches to understanding the psychology, behaviour, and human aspects of passengers to ensure a more accurate and holistic view of their customer base. In essence, passenger segmentation alone will not provide the most robust result to classify airline passengers. A combination of comprehensive methods and innovative tools is preferable.

Stop and think

What are the limitations of using traditional segmentation approaches to sorting airline passengers into specific groups?

10.2 Ageing airline passengers

Although airlines are relatively good at catering for the needs of infants and children, it is only recently that they have started to systematically address the needs of older travellers. Demographic ageing has already become apparent and will increase rapidly in the coming decades. Older air travellers show different behaviour, beliefs, needs, and values compared to younger travellers. In particular, they differ concerning their propensity to fly, travel purpose, destination choice, access modes, airport dwelling time, perception of the travel product, and the use of airport facilities. This demographic has a particularly strong impact on airlines. The over-65s will account for a significantly larger share of the overall mobility market.

The physical and mental conditions of people change with age. Eyesight, hearing, and general physical health decline, mobility decreases, and mental changes such as anxiety may be potential side effects. These changes affect the business and operating model of airlines and the global transport system. Older travellers may be less agile, require mobility aids, take longer to board, be unfamiliar with automated systems, and find it difficult to lift bags into overhead lockers.

Also in this context, it can be observed that not only are people getting older, but behavioural changes can also occur across generations. Experts refer to this as down ageing, i.e., leaving traditional-age roles and changing traditional activities. The result is multi-graphic CVs and travel characteristics. Older people are no longer satisfied with traditional everyday activities but want to reinvent and fulfil themselves. This results in changing behaviour, beliefs, needs, and values; for example, voluntary working, travelling more frequently, and travelling to a greater range of destinations. Age groups are becoming blurred. These older generations change existing structures and create new, different markets and more heterogeneous segments. Researchers and transport experts describe older travellers as the most heterogeneous and complex segment. They will have a significant impact on the business models of airlines, but also on the design of transport systems and the underlying infrastructure.

> **Stop and think**
>
> Why is an understanding of ageing important for airlines?

10.3 Special assistance passengers

Special assistance passengers need additional support when travelling by air. They may need assistance at airport security, in finding the gate, boarding the aircraft, and also during the flight. Typically, special assistance travellers are disabled (e.g. in wheelchairs), are elderly, or have 'hidden' disabilities (such as autism and dementia). Airlines usually have appointed handling agents at airports to aid special assistance travellers.

10.4 Passenger value

Service quality has become a significant issue in the airline industry and is one of the ways in which airlines differentiate themselves and create value for their customers. Major developments in customer service have taken place since the early 1930s, including the introduction of flight attendants, pressurised aircraft, improvements in in-flight catering and entertainment, and innovations in ground service provision. In the new millennium, there is evidence of a renewed focus on product features and service quality, especially in business and first class. The key drivers of service quality are:

- growth in demand for business class travel and premium products;

- technological and digital innovation that enables service features such as individual in-flight entertainment programmes, lie-flat seats, and live service updates;

- a stronger focus on long-haul products by full service network carriers (FSNCs) as a reaction to the declining profitability of short-haul routes due to the rise of low cost carriers (LCCs);

- competitors such as Emirates, Etihad, and Qatar Airways from the Middle East, who specialise in providing superior levels of in-flight service.

New or improved services aim to add customer value. Customer value can be seen as the surplus of customer benefits that occur when examining the utilities that customers experience in the process of consuming a product or service compared to the costs of providing it. Thus, customer value is the subjective value or perceived utility a customer derives from a product or service. Customer value is based on economic utility theory, which assumes that human behaviour seeks to maximise individual utility. Customer **utility** can be used to generate a strategic competitive advantage. An awareness of different customer value dimensions allows direct investments into specific quality or cost factors. **Customer value** is not to be confused with **customer equity**; customer equity defines the value of the customer to the company.

A long-standing debate focuses on hard versus soft service factors (Klein, Linden and Wittmer, 2019). Hard factors are those such as price and ticket flexibility, which are measurable and similar for all customers. Soft factors are often referred to as convenience factors that are subjectively interpreted and experienced by individual customers. Soft factors are often intangible, such as the perceived friendliness of the cabin crew or the quality of in-flight catering, and in the field of airline services, soft factors are increasingly important. The effect of hard and soft factors on customer value can be affected by socio-demographic and contextual demographic variables. For instance, the impact of soft and hard service elements is influenced by the cultural background and previous travel experience of individual passengers.

Service elements have different meanings for different customer segments. First-class travellers show considerably lower price sensitivity in connection with a higher brand affinity and a stronger emphasis on seat comfort. Research has indicated that some customers are willing to pay a premium for green products (Hinnen, Hille and Wittmer, 2017) and to offset the carbon from their flights (Higham, Ellis and Maclaurin, 2019; Wittmer and Wegelin, 2012). Despite this, the problem for customers might be that available information is incomplete, and flyers make assessments of the epistemic reliability of what information they do receive (Higham et al., 2016). Hence, the benefits of carbon offsets are ignored if they do not reinforce travellers' preferred views on the relevant costs and benefits (Becken and Mackey 2017).

Passengers invariably show a higher willingness to pay for safety and ticket flexibility. This can be understood as passengers displaying a high propensity to pay for their wellbeing. Wellbeing is seen as an individualised and subjectively experienced way of being, which is linked to travel-related stress and determined by the behaviour of individual travellers. The higher the travel class, the higher the personal wellbeing.

Airlines are increasingly tailoring more of their products to individual travellers to increase customer value (see Case Study 10.1). Such services come in the form of supplementary services, such as on-board internet access, lounge access, and ground transport services. Passengers pay an additional fee for such services, generating additional ancillary revenue above the basic airfare. This better enables passengers to customise their journey and maximise their value. This trend has been accelerated by developments in information communication technology (ICT) and significant data methods, which enable airlines to identify and respond to individual customer needs. Ancillary revenues have become an increasingly important revenue stream for all airlines, not just LCCs (➤Chapter 8).

Utility: a measure of preferences or benefits of an individual over a set of goods or services.

Customer value: the value a customer places on the product or service received. It is the perceived value the customer receives.

Customer equity: the value of the customer to the company. A loyal customer has a higher value to a company, as the revenue derived from that customer is higher, and the result of lower marketing and sales costs.

Service element: a tangible or intangible aspect of service.

10.5 Passenger retention

Some people repeatedly fly with one airline or alliance because it offers them a high level of personal utility. Airlines seek to support repeated purchasing behaviour by building long-term relationships through **customer relationship marketing (CRM)**. CRM places equal or greater emphasis on maintaining relationships with existing customers as it does on its search for new ones, as it is far cheaper and easier to retain existing relationships than to develop new ones.

Customer relationship marketing (CRM): marketing that aims to create long-term relationships with customers by focusing on customer value and satisfaction rather than quantity of sales.

Some airlines spend considerable time and effort on building customer loyalty through frequent flyer programmes (FFPs). The rationale for operating a FFP is to promote customer retention. Loyal customers have a significant impact on company revenue as their money is captured by the company and denied to their competitors.

Lasting customer relationships are beneficial for airline operators. Customer retention is vital due to increasing competition and developments in ICT, which enable customers to compare the products and prices offered by a range of suppliers. The internationalisation and globalisation of markets, liberalisation, shorter product life cycles, and continuous development of products further accentuate the challenge. This has led airlines to seek to decrease costs and focus on meeting and exceeding the needs of particular customer segments. Focusing on the needs of customers is not an altruistic measure but a business decision based on the knowledge that customer recruitment is more expensive than customer retention. A small increase in customer retention can lead to a significant increase in profits. Over time, airlines have moved away from attempting to satisfy every single customer to prioritising the most valuable customers and influencing the behaviour of less valuable customers to convert them into loyal ones.

An essential approach to enhancing customer retention is a customer loyalty programme. Customer loyalty is central to relationship marketing and takes into account how companies can benefit from loyal customers by increasing customer profitability and lower marketing costs. Companies in different industries have established loyalty programmes. The classic examples are retail companies; customers collect points with every purchase, which subsequently can be used to purchase discounted products. The mechanism of the loyalty programme in the airline industry is similar. Passengers collect mileage points with every flight which can later be used to purchase flights, upgrades, or airline-branded products. Points can also be collected through the use of linked credit cards. A customer can attain different status levels depending on the number of mileage points collected. The higher the status level, the more benefits the customer receives. The benefits may include the use of a priority check-in, priority security line, priority boarding, priority standby for fully booked flights, a more personalised service, access to higher-class lounges, limousine services, higher baggage allowance, and upgrades. By having different tiers of membership, airlines engage in **aspirational marketing** as benefits increase with each status level. Individual status levels are often named after rare jewels (e.g., sapphire, ruby, emerald, or diamond) or precious metals (such as silver, gold, and platinum) to suggest exclusivity and value.

Aspirational marketing: marketing that creates a desire among consumers to obtain an exclusive or luxury product or service that, in reality, few can afford.

THE USE OF PERSONAS FOR SEGMENTATION IN THE LUFTHANSA GROUP

The Lufthansa Group has introduced six different "Personas," fictitious people who act as representatives for a specific target group. Personas help staff to put themselves in the customers' shoes and thereby identify possible pain points in the customer journey. The aim is to achieve customer-centricity, i.e., to put the customers' needs at the centre of attention. The personas have been developed from detailed market research on needs-oriented and psychographic factors, focusing on the emotional drivers for travel behaviour. Up to 80 per cent of all travellers are covered by these six personas.

The central insight is that an individual might be more than one customer. For instance, Dr Smith might travel to Brussels on business during the week and then fly to Mallorca with her family at the weekend. For the Brussels trip, what she wants most is a quick air travel experience that is as smooth and efficient as possible; for Mallorca, she is looking for all the help she can get to make it a stress-free flight. Depending on the travel context, customers move from one role to another.

To cater for these needs, the Lufthansa Group has introduced Personas such as the *Care Seeking Family*, the *Efficiency Seeker* or the *Individuality Seeker*. The "*Care Seeking Family*" wants a stress-free holiday flight. The *Efficiency Seeker* wants to get from A to B as quickly as possible, looking for consistency in service and punctuality. Finally, the *Individuality Seeker* wants to be as flexible and independent as possible and will readily use new technologies to manage the journey in order to have a seamless experience.

Customer retention programmes such as loyalty cards and membership clubs have three main targets which collectively have a positive impact on customer retention and profit:

1 *Customer selection.* A loyalty programme enables a company to build more detailed customer segments. For example, access to a loyalty programme can be limited to a specific customer group. It is also possible to specifically address specific customer segments. By creating incentives such as a status programme, some customers are selected and treated better than others. With a precisely structured segmentation, customer satisfaction, customer value, and retention can be enhanced and, furthermore, customer information can be improved. Loyalty programmes enable a firm to collect vast amounts of data on its passengers, such as flight frequency and purchasing behaviour. The airline can simultaneously profit from increased market awareness, which it can utilise to develop a more targeted marketing plan and individualised services. Gaining this data is of utmost importance to increase **cross-buying activities** and decrease price sensitivity. Information management is vital for successful customer relationship management as it helps to segment customers and ultimately increase customer value and loyalty.

Cross-buying activity: where customers buy different services from the same provider.

2 *Interaction and integration through dialogue.* Loyalty programmes require regular contact and interaction with members.

3 *Image improvement and strengthening of identification.* If a customer feels directly and individually targeted by an airline, a particular additional emotional value is generated. Customer programmes can be developed either in a **company-specific** or

Company-specific: the benefits go to the company rather than to the alliance.

Company-overarching: the benefits are distributed among the programme partners.

company-overarching way. Company-overarching programmes are, for instance, programmes in partnership with credit card companies or retail chains where miles can be generated through transactions or purchases in other industries. Such programmes distribute the revenue between the programme partners rather than being retained wholly by the company.

!

Stop and think

What are the benefits of FFPs for airlines and passengers?

10.6 Airline alliances and loyalty programmes

Airline alliances are an essential factor in passenger loyalty and a vital feature of loyalty programmes. Alliances enable their partner airlines to offer members of their individual loyalty programmes the benefit of being able to collect and redeem frequent flyer points with other carriers in the alliance in addition to granting them access to a broader range of benefits that increase customer value, such as more destinations, improved flight connections, enhanced flight frequencies, and access to more airport lounges. Nevertheless, each airline within an alliance retains and administers its own FFP and selects the level of reward and bonuses that it offers. This can make aligning the benefits and rewards of different FFP memberships within one alliance challenging.

Some airlines, for example, may provide passengers who hold a ticket with their airline with some form of unique service (such as lounge access for a high-status member flying economy class) that is not available to passengers with equivalent FFP status holding tickets issued by a partner airline. This is because the frequent flyer points needed to gain a particular status level with different airlines within an alliance vary. At the time of writing, a Lufthansa passenger needs to collect 100,000 status miles points to become a gold member with Miles & More, whereas an Aegean Airlines passenger needs to collect only 20,000 status miles points to become a gold member of the Aegean Miles+Bonus programme. As a consequence, a passenger might hold an Aegean Miles+Bonus Gold Card despite flying mostly with Lufthansa. This fact has led airlines within an alliance to differentiate benefits for passengers based on their FFP membership and status level.

As Chapter 8 explains, airlines enter alliances to gain increased economies of scale and scope, access to a broader range of markets, and opportunities to increase customer benefits and utility. An alliance aims to allow customers to benefit from the combined geographic coverage and service of all member airlines. Three alliances – Star Alliance, oneworld, and SkyTeam – currently dominate the passenger market (see Table 10.2 and Case Study 10.2). Although FSNCs were the first to enter into formal alliance arrangements, they are not the only airlines who seek to create loyalty benefits for customers. Several LCCs, including Southwest (US), GOL (Brazil), and AirAsia (Malaysia), operate membership or loyalty programmes which offer benefits such as priority boarding or a higher baggage allowance. In addition to the three major alliances, there are also three new regional alliances – Vanilla

Alliance (formed in 2015 to increase connectivity in the Indian Ocean), the Chinese U-FLY Alliance (formed in 2016) and the Value Alliance (formed in 2016 by airlines in East Asia and Australia).

Stop and think

Is there an optimum size for an airline alliance, and what issues does alliance membership pose for individual carriers?

!

Table 10.2 Comparison of main airline alliances, 2019

	Star Alliance	oneworld	SkyTeam
Formed	May 1997	February 1999	June 2000
Members	28	13 (+30 affiliated)	19
Daily flights	18,800+	14,000	14,500+
Destinations	1,300+	1,100+	1,150+
Countries served	193	180+	175+
Annual passengers (million)	757	536	630+
Employees	443,703	almost 400,000	459,781
Aircraft fleet	5,046	3,500+	4,467
Lounges	1,000+	650	750
Total revenue (US$)	177.24 billion	135 billion	156 billion (2017)

Source: Individual alliance webpages

CASE STUDY 10.2

A COMPARISON OF THE MAJOR AIRLINE ALLIANCES (2019)

Star Alliance

Star Alliance is the world's largest global airline alliance. It was founded in 1997 by Air Canada, Lufthansa, Scandinavian Airlines, Thai Airways, and United Airlines. New members have since joined the alliance, and 28 member carriers currently operate at over 1,300 different airports within 193 countries. Star Alliance categorises its frequent flyer customers into silver, gold, and (depending on the issuing airline) platinum or honorary status tiers. This is in addition to the status level that is held with an individual airline's FFP.

Star Alliance Silver status: After reaching the premium level of one of the different airline members, the frequent flyer receives Star Alliance Silver status. This status includes priority waitlisting and a guaranteed seat reservation if a place becomes available on a fully booked flight. Passengers also have priority standby on the next scheduled flight in the event of missing their original flight.

Star Alliance Gold status: Gold status cardholders receive the same benefits as the Silver status members plus five additional benefits. The cardholder receives access to all Star Alliance airport lounges worldwide, regardless of the class of travel. Priority check-in is permitted at all airports and cardholders receive priority boarding and an additional 20 kg baggage allowance. Bags belonging to Gold card members get priority handling and are among the first to be unloaded.

oneworld

The oneworld alliance was founded in 1999 by American Airlines, British Airways, Cathay Pacific, Canadian Airlines, and Qantas. It has 13 airlines and 30 affiliated partners who collectively serve over 1,100 destinations in 180 countries. oneworld offers different tier benefits to its customers. oneworld member airlines work together to deliver a superior, seamless travel experience consistently, with special privileges and rewards for frequent flyers, including earning and redeeming miles and points across the entire alliance network. Some of the status benefits are intangible, unlike direct discount schemes such as mileage points:

oneworld Ruby privileges: The lowest tier status is awarded when a customer reaches the first premium level of a members' FFP. In addition to the benefits afforded by the member airline, three oneworld privileges exist. These are access to business class priority check-in; preferred or pre-reserved seating; and priority standby on fully booked flights.

oneworld Sapphire privileges: A Sapphire member receives Ruby benefits plus additional privileges. Sapphire members can access business class lounges at every airport, even if they are flying in economy class and they receive priority boarding and an additional baggage allowance.

oneworld Emerald privileges: The benefits in the Emerald tier status include those of the Ruby and Sapphire levels and two additional privileges. If first-class lounges are available at an airport, cardholders may use them regardless of the class they are flying in. Emerald status cardholders are permitted to check in at the first-class priority check-in desks, can access fast-track security lanes, and can receive an additional baggage allowance.

SkyTeam

SkyTeam was formed in June 2000 by Aeroméxico, Air France, Delta Air Lines, and Korean Air and has its headquarters in Amsterdam. As of 2019, SkyTeam has 19 member airlines. SkyTeam offers different status levels and benefits:

SkyTeam Elite: Elite status customers benefit from an extra baggage allowance, priority check-in, priority boarding, preferred seating, and priority standby.

Sky Team Elite Plus: Elite Plus offers three additional benefits. Members have access to exclusive member lounges and may invite a guest to accompany them. They are guaranteed an economy class seat on every long-haul flight if they book more than 24 hours in advance of departure, and their luggage receives priority handling.

10.7 Challenges of frequent flyer programmes: induced disloyalty

In response to the rapid accumulation of frequent flyer miles and the legacy of unredeemed miles, some airlines have made the terms and conditions of their loyalty programmes more restrictive. As a consequence, the benefits and status associated with FFPs have, in many cases, decreased since the early 2000s as airlines have switched from using a distance-based metric (how far a passenger flies) to award points, which is a revenue-based one (how much they pay for their ticket). Although airlines can change the rules of their loyalty programme by reducing the value of mileage points, this might alienate previously loyal customers, and carriers need to be aware of the effect of customer reactions. It is important to note that many frequent flyers are members of more than one programme.

Despite the anticipated benefits of operating an FFP, research has revealed that FFPs can be less successful in creating long-term loyalty than expected. FFPs are expensive to administer, and passenger dissatisfaction with loyalty programmes can evoke negative publicity. This can range from low ratings and non-recommendation to switching to other programmes. There is an ongoing debate into the overall effect of loyalty programmes, and the danger is that loyalty programmes may destroy customer value over time, rather than create it.

Stop and think

How might FFPs evolve in the future?

Key points

- Only airlines that create long-term value for customers will be able to compete in the marketplace successfully.

- Airlines must identify different customer segments according to different criteria and design products to satisfy their needs.

- Airlines must maximise customer value for each passenger by innovating and offering services that meet or exceed customer expectations.

- Airlines must retain customers, and many use loyalty programmes to increase retention and engender loyalty.

References and further reading

Becken, S. and Mackey, B. (2017). What role for offsetting aviation greenhouse gas emissions in a deep-cut carbon world? *Journal of Air Transport Management*, 63, pp. 71–83.

Boetsch, T., Bieger, T. and Wittmer, A. (2011). A customer-value framework for analyzing airline services. *Transportation Journal*, 50(3), pp. 251–270.

Hapsari, R., Clemes, M. D. and Dean, D. (2017). The impact of service quality, customer engagement and selected marketing constructs on airline passenger loyalty. *International Journal of Quality and Service Sciences*, 9(1), pp. 21–40.

Higham, J., Cohen, S. A., Cavaliere, C. T., Reis, A. and Finkler, W. (2016). Climate change, tourist air travel and radical emissions reduction. *Journal of Cleaner Production*, 111, pp. 336–347.

Higham, J., Ellis, E. and Maclaurin, J. (2019). Tourist aviation emissions: a problem of collective action. *Journal of Travel Research*, 58(4), pp. 535–548.

Hinnen, G., Hille, S. L. and Wittmer, A. (2017). Willingness to pay for green products in air travel: ready for take-off? *Business Strategy and the Environment*, 26(2), pp. 197–208.

Klein, M., Linden, E. and Wittmer, A. (2019). Influence of marketing instruments on consumer behavior in the process of purchasing leisure flight Tickets. *Marketing Review St. Gallen*, 3, pp. 30–38.

Wittmer, A. and Wegelin, L. (2012) Influence of airlines' environmental activities on passengers. *Journal of Air Transport Studies*, 3(2), pp. 73–99.

CHAPTER 11

12 Airline scheduling and disruption management

Cheng-Lung Wu and Stephen J Maher

LEARNING OBJECTIVES

- To understand the principles of airline scheduling including schedule generation, fleet assignment, aircraft routing, and crew rostering.

- To appreciate the complexity of airline scheduling and optimisation.

- To recognise the role of operational uncertainties and their impact on airlines.

- To assess options in airline recovery and disruption management.

11.0 Introduction

This chapter examines airline schedule planning and includes considerations of: airline scheduling (procedures and methods), airline operations, disruption management, and schedule recovery. The chapter introduces the major elements of scheduling and the mathematical models that underpin them. The chapter also addresses operational uncertainties and highlights the influence of scheduling practices on managing schedule operations. Disruption management is introduced to demonstrate how schedule disruptions may occur, how airline schedules are recovered, and how disruption management can inform schedule planning through feedback.

11.1 Airline schedule planning and resource utilisation

Tasks in schedule planning

The task of airline schedule planning is essentially equivalent to resource allocation and management with a strong focus on the optimisation of resource utilisation. Airline schedule planning comprises four main tasks, which are often conducted sequentially. These are: schedule generation, fleet assignment, aircraft routing and crew rostering. The aim of the schedule generation process is to design a timetable that is competitive, can meet potential travel demands (in terms of departure times, flight frequency and origin/destination airports) and can recover delays. The task of fleet assignment is to determine which type of aircraft should fly a particular **sector** to maximise revenue.

Sector: (also known as a route) is a single flight that connects an origin and destination airport (an OD pair).

The task of crew scheduling is then to assign individual crew members to flying duties in accordance with their qualifications and working hour limitations. The aim of crew scheduling is to maximise resource efficiency and utilisation while minimising operational expenses (most notably crew expenses since labour is often the second largest cost to an airline, after fuel). It must also satisfy legal requirements of crew competence and minimum crew numbers for each aircraft type.

Resource utilisation

Efficient resource utilisation is the goal of airline scheduling. Aircraft are expensive assets (►Chapter 12), and airline crew (particularly pilots) are highly skilled and costly to employ (see Chapter 18). Depending on employment conditions and countries, an A380 captain can receive an employment package worth more than US$250,000 per annum. Given the low net profit margins of global airlines, resource utilisation strongly influences airline scheduling and mathematical models that assist in optimising the tasks of airline scheduling are often employed.

Resource synchronisation

A critical element in airline scheduling is the synchronisation of resources. In this context, synchronisation means the pairing or matching of two or more resources that cannot operate independently. For a given timetable, if an airline operates different fleets of aircraft, then flights are assigned to different fleets to best utilise aircraft capacity and meet flight demand. The result of fleet assignment is then matched with **routings** coming from aircraft route planning to synchronise flights with individual aircraft. Crew are then paired with these routings to minimise equipment change during operations and crewing costs. Rosters for individual crew members are then synchronised with those paired flights (called crew pairings) as well as aircraft routings. During flight operations, synchronisation also extends to passenger itineraries that pair with various flights in the network.

Routing: a series of connected flights that are assigned to an aircraft.

Given the nature of resource synchronisation and the pursuit of optimisation, the task of airline scheduling becomes extremely complex. It is because of the need for synchronisation and the benefit of cost minimisation and profit maximisation that planning efficiency can occur at the expense of operational performance. Over optimisation in the scheduling process can make disruption management of daily operations a complex task. Mathematical models are used to plan for disruptions and aid schedule recovery.

11.2 Flight schedule generation and travel demand

Forecasting consumer demand is inherently challenging (►Chapters 2 and 5). Airlines publish their flight schedules one season in advance, although ad-hoc changes to flight schedules can happen during short-term planning. Medium- or long-term demand forecasts are predominately used for route development and fleet planning purposes and may contain forecast errors.

Forecasting travel demands for individual sectors is important. The aim of flight schedule generation is to create a schedule that is appealing to potential travellers while balancing the availability of the operator's aircraft capacity. The travel purpose of a passenger can influence the preference of flight choices (see ►Chapter 10), and this in turn determines the demand for a particular sector along with service and product features and related customer expectations. Business travellers tend to leave early in the morning and return in the evening for domestic trips, whereas leisure travellers are more flexible with respect to departure times and often seek cheaper tickets. Therefore, airlines typically provide more flights for business travellers during their preferred travel hours and charge a premium for those flights. These flights typically depart during morning and evening peak hours. In contrast, flights scheduled during off-peak hours tend to be cheaper and attract more leisure or less time-sensitive passengers.

Schedule generation (timetabling)

Apart from determining the departure times of flights between an OD pair, the other critical element in schedule generation is the flight frequency for each sector. The 'rule-of-thumb' in determining flight frequency is that the higher the frequency of a sector, the more appealing the airline (and its flights) will be to travellers, especially business passengers. This is primarily about market share; the higher the market exposure, the higher the likelihood of achieving a bigger market share. Flight frequency can be calculated using Equation 11.1:

$$frequency = \frac{D_{ij}}{C_k \cdot \rho.}$$

(11.1)

Where D_{ij} is the forecast demand between airport i and j; C_k is the capacity of aircraft type k; and ρ is the assumed load factor of the sector (often set between 75 per cent and 80 per cent).

> **!**
>
> ## Stop and think
>
> In a hypothetical situation, the demand for travel between Sydney and Hong Kong is 1,000 passengers per day in both directions. An airline operates fleets of A380s (500 seats), A330s (280 seats), and B787s (240 seats). When an A380 is used, the frequency between SYD and HKG is one per day in each direction. A330s can do two flights per day, offering 560 seats each way and if B787 is utilised, the frequency can be increased to three services per day each way (720 seats in total). How frequently should the airline fly the SYD-HKG sector, and with which fleet?
>
> The answer depends on the scheduling strategy of the airline and the limitations of the fleet size. Assuming all three fleets are available to operate the sector, the use of the A380 offers savings by conducting a single flight to meet demand. Hence, the unit cost can be lowered and profits increased.
>
> This strategy consolidates demand on one flight. The lower frequency may limit passenger choices in terms of departure times during any booking day. Often, this type of strategy is used by an airline to 'feed' traffic to a destination hub airport such as HKG. Inbound traffic is then fed to a partner airline based at HKG, which can provide services to other destinations beyond HKG, often through code-sharing (➤Chapter 8).
>
> If smaller aircraft are used, the airline can offer a higher frequency for this sector; A330 and B787 can offer two and three flights per day respectively. This strategy maximises the potential market share due to higher 'exposure' to the market, and it is also more convenient for passengers in choosing departure times; ideal to cover both leisure passengers who are less time sensitive and more cost sensitive, and business travellers who are more time-sensitive in choosing departure/arrival times (➤Chapter 9).

Schedule generation: also known as timetabling. This is a process where an airline tries to use limited aircraft capacity to meet travel demand while staying competitive.

Season: there are two seasons in airline scheduling, known as the Northern Summer Season (which commences on the last Sunday in March) and the Northern Winter Season (which commences on the last Sunday in October).

Schedule generation is often conducted about 8–10 months before a new **season** to facilitate early bookings. However, the majority of bookings usually occur three months before the departure date.

11.3 Fleet assignment and aircraft routing

Fleet assignment

After a timetable is drafted, an airline needs to allocate its available aircraft fleets to meet potential demand of each flight in the draft timetable. By using aircraft with different capacities, the flight frequency can be adjusted in order to vary the supply of aircraft capacity in an airline network. Large aircraft are often used on *trunk routes* with a lower frequency where the traffic volume is large, while narrow body aircraft are deployed where the volume is 'thinner' or where frequency is more valuable in the market. Hence, the task of fleet assignment is to match uncertain demand with a fixed supply of aircraft capacity, and to maximise potential profits. The uncertainty of demand and fixed aircraft capacity makes **fleet assignment** an inherently challenging problem (see Example 11.1). It is important to note that the majority of low cost carriers (LCCs) operate a single fleet and therefore any aircraft can be assigned to any route. Although operating a single fleet offers a number of financial benefits (►Chapter 8), it reduces scheduling flexibility, particularly if individual aircraft in the fleet have different cabin configurations.

Example 11.1

Optimal fleet assignment

In a hypothetical case, the demand forecast of a sector is 220 passengers per day in each direction. Since demand is uncertain, the forecast is best represented as a random variable following a normal distribution with mean value of 220 passengers and a standard deviation of 25, i.e. N (220, 25). The airline has two fleets: B737 (180 seats) and B787 (240 seats). If the airline uses the B737, then there will be excess demand. This means demand will spill: on average, 40 passengers will be turned away. If the airline chooses to use the B787, then there will be no excess demand, but the average seat factor will be lower at 92 per cent. The question is, therefore, which is the optimal fleet assignment for this sector?

The operating cost of a large aircraft is higher and the use of a B787 may result in a lower seat load factor and less revenue to the airline compared with using a B737. The B737 has lower operating costs and a higher seat load factor, with supply being less than demand, and the airline can expect a higher yield by charging more.

For the whole network, optimal aircraft types must be assigned to each sector to maximise total profits of operating a schedule. Hence, the optimal choice of fleet for our hypothetical case can be a B737 or B787 depending on the fleet assignment result for the whole network.

The Fleet Assignment Model (FAM) is often formulated as an integer programme that aims to minimise the operating costs of the network and maximise profits as shown in Equation 11.2.

$$\min C = \sum_{j \in J} \sum_{i \in F} c_{i,j} x_{i,j} \qquad (11.2)$$

where C represents the total operating costs of the network; $c_{i,j}$ is the total operating costs if flight i is assigned to fleet type j; $x_{i,j}$ is the binary decision variable of assigning flight i to fleet j. The total operating costs, $c_{i,j}$, include the physical operating cost of flying sector i–j, the expected spill cost of adopting fleet j, and the **recapture** (offsetting the **spill** costs) of this sector due to adopting fleet j.

Fleet assignment: the optimisation process that aims to maximise potential revenue by allocating different fleets of aircraft to undertake flights with different levels of demand. Flight demand is affected by the time of day, day of week, the nature of the origin-destination market (more leisure, business, or mixed market), and the network demand that is generated from connecting flights via hubs. This can occur on a destination/sector level but also by time of day, with larger aircraft rostered to daytime flights.

Recapture: if a spilled passenger chooses to buy a more expensive ticket on the same flight or change departure time/date but remains with the same airline, then this passenger is 'recaptured' and there is no revenue loss for an airline.

Spill: when a passenger wants to book a particular flight but cannot get a ticket. When this happens, this passenger is 'spilled'. This represents a potential loss of revenue for an airline and dissatisfaction and inconvenience for the passenger.

There are three sets of constraints when optimising the fleet assignment model:

- Flight coverage: each flight must be assigned to one aircraft type.

- Aircraft flow balance: the total number of inbound aircraft at an airport must equal the total number of outbound aircraft plus any other aircraft remaining on the ground.

- Fleet size: the number of aircraft used must be less than the total fleet size to account for aircraft which are grounded for maintenance checks or other operational reasons.

Aircraft routing

The FAM partitions the flight schedule into sub-timetables, one for each fleet. An Aircraft Routing (AR) problem is then solved on each of these sub-timetables to connect flights into routings. Constraints in the aircraft routing problem include:

- flow balance: the destination airport for a flight must be the departure airport for the subsequent flight by the same aircraft;

- flight coverage: one aircraft must be assigned to each flight;

- the availability of desired slots (➤Chapter 6) at all the destination airports.

The aircraft routing problem is given by Equation 11.3. The AR problem is an important component of the sequential planning process. The flight partition related to fleet f is denoted by N^f, with each contained flight indexed by j. Since the objective of this problem is to identify a set of aircraft routings, each decision variable (y_p) identifies a feasible routing. The set of all feasible routings for fleet f is given by p^f and indexed by p. The parameters a_{jp} equal 1 to identify whether flight $j \in N^f$ is contained in routing $j \in P^f$.

$$\min \quad \sum_{p \in P} c_p y_p$$
$$\text{subject to:} \sum_{p \in P} a_{jp} y_p = 1 \quad \forall j \in N^f , \quad (11.3)$$
$$y_p \in \{0,1\} \quad \forall p \in P^f$$

Since all aircraft are of the same fleet for each individual aircraft routing model, the cost of flying each aircraft is almost identical. However, there are alternative costs that can be used in the objective function to achieve different optimisation goals. For example, the objective function can be minimising the 'cost' of routing (if certain flight connections are cheaper or more expensive, e.g. direct flights are desired but connections with long transfer times are not), or minimising the requirement of fleet size by creating tight connections to reduce turnaround times and increase aircraft utilisation.

It is also possible to set $c_p = 0$, $\forall p \in P$ and solve it as a feasibility problem. However, it could still be advantageous to set c_p to some small costs to ensure the number of selected routes, given by $y_p = 1$, is minimised. The first set of constraints ensures that each flight within the schedule partition for fleet f is assigned to exactly one aircraft. Finally, the aircraft routing variables (y_p) are binary, which means that they can only take the values 0 (not selected in the final solution) or 1 (selected in the final solution).

The outputs of the AR model are routings for a particular fleet. An example of a domestic B737 routing for a one-day duty in Australia incorporating Sydney (SYD), Brisbane (BNE), Cairns (CNS) and Melbourne (MEL) is: FLT025 (SYD-BNE) – FLT026 (BNE-SYD) – FLT085 (SYD-CNS) – FLT076 (CNS-BNE) – FLT046 (BNE-MEL). The aircraft overnights at MEL after finishing the routing. Routings in international operations tend to be longer and may span several days.

AR is commonly performed two to three months in advance of operations. The results are then passed to the crew planning team for crew scheduling. At this point routings are specific to a particular fleet but not to any specific aircraft within that fleet. The job of assigning a routing to a particular aircraft on a particular day is called Aircraft Tail Assignment.

Aircraft Tail Assignment

Each aircraft has a tail number, which is its unique identification or registration number. When routings are assigned to individual aircraft, tail numbers are used to identify each aircraft and its assignments. This process is called Tail Assignment (TA).

The objective of TA is to assign routings to each individual aircraft and to meet the requirement for aircraft maintenance. Depending on the usage of an aircraft, different categories of maintenance activity are scheduled to maintain airworthiness and meet legal safety requirements. Major maintenance checks such as C Checks can take weeks to complete and require an aircraft to be taken out of service. Routine checks such as defect corrections and small A Checks can be performed overnight or between flights. Some maintenance tasks are due by calendar days, while others are required after a certain number of flying hours or after a specified number of take-off and landing cycles.

In TA, schedulers need to take into account the maintenance history of individual aircraft and also the projected maintenance activities that need to be carried out over the next one or two months. The task of TA is then to assign routings to specific aircraft so the aircraft can arrive at a specific maintenance base at the right time, and ideally with some flexibility in the routing schedule. Some airlines will undertake basic checks themselves, whilst others are performed by third parties at a maintenance base, which is not always the same as the airline's home base.

The challenge in TA is that if there is no option to route an aircraft to a maintenance base before a key maintenance task is due, then either the routings must be modified (revising the solution of AR), or the aircraft must be taken out of service because it is not able to meet the maintenance requirements. Not being able to route an aircraft to a maintenance base in time can be very costly to an airline. However, if an aircraft is brought to a maintenance base too early, then valuable time is 'wasted' and the aircraft may receive more maintenance than needed during its lifespan, costing more to an airline; see Example 11.2.

Example 11.2
Tail Assignment buffer planning

Compare the following three hypothetical TA routings and identify the optimal TA choice:

Route 1: FLT025 (SYD-BNE) – FLT026 (BNE-SYD) – FLT085 (SYD-CNS) – FLT076 (CNS-BNE) – FLT046 (BNE-MEL), arriving in Melbourne at 10 pm for a scheduled A check (the most basic of the four checks) from 11 pm to 7 am the next day.

Route 2: FLT025 (SYD-BNE) – FLT026 (BNE-SYD) – FLT085 (SYD-CNS) – FLT120 (CNS-MEL), arriving in Melbourne at 4 pm for a scheduled A check from 11 pm to 7 am the next day.

Route 3: FLT025 (SYD-BNE) – FLT026 (BNE-SYD) – FLT090 (SYD-MEL), arriving in Melbourne at 2 pm for a scheduled A check from 11 pm to 7 am the next day.

The optimal choice may seem to be Route 1, because the aircraft arrives at MEL just in time for the A check. However, schedulers may opt for Route 2 or even Route 3 because if any delays happen to earlier flights in Route 1 during the day of operation, then the arrival in MEL may be delayed. For a late arrival to the maintenance base, the available maintenance time is reduced, causing maintenance tasks to be delayed or 'skipped' to the next maintenance opportunity. If critical maintenance tasks cannot be finished during the planned maintenance slot, it may delay the morning operation or disrupt other maintenance tasks in MEL.

However, bringing the aircraft back to base earlier can also be costly. Route 3 contains two fewer flights than Route 1 and may leave the aircraft idle in MEL for nine hours. This idle time can be seen as schedule buffer time in TA with an opportunity cost of earning extra revenue.

Apart from maintenance requirements, some airlines also require aircraft to experience balanced wear and tear from different operating conditions, such as weather and landing

Cycle: one complete take-off and landing

cycles. Too much exposure to extreme (cold/hot) weather conditions may cause certain parts of an aircraft to wear more quickly, and hence require more frequent maintenance. Other operating conditions such as short flights with a high number of landing cycles may require the airline to service the landing gear more frequently. Hence, schedulers rotate routings among aircraft in TA, so over the course of a year each aircraft will encounter most, if not all, operating environments to balance the wear and tear among their fleet.

There are also aircraft specific restrictions that must be considered when planning the fleet assignment, aircraft routes and the tail assignment. Most often aircraft specific restrictions are handled in the tail assignment, but such restrictions may not be satisfied without proper consideration in the fleet assignment or aircraft routing. Aircraft characteristics that restrict the allowable flights include defective reverse thrusters preventing an aircraft from landing on short runways, not enough fuel tanks to complete flights over a certain length and lacking **ETOPS** certification, which forbids an aircraft from flying over water. There are many different individual aircraft characteristics that need to be accounted for in the optimisation of the aircraft routing and tail assignment, increasing the complexity of these planning and scheduling models. This increased complexity highlights the need for sophisticated software solutions to solve these optimisation challenges.

11.4 Crew scheduling

Crew planning

After AR, flights are synchronised with aircraft. The task of crew planning is then to design a crewing schedule that is able to match crewing requirements of different fleets with routings from AR. Typically the airline crewing problem is broken down into two sub-problems, namely Crew Pairing (CP) and Crew Rostering (CR). CP is conducted first, and the outputs of CP (called **pairings**) are used in CR – building rosters for individual crew members.

Pairings must meet legal crewing requirements to ensure safe operation of aircraft. Typically, these requirements regulate the working hours of pilots and cabin crew. For instance, the '8-in-24 rule' mandates that a pilot cannot fly more than 8 hours within any 24-hour period. In addition to legal mandates, crew bargaining agreements, if applicable, between an airline and its crew unions (➤Chapter 19) also impose further conditions on crewing and are at least equal to and, in most cases, stricter than those conditions imposed by government authorities. For instance, the cap imposed by many aviation authorities on the total flying hours of a pilot per annum is about 1,000 hours, but some crew bargaining agreements impose an 800-hour yearly cap; a productivity reduction (hence, cost increase to airline businesses) of at least 20 per cent.

Minimising pairing costs is important for airline profitability. Given the complex crewing conditions and potential choices in building pairings, CP is a difficult mathematical problem. A typical model form is shown by Equation 11.4:

$$\min C = \sum_{j \in P} c_j x_j \qquad (11.4)$$

where c_j is the cost of choosing pairing x_j, among all possible pairing candidates in set $P (\forall j \in P)$. The only set of constraints for this CP model is the flight coverage in which each flight must be 'covered' only once in the pairing result. This model form is elegant with few constraints. However, the set of potential candidate pairings, $P (\forall j \in P)$, can grow exponentially when a network gets larger and more complex (see Example 11.3).

ETOPS (Extended Range Twin-engine Operational Performance Standards): allows certified twin-engine aircraft to perform long-range flights that are a distance away from a suitable diversion airport. ETOPS-180 permits twin-engine aircraft to fly up to 180 minutes single-engine flying time away from a suitable diversion airfield.

Pairing: or 'tour of duties', are a series of flights that are: connected for a single crew member to conduct, start and end at the same crew base, and meet crewing conditions. A pairing can span multiple days but is often capped with a maximum time away from base.

Example 11.3

Determining the length of a pairing

Compare these three hypothetical pairings for pilots based at Pudong International Airport (PVG), Shanghai and flying to Taiwan (TPE), Singapore (SIN), Incheon, Seoul (ICN), Hong Kong (HKG):

Pairing 1: {Day-1: (PVG-TPE) – (TPE-PVG) – (PVG-SIN)} – <overnight at SIN> – {Day-2: (SIN-PVG) – (PVG-ICN) – (ICN-PVG)}; flying hours: 8 hours on day-1 and 8 hours on day-2.

Pairing 2: {Day-1: (PVG-TPE) – (TPE-PVG) – (PVG-HKG) – (HKG-PVG)}; flying hours: 7 hours on day-1.

Pairing 3: {Day-1: (PVG-TPE) – (TPE-PVG)}; flying hours: 3 hours on day-1.

All three pairings start and finish at PVG. Pairing 1 is a two-day pairing with an overnight stay at Singapore, while Pairings 2 and 3 are both one-day pairings. From the viewpoint of crew productivity with the imposed 8-in-24 rule, Pairing 1 is the most productive and efficient pairing. Pairing 2 is less productive, while Pairing 3 is the least productive. So, which one is better?

From the 'cost' perspective, for one-day pairings such as P2 and P3, the cost of crewing is mostly from flying hours. Hence, it's ideal to get these pairings as close as possible to the 8 hour daily flying-hour cap. P2 is superior to P3. But P1 might be the best option because it is both long and reaches the daily hour cap for both days of the pairing.

Apart from the costs of flying hours, P1 also involves the overnight expenses of crew in Singapore such as accommodation costs, ground transport to/from the airport and a living allowance. These expenses can be significant and it motivated some full service network carriers to establish their own hotel chains in the 1980s and 1990s.

From the cost minimisation perspective, P1 is more expensive than the combination of two P2-type pairings, if feasible, so it is a better choice to break down P1 and replace it with two P2-type one-day pairings. However, this is not always feasible in pairing optimisation since some flight times are longer and it is not always possible to bring crew back to base on the same day.

Credit hours: the total number of pay hours that a crew member may be compensated for. Apart from the flying time, credit hours also include synthetic hours within a duty day of a pairing. Hence, the credit hour is always longer than the flying hour for any pairing.

Premium: the extra cost that an airline pays its crew for conducting duties. Premium is calculated as the percentage of the hour difference between credit hours and actual flying hours compared with the flying hours alone.

Synthetic hours: non-flying work hours such as sign on/off times, downtime between flights and rest times within a duty day.

For crew salaries, many airlines pay crew not only by flying hours, but also by **credit hours**. Cabin crew remuneration may comprise a basic salary based on flying hours and experience which (particularly in the case of LCCs) may be supplemented with commission from on-board sales. An airline pays a **premium** for each pairing depending on how many **synthetic hours** are included in each pairing; the more synthetic hours, the higher the premium and the more expensive a pairing becomes. The impact of premium on crew pairings is demonstrated by Example 11.4.

Example 11.4

Pairing cost calculation – the impact of premium

Following on from Example 11.3 we can calculate the cost of each pairing by the following three cost elements: flying time ($100/hour for pilots and $50/hour for crew), non-flying time ($25/hour for all crew), and overnight hotel costs. We calculate the costs of pairings for an A320 with two pilots and four cabin crew.

All turnarounds between flights are assumed to take one hour and the sign on/off times before starting and finishing a daily duty is one hour. For a set of six crew members for A320 operations, six hotel rooms are required per night at Singapore (assuming no room sharing), costing US$1,200 per night. If this particular flight has a daily frequency between PVG and SIN, then the total accommodation bill for the crew of the PVG-SIN flight to the airline is US$438,000 a year for this particular flight alone. The cost breakdown of each pairing is provided as follows.

Cost comparison of example pairings

	Flying	Non-flying	Hotel	Total costs	Premium
P1	$6,400	$900	$1,200	$8,500	38%
P2	$2,800	$600	$0	$3,400	57%
P3	$1,200	$300	$0	$1,500	67%

From a cost perspective, P1 is the most expensive choice and P3 is the cheapest. However, from a premium perspective, P3 is the least efficient (67 per cent premium) because the flying time (3 hours) is relatively short when compared with non-flying time (2 hours). On the contrary, P1 is the most efficient pairing with a 38 per cent premium only. Typically, airlines prefer pairings with low premiums because these pairings end up cheaper without paying for excessive synthetic hours.

High premiums are sometimes unavoidable, especially for networks that have many short sectors. While an airline can utilise a narrow-body jet for 12 hours a day, a crew pairing can cover only half of the aircraft routing. Accounting for the turnaround times between flights of a domestic pairing causes the premium to increase; domestic pairings typically have premiums ranging between 45 per cent and 65 per cent, depending on sector lengths. In contrast, long-haul flights tend to have longer flight times, so premiums are typically low, ranging between 15 per cent and 30 per cent.

The outputs from CP are lists of pairings for each fleet type. Some pairings are fleet specific, while others can be more flexible. Regulations governing type rating mandate that a pilot can hold only one aircraft type certification at any one time, so an A330 pilot can only fly an A330 and not an A320, even though the pilot may have previously held an A320 certification. However, if a cabin crew member is qualified to operate both the A319 and A320, they can accept duties from both fleets. This can increase crewing flexibility and reduce crewing costs, although the training itself will be a cost to the airline. Dual qualified cabin

Deadheading: transporting flightdeck or cabin crew as passengers so they can operate the aircraft on its return journey.

crew are not required by very large airlines where the number of available crew qualified for a particular aircraft type is large enough to provide flexibility and coverage.

To ensure that sufficient crew are positioned to operate the return sectors of long-haul flights, it may be necessary for airlines to 'deadhead' crew. While this ensures that the return flight can be operated, the deadheading crew may take up space (often in premium cabins) that could otherwise be occupied by revenue generating passengers.

Crew rostering

Crew rostering: is the task of assigning named crew members to individual flights so the skill requirements and crew size for a particular aircraft are met.

Crew pairings are anonymous and not crew member specific, but they are fleet type specific. Therefore, in terms of resource allocation, crew are now synchronised with aircraft (AR) and flights (FAM). The task of assigning pairings to individual members of the crew is called **crew rostering** (CR). The goal of rostering is to ensure that the employment conditions of crew are met including training days, annual leave entitlements, *ad hoc* leave requests, flying duties (and annual hour caps), and non-flying duties (such as stand-by duties). Some qualified crew members (both pilots and cabin crew) also carry out management and training responsibilities. Their block hours are typically much lower than a fully operational crew member. The other key aspect of CR is to ensure that each flight has enough qualified crew aboard. The required qualifications vary by national authority. For airlines registered in the UAE, for example, the General Civil Aviation Authority mandates that at least one native Arabic speaker is on board every flight. The goal of CR is to ensure that all flights are adequately resourced.

Rostering conditions can be complex, and crew bargaining agreements may impose specific crewing conditions such as the maximum number of hours they can be on duty per day and per week (per seven days), days off between duties, rest days after duties that cross the International Date Line or multiple time zones, and a minimum guaranteed number of working hours per roster period. These crewing conditions and the cap of credit hours or flying hours make the roster problem challenging. The common CR equation is:

$$\min C = \sum_{j \in R} c_j x_j$$

(11.5)

where c_j is the cost of choosing roster x_j, among all possible roster candidates in set $R(\forall j \in R)$. The only set of constraints for this CR model is pairing coverage. In other words, each pairing must be covered by enough crew rosters on each operational day with the right skill mix. The objective of CR optimisation can be the minimisation of crew employment, that is, the minimum crew (base) size, the balance of working hours per roster period among crew, and the maximisation of crew productivity (in terms of flying or credit hours).

Roster period: is the period of time that a crew roster spans. This is typically four weeks for most western airlines. Crew members are also paid according to roster periods.

As the total flying hours of pilots are capped (restricted by law), it is unwise to oblige them to fly too many hours early in the year, as they will have only a small number of hours remaining from the cap later in the same year, while an airline is obliged to pay the pilot at least the minimum flying hours for each roster period. If there are 12 **roster periods** in a year, after taking off six weeks of annual leave, there are only 11.5 roster periods remaining. If the total hour cap is 800 hours per annum, then on average, a pilot flies less than 80 hours per roster period. If we discount this figure by considering other non-flying duties and anticipated sick leave, then at best a pilot can fly about 70–75 hours per roster period. If the average length of a one-day pairing is seven flying hours, then on average a pilot will work about 10 days per roster period; the industry norm is about 10–15 working days for a pilot for each roster period. Similar calculations can be made for cabin crew productivity.

Results of CR are rosters for individual crew members that incorporate individual crew requests such as annual leave. Ultimately, all individual rosters must cover each flight and each aircraft operation in the timetable with the right skill mix. Finally, an adequate number of reserve crew is also required at each crew base to respond to unforeseen circumstances such as crew member illness and flight delays.

11.5 Operational uncertainties and disruption management

Uncertainty is observed in daily airline operations and often causes flight delays and cancellations. **Disruption management** is a decision-making process employed by airlines to address these uncertainties and minimise their impact.

The previous sections describe the airline planning process that is routinely undertaken to efficiently allocate the available resources, such as crew and aircraft. Schedule planning is typically conducted under the expectation that the day of operations will be performed without any **schedule perturbations**; however unlikely this may be. For example, an aircraft routing may be designed with very short turnaround times between each pair of flights. While this is an efficient use of expensive resources, a delay on one flight is likely to cause further delays on all subsequent flights. **Delay propagation** results in higher than expected operational costs and delay propagation is a common (if not daily) occurrence.

Since planning solutions designed with a focus on efficiency are susceptible to significant cost impacts from schedule perturbations, operations controllers are employed to ensure that daily operations are executed close to plan. The process undertaken by operations controllers to achieve this goal in the presence of schedule perturbations is called **schedule recovery**. This is a *reactive* form of disruption management. It is possible, however, to consider schedule perturbations during the airline planning process to reduce their prevalence or impact. This involves using techniques of **robust planning**. Such methods are described as a *proactive* form of disruption management.

Proactive disruption management

The main objective of proactive disruption management is to avoid or reduce the potential impacts of schedule perturbations on the day of operations. These may include:

- increasing aircraft turnaround times to provide a greater buffer against delays;
- minimising the impact of propagated delays;
- introducing aircraft swapping opportunities.

The first approach is conservative and does not take into account any flight or time-of-day specific aspects. The application of this robustness technique involves planning for all aircraft to remain on the ground for longer time periods. For example, an aircraft routing

Disruption management: the process and actions taken by an airline to minimise the costs resulting from operational disruptions.

Schedule perturbation: a change to scheduled departure or arrival times during the day of operation. Possible causes include bad weather, late arriving passengers, industrial action by employees or unplanned maintenance.

Delay propagation: intensification of delays during the course of an operating day.

Schedule recovery: reactive interventions undertaken by an airline to return operations to normal following schedule perturbations.

Robust planning: proactive approaches used during the schedule planning stages to avoid or minimise the potential impact of schedule perturbations.

with flights: FLT025 (SYD-BNE) – FLT026 (BNE-SYD) – FLT085 (SYD-CNS) may be formed such that a minimum turnaround time of one hour is scheduled in between each pair of flights. Assuming the aircraft requires a turnaround time of 40 minutes, this allows for a maximum delay of 20 minutes on each flight without causing departure delays, a 20-minute buffer.

If historical records show that departure of FLT025 (SYD-BNE) is regularly delayed by 30 minutes and FLT026 (BNE-SYD) is rarely delayed, then a uniform turnaround time does not adequately match the expected delays. Setting the minimum turnaround time to 90 minutes for all flights will avoid delays for FLT025, but it is unnecessary for other services, such as those following FLT025. A drawback of this approach is that it reduces aircraft utilisation which may result in the need for additional aircraft to operate the flight schedule.

The second approach increases the ground time for aircraft, but only between flights where there is an expectation of delay propagating onto subsequent flights. This expectation is computed by reviewing historical delay data to calculate the probability of delay propagation for every possible pair of connected flights. Assuming the routing above is selected, the expected 30-minute delay for flight FLT025 (SYD-BNE) will impact the on-time performance of flight FLT026 (BNE-SYD). Hence, it is better to construct a routing where a turnaround time of at least 70 minutes is scheduled for FLT025 (SYD-BNE). By taking into account historical delay data, turnaround buffer times can be more efficiently and effectively allocated to improve the utilisation of the fleet and contain delays to a defined level.

The final technique considers aircraft swapping opportunities. This is possible when at least two aircraft of the same type are planned to be on the ground at a particular airport at the same time. In such a situation, if one aircraft is delayed, it is possible to substitute another aircraft to perform the flights that were originally scheduled for the delayed one. Aircraft swapping is a valuable management technique that is available to operations controllers to minimise the impact of schedule perturbations. Increasing the prevalence of swapping opportunities at the scheduling stage aids schedule recovery, leading to fewer delays and lower operating costs (see Example 11.5). However, this may not be desirable or practical for low cost carriers who aim to operate their fleets with as little ground time as possible.

Example 11.5
Aircraft swapping and schedule recovery

The inclusion of aircraft swapping opportunities is one of the few robust planning approaches already widely employed by airlines. A major reason is that increasing the number of swapping opportunities in aircraft routing comes at very little additional cost to an airline.

The existence of swapping opportunities provides operations controllers with an opportunity to minimise the impact of flight delays, in particular delay propagation in a network. For example, a swapping opportunity exists for aircraft TA011, arriving in SYD at 1200 (operating flight FLT057) and departing at 1415 (flight FLT070), and aircraft TA027, arriving in SYD at 1230 (flight FLT059) and departing at 1445 (flight FLT072). A severe delay of two hours on flight FLT057 will prevent aircraft TA011 from operating FLT070 on time, potentially propagating delays onto subsequent flights. However, because this opportunity exists, TA027 can be swapped to operate FLT070 and continue the remaining routing assigned to TA011. TA011 is then used to operate FLT072 and the remaining routing originally assigned to TA027. An advantage of this swap is that while flight FLT057 (by TA011) is delayed, this delay does not propagate onto any other flights in the same routing originally assigned to TA011.

Increasing aircraft swapping opportunities in the planning stage is an approach used to improve the recoverability of an airline schedule. By increasing swapping opportunities, a greater number of options are available for the operations controllers in the event of a schedule perturbation.

Reactive disruption management

Reactive disruption management is employed when an event causes a schedule perturbation and prevents the original aircraft routings, crew pairings, and even passenger itineraries from being operated as planned. Many different techniques are available to the operations control centre to *recover* from a schedule perturbation. Such techniques include:

- delaying or cancelling flights;

- rerouting aircraft and/or crew to operate a different sequence of flights;

- using additional (reserve) crew to operate flights to avoid regular crew exceeding work limits;

- transporting crew as passengers ('deadheading' crew) to reposition them to operate flights out of different airports;

- chartering additional aircraft under ACMI leases (➤Chapter 12).

The employment of these recovery techniques by the operations control centre depends on the nature of the schedule perturbation. Since there are many interconnected resources in an airline's operations, there are conflicting objectives during the recovery process. Primarily these recovery techniques are employed to minimise the additional costs to the airline; these include additional crew costs and lost revenues. However, actions such as delaying and cancelling flights have a significant impact on passenger itineraries and service satisfaction, so they must be carefully evaluated in the recovery process. Evaluating the direct and indirect costs of recovery actions is a critical consideration during disruption management, and some airlines may employ automated decision support systems (see Example 11.6).

Example 11.6
Automated decision support systems

The expansion of airline networks and services during the 1980s and 1990s prompted an interest in automated decision support systems. During this period, irregular operations were handled within airline operations control centres by practitioners basing recovery decisions primarily on their experience and intuition. To address the rising incidents of irregular operations, major US airlines invested in automated systems. Initially these systems collected and displayed operational information and data. This complemented the work performed by the operations controllers by providing up-to-date information about the airline's operations.

The benefits of providing this information were soon apparent, with significant reductions in delays. Over time, automated decision support systems were able to provide operations controllers with suggestions for cancellations, possible new aircraft routings, and crew pairings. These systems have evolved into collaborative decision-making tools. In time, fully automated airline recovery systems will become available.

Given the complexity of the airline recovery problem, it is common to focus on each key resource in isolation. Specifically, the airline recovery process involves:

- constructing an updated flight schedule;
- rerouting aircraft to operate the schedule;
- allocating crew to the rerouted aircraft schedule;
- constructing new itineraries to ensure disrupted passengers arrive at their intended destinations.

Disruption scenarios

In order to practice their response to disruption and become more efficient, airlines regularly simulate disruptive events. Often such exercises are performed just before a period of anticipated disruption. This enables the airline to refresh its contingency plans and airline staff to understand the company's procedures and their role in recovering disrupted schedules in order to deliver the best possible service. Airlines will also review their performance after a period of disruption to revise their action plans and enable them to continuously improve their operation.

Key points

- Scheduling (timetabling) reflects market competition, potential travel demand and the nature of an airline network (hubbing or non-hubbing).

- Fleet assignment maximises airline profits by minimising the risk associated with the use of fleets that often have various sizes and capacity, while travel demand is uncertain.

- Aircraft routing provides the backbone of airline operations by synchronising flights with aircraft. This synchronisation is extended to crewing at a later stage of scheduling.

- Airline crewing is expensive and a fundamentally hard mathematical problem due to complex crewing conditions and regulatory requirements.

- Schedule recovery tactics and strategies are developed to recover disrupted schedules and operations in order to maintain airline service levels and transport passengers to their destinations amid unpredictable operational disruptions.

Further reading

Barnhart, C., Cohn, A. M., Klabjan, D., Nemhauser, G. L. and Vance, P. (2003). Airline crew scheduling. In: R.W. Hall, ed., *Handbook of transportation science*. Norwell, MA: Kluwer, p. 560.

Clausen, J., Larsen, A., Larsen, J. and Rezanova, N.J. (2010). Disruption management in the airline industry: concepts, models and methods. *Computers & Operations Research*, 37(5), pp. 809–821.

Cohn, A. and Barnhart, C. (2003). Improving crew scheduling by incorporating key maintenance routing decisions. *Operations Research*, 51(3), pp. 387–398.

Dunbar, M., Froyland, G. and Wu, C. L. (2012). Robust airline schedule planning: Minimizing propagated delay in an integrated routing and crewing framework. *Transportation Science*, 46(2), pp. 204–216.

Froyland, G., Maher, S. and Wu, C. L. (2014). The recoverable robust tail assignment problem. *Transportation Science*, 48(3), pp. 351–372.

Gronkvist, M. (2005). *The tail assignment problem*. PhD Thesis, Chalmers University of Technology and Göteborg University.

Wu, C. L. (2010). *Airline operations and delay management*. Aldershot: Ashgate.

CHAPTER 12

Airline finance and financial management
Joe Kelly

LEARNING OBJECTIVES

- To understand how airlines finance aircraft.
- To consider the relative merits of different sources of finance.
- To interpret airline financial statements.
- To recognise key financial performance indicators.
- To understand how airlines manage their exposure to financial risk.
- To understand the causes and consequences of insolvency, bankruptcy and liquidation.

12.0 Introduction

The availability of appropriate levels of finance and effective financial management are important components of any successful airline operation. Airlines need to be able to access significant amounts of capital to commence operations, expand their fleet of aircraft, invest in new aircraft, develop new in-flight products (such as new seats or aircraft interiors) and serve new markets. Airlines that are not able to adequately finance and sustain their operations are vulnerable to being taken over by other carriers or ceasing to operate.

This chapter aims to highlight selected aspects of airline financing and financial management. It introduces the different sources of finance available to airlines, with a particular focus on how airlines fund the acquisition of aircraft. It also describes the presentation and content of airline financial statements and discusses financial ratios, selected financial risk management strategies and the consequences of financial failure and how that can differ depending upon the country of operation.

12.1 Sources of airline finance

One of the most significant aspects of airline finance concerns the acquisition of aircraft. In 2018, 1,839 new commercial jet aircraft were delivered by Airbus and Boeing. Boeing (2019a) forecasts 44,000 new aircraft will be required by 2038 at a value of $16 trillion as the global fleet grows from 25,830 to 50,660.

List price: the manufacturer's published price of an aircraft.

The costs associated with acquiring aircraft pose a considerable challenge for airlines. The 2018 **list price** for a new narrow-body A320neo is US$110.6 million (Airbus 2018), while the list price for a wide-body Airbus A380–800 was US$445.6 million (Airbus 2018). Although bulk purchase discounts mean airlines rarely pay the full manufacturers' list price, carriers still need to source large amounts of capital to finance the acquisition of these assets. Even second-hand aircraft can retain a relatively high resale or residual value, and airlines will need to source sufficient finance to cover their acquisition.

Worldwide, the annual financing requirement for aircraft is predicted to increase from US$126 billion in 2018 to US$181 billion by 2023 (Boeing, 2019b). For an airline such as easyJet, which is planning to acquire 152 new Airbus aircraft by 2022 at an average list price of US$92.4 million (easyJet, 2018a), the importance of informed aircraft finance decision-making is clear.

Aircraft financing: funding mechanisms for the purchase and, in some cases, operation of aircraft.

Liquidity: the ability of an airline to convert its assets into cash relatively quickly.

Solvency: the ability of an airline to meet its liabilities as they become due.

There are several options available to airlines when financing aircraft. A range of **aircraft financing** sources can be used to fund the transaction depending on the type of finance that is required and the duration of any loan. These options include: buying the aircraft using retained earnings, borrowing from banks, export credit loans, additional equity finance, manufacturer support, Islamic finance or leasing. The final decision will depend in part on the availability of funding sources and the alignment of the funding sources and funding costs with the adopted business model (➤Chapter 8). The business model will guide the selection process by clarifying the levels of cost that can be incurred while remaining profitable and the levels of debt (gearing) that can be incurred without endangering the business. Furthermore, the funding used can vary significantly from region to region.

Overdraft: an authorised deficit in a bank account caused by withdrawing more money than the account holds.

Cash

Cash flow: the net amount in cash terms being received and spent by an airline.

Using cash is arguably the most straightforward method, as aircraft are paid for by using internal funds that are available to the business such as retained earnings from previous years. While using internal funds means that the airline doesn't have to take out a loan or find alternative sources of finance, it does reduce the airline's **liquidity** (and potentially also its **solvency**). Airlines do use internally generated funds (cash) to partially fund the acquisition of aircraft, blending it with other sources to achieve the most effective combination of equity/debt funding.

Interest: the amount charged by a lender to a borrower for providing them with finance.

Overdrafts

The most obvious source of short-term finance is an **overdraft**. These facilities are designed to provide temporary funding to cover a shortfall in **cash flow**. Overdrafts are typically agreed with commercial banks and often attract a higher rate of **interest** than other loans.

Airlines, however, would not normally use short-term facilities such as overdrafts to fund long-term assets such as the purchase of aircraft. Overdrafts would normally be used to smooth a short-term cash flow demand.

Bank loans

A major source of long-term finance is a bank loan. Typical bank loans might be for a period of up to seven years and provide a source of capital that can be used to fund the purchase of new aircraft or other investments. Loans are set for a fixed period and the capital is paid back over the life of the loan, normally on a monthly or quarterly basis. Features of bank loans include the following:

- *Regular repayments.* These include a **fixed rate** or **variable rates** of interest and have to be made irrespective of whether the company is making a profit.

- *Security.* The lender will normally require security for the loan, which is typically the aircraft. This means the aircraft can be repossessed if repayments are not made on time. This is known as a 'secured loan'.

- *Covenants.* There will normally be conditions included in the loan agreement which are referred to as 'covenants'. These frequently include:

  - a limit on future **gearing** (see Section 12.3) to agreed levels;

  - a minimum level of profit to debt levels to ensure repayment capability;

  - a minimum level of interest cover to ensure future interest serviceability.

> **Fixed rate:** the interest that is charged does not alter during the loan period.
>
> **Variable rate:** the interest that is charged changes during the loan period and is tied to a market benchmark rate.
>
> **Gearing:** a financial ratio that compares a company's debt against its equity.

In addition to financing new aircraft, airlines may also be able to refinance their existing aircraft, which would increase debt but by doing so free up cash that can be reinvested elsewhere in the business. Airlines will normally try to ensure they have a staggered programme of loans and other funding renewals so they minimise the risk of having to refinance significant amounts of funding in difficult market conditions.

As well as obtaining secured loans, it may be possible for an airline to obtain an unsecured loan (known as a 'corporate facility'), which can provide funding and available resources (i.e. not used but available) to support the solvency needs of the business. However, with no security (i.e. assets that could be sold in the case of default), the loan is considered riskier, and a higher rate of interest will be charged as a consequence. Unsecured loans are relatively uncommon unless the airline operator is well established with a high level of **creditworthiness** and reliable proven cash flow history.

Asset-backed financing is, however, not without its risks, as the value of aircraft can vary quite significantly depending upon the economic outlook and the supply and demand of aircraft. 2009 was not a good year to be selling aircraft, as their value dropped by as much as 20 per cent between 2008 and 2009.

> **Creditworthiness:** an assessment of the likelihood of a borrower being able to repay a loan.

Export credit finance

These are loans guaranteed by national governments or their appointed agents to support the sales of their country's domestic aircraft manufacturing industry overseas. Export credit loans are particularly important during periods of economic downturn, when commercial banks normally reduce their lending. They might cover 85 per cent of an aircraft's value, with the remaining 15 per cent being financed by the airline through other sources (such as the ones already mentioned in this chapter).

Equity finance

Equity represents the shareholders' funds in the airline. Raising additional equity finance can be an appropriate way of obtaining additional finance. Its principal advantage is that it is less risky than debt, as there is no obligation to pay interest. Its principal disadvantage is that it can dilute the existing shareholders' ownership and earnings. This form of long-term financing involves privatised airlines raising cash by issuing the airline's shares to investors. Several methods of raising equity finance are used by airlines. These include new share issues in the form of initial public offerings and rights issues. In a new share issue, an airline will sell a proportion of the ownership of the airline in the form of shares to investors. Shareholders can be individuals, companies or institutions such as pension schemes, governments or **sovereign wealth funds**. Airlines that are listed on a stock exchange can raise additional new capital through a rights issue in which new shares in the company are offered to existing investors in proportion to their existing shareholding.

Sovereign wealth fund: a state-owned investment fund.

While many state-owned airlines have been privatised by selling shares to investors, the equity of some airlines remains in full or partial government ownership. Some countries and regions also have strict rules governing the foreign ownership of airlines and restrict the proportion of shares that can be held by foreign investors (see Table 12.1).

Table 12.1 Maximum permitted foreign ownership of airlines in selected countries and regions

Country/region	Maximum % foreign ownership permitted
Chile	100
South Korea	50
European Union	49.9
Kenya	49
Australia	49
China	35
US	25
Brazil	20

Manufacturer support

In order to secure a sale, an aircraft manufacturer (with or without government assistance) might agree to lease an aircraft to an airline operator or guarantee a residual value for the airframe at the end of a defined loan period.

Islamic finance

In order to be compatible with Islamic principles, aircraft finance packages can be arranged in such a way that no interest is paid on a loan.

Leasing

Aircraft leasing refers to the renting of aircraft on a contract basis, allowing airline companies and other aircraft operators to use the aircraft on a lease without having to purchase them outright. The leasing companies (the lessor) can also provide to airlines (the lessee) the flexibility of short- or longer-term contracts.

Leasing an aircraft from a third-party provider offers the same (or in some cases more) benefits as full ownership but without the initial high capital investment. While leasing allows airlines to add additional aircraft to their fleet without some of the risks associated with ownership, it also poses a number of challenges. There are two main types of lease, an **operating lease** and a **finance lease**. Operating leases are the more common of the two.

Operating lease

Under the terms of an operating lease, an airline hires an aircraft for a defined period of time. An operating lease might last between 3 and 10 years (although wide-body aircraft leases can be longer). This is considerably shorter than the aircraft's useful life, which is usually considered to be in the region of 20–25 years. Airlines pay the aircraft's owner (the lessor) a fixed sum to operate the aircraft for the duration of the contract.

The cost of an operating lease is the sum of three factors:

1. *The leasing cost.* This is typically set at 1 per cent of the new aircraft cost per month, but this can vary according to the supply and demand of particular aircraft.

2. *Maintenance reserves.* These are paid to the lessor to cover future maintenance activities on the aircraft.

3. *Security deposit.* This is usually equivalent to two to three months of lease payments and is returned to the airline at the end of the lease.

In addition to paying to hire the aircraft, the lessee pays all fuel, air traffic control (ATC) and airport charges. If required, the costs of repainting a leased aircraft in the lessee's corporate livery can be negotiated separately.

Operating leases differ not only in terms of their duration but also in the nature of the lease, and as such they can be classified as one of three types:

Aircraft lease: a contract agreed between an owner (the lessor) and an operator (the lessee) for the right to operate an aircraft.

Operating lease: risks and rewards related to asset ownership remain with the lessor and the asset is returned by the lessee after using it for the agreed lease term.

Finance lease: substantially all of the risks and rewards of ownership are transferred to the lessee.

- *Dry lease*. The airline hires the aircraft (without crew, maintenance or insurance) from a leasing company such as GE Capital Aviation Services (GECAS) or International Lease Finance Corporation (ILFC) but uses its own flight and cabin crew to operate it. This is the most common form of operating lease.

- *Damp lease*. The airline hires the aircraft plus the flightcrew and maintenance needed to operate it. Cabin crew are provided by the lessee.

- *Wet lease (also called an ACMI lease)*. A leasing arrangement in which the owner of the aircraft (the lessor) provides the aircraft, crew (pilots and cabin crew), maintenance and insurance (ACMI). Wet leases typically last for at least a month (a one-off flight would usually be termed a 'charter service'). They are the most comprehensive (and thus expensive) of the three types and are often used to provide a short-term increase in capacity or meet a shortfall in capacity due to mechanical or technical problems with other aircraft in the airline's fleet. Wet leases might also be used to enable an airline to operate into a third country where the lessee is forbidden from operating for safety or political reasons.

Operating leases are tailored to the needs of individual operators and help preserve an airline's liquidity. They are often employed to provide a short-term increase in capacity during periods of peak demand, to launch new routes, to fulfil a short-term contract or to provide interim capacity during periods of scheduled maintenance to the usual fleet. In addition, as the aircraft are owned by the lessor, they do not necessarily need to appear on the balance sheet, which can improve an airline's assessed asset performance measures. Potential drawbacks are that they are expensive and, with ACMI leases, the airline can lose control over the level of in-flight service provision which may adversely affect the airline's reputation.

Operating leases are often used by new entrant airlines that lack the capital required to purchase aircraft outright or the financial history to secure favourable loan terms. They are particularly attractive to airlines that need flexibility and do not want to be committed to long-term aircraft ownership and uncertain future residual and/or re-sale values. Leases may also be used by airlines as an interim measure to cover short-term capacity shortfalls while waiting for new aircraft to be delivered, or when there is a long backlog of orders and leasing is the only way to acquire use of a particular aircraft.

Stop and think

Under what conditions might airlines use ACMI leases?

Finance lease

A finance lease for an aircraft normally offers the lessee the option of purchasing the aircraft at the end of the lease. A financing lease is attractive when the cost of borrowing is lower than the airline's required return on capital. Unlike an operating lease, the airline can gain ownership of the aircraft once the loan is paid and comes to an end. Airlines may take out a combination of operating and finance leases to cover their fleet financing requirements (see Case Study 12.1).

EASYJET'S AIRCRAFT OWNERSHIP AND LEASING STRATEGY

UK-based easyJet is Europe's second largest low cost carrier (LCC). In 2018, it carried 88.5 million passengers on over 979 different routes to more than 156 destinations in Europe, North Africa and the Middle East using a fleet of 315 A320 family aircraft (comprising 156-seat A319s and 180–186 seat A320s). Of the 315 aircraft, 220 were owned, 90 were on dry operating leases and 5 were on finance leases, an owned-to-leased split of 70:30. The terms of the initial operating leases ranged from 4 to 12 years and aircraft with a net book value (net book value means original purchase value plus any additions less depreciation) of £591 million were mortgaged to lenders as security. As part of the operating lease agreements, easyJet is contractually obliged to perform maintenance on the aircraft. The finance leases incur interest at both fixed and variable rates linked to LIBOR (London Interbank Offered Rate, i.e. the rate banks charge each other for short-term loans, used as a basis for calculating UK interest rates).

Leasing enables easyJet to remove older aircraft from the fleet, protect the residual values of the aircraft and benefit from the flexibility of being able to increase or downsize the number of aircraft in the fleet to as few as 316 or as many as 385 by 2022 according to economic conditions. The new aircraft are also bringing substantial operating cost savings. The new generation A321 are 19–22 per cent cheaper than the A319 (easyJet, 2019). The new aircraft that are due to enter the fleet between 2019 and 2022 are expected to be funded through a combination of cash flow, sale and leaseback transactions and debt (easyJet, 2018a).

Sale and leaseback agreements

In addition to conventional operating and finance leases, airlines may choose to sell some or all of their aircraft (whether new or mid-life) to a third party and then immediately lease the aircraft back under an operating or finance lease. The operation of the aircraft is not interrupted and the arrangement releases capital which the airline can reinvest in other areas of the business. There are a number of reasons why airlines might engage in sale and leaseback agreements:

- They can be used as a method of financing fleet expansion, as new aircraft can be used as security for a loan.

- They can release the value in an aircraft so that it can be used in other areas of the business.

- They can increase an airline's fleet flexibility.

- They can, provided strict criteria are met, improve an airline's balance sheet by removing the aircraft and its associated debt from it.

- There may be tax advantages associated with leasing rather than owning an aircraft.

- The risk of a reduction in an aircraft's second-hand value is transferred to the new owner.

Leasing trends

In 1990, fewer than 6 per cent of the world's Western-built jet aircraft were leased. This has increased steadily during the intervening years to the point where almost 50 per cent of the world's aircraft are now leased (Tozer-Pennington, 2019). In 2002, there were fewer than 100 leasing companies, with the top two lessors maintaining more than 40 per cent market share (by aircraft). Today, there are more than 150. The top 10 leasing companies (by numbers of aircraft) have 5,593 aircraft in their combined portfolios (Table 12.2). While some consolidation has taken place, it has been more than offset by new participants, leading to an increasingly diverse market

Table 12.2 Top 10 aircraft leasing companies (by number of aircraft), 2019

Rank	Company	Total aircraft portfolio (excluding orders)	Estimated portfolio value (US$ million)
1	GECAS	1,229	23,602
2	AerCap	1,056	32,975
3	Avolon	521	18,725
4	BBAM	498	20,499
5	Nordic Aviation Capital	471	6,285
6	SMBC Aviation Capital	422	15,723
7	ICBC Leasing	377	15,448
8	DAE Capital	352	10,257
9	Air Lease Corporation	336	14,559
10	BOC Aviation	331	14,051

Source: Data derived from Tozer-Pennington (2019, p. 16)

Purchase or lease?

A key decision for airline management is whether to buy or lease aircraft. Table 12.3 identifies the potential advantages and disadvantages of leasing as a form of aircraft acquisition. In order to come to an informed decision about which approach is most appropriate, airline managers need to consider their business model, assess the current and likely future external economic and competitive environment in which they operate and evaluate whether leasing or ownership will enable them to deliver their future business development objectives. A number of airlines, including easyJet (see Case Study 12.1), operate both owned and leased aircraft, as this offers both the security of owning individual assets and also the flexibility to respond to changing market conditions and new business opportunities.

Table 12.3 The relative merits of leasing

Potential advantages	Potential disadvantages
• Reduced capital investment	• More expensive than ownership
• Possible earlier aircraft delivery	• Exposure to interest rate fluctuations (if variable rates are chosen)
• Improved balance sheets in the short term	• Aircraft may be repossessed
• Flexible payment terms	• Airline cannot access aircraft equity (operating leases only)
• Possible tax advantages	• Ongoing maintenance/service charges
• Provide short-term or interim capacity	• Return conditions can be strict
• Flexible entry and exit terms	• Early return or exit fees may apply
• Risk of resale value transferred to lessor (operating lease only)	• Restrictions may be placed on where and how the asset can be used
• Option to purchase aircraft at the end of the lease (finance leases only)	• Vulnerable to foreign currency exchange rate fluctuations (if the lease is denominated in US dollars but the earning capability of the airline is in non-US dollars)
• Services of an aircraft management company	• More expensive than ownership

Stop and think

!

Why have leases become an increasingly important source of financing aircraft in recent years?

New or old?

A further decision is whether to acquire new aircraft or second-hand aircraft that have been previously utilised by other operators and at what stage of an aircraft's life it should be replaced. New aircraft are initially more expensive to acquire, but they are also more fuel-efficient, are more environmentally friendly and have lower operating and maintenance costs than older aircraft. This means that they are cheaper to operate in the medium to long term. Newer aircraft are often also assumed to be safer. An airline may also need to consider its brand reputation when evaluating whether to acquire new or second-hand aircraft, as the average age of the fleet may be used by customers to evaluate the safety, comfort, financial health and vitality of an airline when choosing which carrier to use. The airline's business model will again guide these decisions (➤Chapter 8).

Stop and think

Why do airlines require access to different sources of finance, and what are the benefits and limitations associated with the various types of finance?

12.2 Financial statements

Irrespective of the composition of their aircraft fleet and the business model they adhere to, all airlines must prepare annual financial statements that include details of:

- the airline's profitability in its income statement (also known as a profit and loss account or, more recently, the statement of profit and loss and other comprehensive income);

- its statement of financial position (or balance sheet) which details a company's assets, liabilities and equity at the year-end);

- its statement of cash flows (or cash flow statement), which details the movement of cash into and out of the business for the year.

These statements give an indication of the financial performance of the airline to its stakeholders, such as regulators, current and potential investors, employees and customers.

Although considerable variation exists in the format and presentation of airline financial statements, all contain common elements as they need to convey details of an airline's income in terms of its profit or loss, its financial position (balance sheet) indicating its assets and liabilities, and its cash flow. In order to illustrate key points, this section uses extracts from easyJet's 2018 annual report and accounts as an example of how financial information can be presented.

Airlines that are parent companies produce consolidated financial statements which provide details of all the combined company's assets, liabilities and operating accounts, including those of the parent company and any subsidiaries in which it has a controlling interest. In the case of easyJet, the consolidated financial statement includes details of easyJet plc and its subsidiaries.

Income statement

An airline's income statement is a key document that details the financial performance of the company during a specific time period, usually a year. The income statement shows an airline's revenue and expenses and any profits or losses it has incurred. It will also provide an indication if these profits have been retained or distributed to shareholders. The format and presentation of airline income statements varies according to the complexity of the business that is being reported, but typical headings include the following:

- *Total revenue.* This is derived from ticket sales, carriage of cargo and ancillary sources such as paid-for in-flight catering, hold baggage, priority seating, car rentals and hotel bookings. Low-cost airlines in particular derive a significant proportion of their revenue from ancillary sales (➤Chapter 8).

- *Total expenses.* Expenses incurred in the process of earning revenue include fuel (which can account for almost half of an airline's total expenditure), crew salaries and expenses, interest, insurance, maintenance, **depreciation** (see Case Study 12.2), airport expenses, navigation and ATC fees, passenger service fees, marketing and other costs.

> **Depreciation:** the reduction in value of a tangible asset over time.

- *EBITDAR – earnings before interest, taxes, depreciation, **amortisation** and rent.* This can be seen as an indication of the underlying earnings of the company before these charges, which generally, but not always, offers a close approximation to the cash flow being generated by the company.

> **Amortisation:** the reduction in value of an intangible asset over time.

- *Operating profit/loss.* This describes the profit or loss that is earned before interest and tax is deducted.

- *Profit before tax.* This is a measure of profitability that shows an airline's profits before deducting the costs of any taxes.

An example of an income statement is provided in Table 12.4.

CASE STUDY 12.2

AIRCRAFT DEPRECIATION

Depreciation describes the reduction in value of an asset over time. The reduction is based upon estimated useful life of an aircraft, which can and does vary by airline, and the intensity with which the asset is used. An aircraft acquired for US$101 million initially appears on the balance sheet as an asset of US$101 million. Twenty years later, it is sold for scrap at US$1 million. Over the 20 years, US$100 million of value has been used to generate revenue and profits. Most airlines use a straight line (constant) rate to depreciate their aircraft. The aircraft in this example therefore depreciates at US$5 million each year (100 ÷ 20 years). The figure in the balance sheet is reduced by US$5 million each year, and an expense of US$5 million is charged to the income statement. This reflects the decline in value of the asset and offsets the income generated each year by the aircraft against the cost of its depreciation. If an aircraft operates for only 15 years, the depreciation becomes US$6.7 million a year (US$100 million ÷ 15 years) but if it operates for 25 years, the expense would be US$4 million each year (US$100 million ÷ 25 years). Notably different policies are adopted by different airlines.

easyJet's 2018 consolidated income statement shows the performance for the year ended 30 September 2018 with a prior year comparison (Table 12.3). In 2018 the airline generated revenues of £5,898 million, of which 79.5 per cent came from seat sales and 20.5 per cent from non-seat revenue. Year on year easyJet's revenues increased by 16.8 per cent and costs by 16.7 per cent, and a significant factor in those increases was the acquisition of Air Berlin's

Tegel operations. The company spent £1,184 million on fuel and another £1,649 million on airports and ground handling and generated £836 million EBITDAR at 14.2 per cent slightly up on the 14 per cent achieved in the prior year. The company had a profit before tax of £445 million up £60 million and paid tax of £87 million on those profits.

Table 12.4 easyJet's consolidated income statement, 2018

Year ending	30-Sep-18	30-Sep-17
	£ million	*£ million*
Passenger revenue	4,688	4,061
Ancillary revenue	1,210	986
Total revenue	**5,898**	**5,047**
Fuel	(1,184)	(1,062)
Airports and ground handling	(1,649)	(1,465)
Crew	(761)	(645)
Navigation	(400)	(381)
Maintenance	(335)	(274)
Selling and marketing	(143)	(122)
Other costs	(590)	(389)
EBITDAR	**836**	**709**
Aircraft dry leasing	(162)	(110)
Depreciation	(199)	(181)
Amortisation of intangible assets	(15)	(14)
Operating profit/(loss)	**460**	**404**
Net finance (charges)/income	(15)	(19)
Profit/(loss) before tax	**445**	**385**
Tax (charge)/credit	(87)	(80)
Profit/(loss) for the year	**358**	**305**
Earnings per share, pence		
Basic	90.9	77.4
Diluted	90.2	76.8

Source: Reproduced from easyJet (2018a, p. 118)

Consolidated statement of financial position (balance sheet)

The statement of an airline's financial position (or balance sheet) summarises what the airline owns in term of assets and what it owes in terms of liabilities as at the year-end date. Although the format of balance sheets varies by airline, assets and liabilities are typically defined as being current (meaning they will be realised within 12 months) and non-current (meaning they will not be converted into cash within 12 months). Examples of current assets are cash and deposits, while non-current assets include property, plant and equipment, including any aircraft an airline owns. Current liabilities include borrowings and tax liabilities, while non-current liabilities include maintenance provision and deferred tax liabilities. The easyJet 2018 consolidated statement of financial position appears in Table 12.5.

Table 12.5 easyJet's consolidated statement of financial position, 2018

Year ending	30-Sep-18	30-Sep-17
	£ million	£ million
Goodwill	365	365
Other intangible assets	181	179
Property, plant and equipment	4,140	3,525
Derivative financial instruments	175	87
Restricted cash	11	7
Other non-current assets	122	74
Non-current assets	**4,994**	**4,237**
Trade and other receivables	408	275
Derivative financial instruments	220	131
Money market deposits	348	617
Cash and cash equivalents	1,025	711
Current assets	**2,001**	**1,734**
Current liabilities	(1,023)	(714)
Trade and other payables	(877)	(727)
Unearned revenue	(9)	(8)
Borrowings	(24)	(82)
Derivative financial instruments	(9)	(35)
Current tax payable	(118)	(104)
Provisions for liabilities and charges	**(2,060)**	**(1,670)**
Net current (liabilities)/assets	**(59)**	**(64)**
Borrowings	(968)	(963)
Derivative financial instruments	(7)	(44)
Non-current deferred income	(18)	(25)
Provisions for liabilities and charges	(335)	(218)
Deferred tax	(348)	(249)
Non-current liabilities	(1,676)	(1,499)

Continued

Table 12.5 Continued

Year ending	30-Sep-18	30-Sep-17
	£ million	£ million
Net assets	3,259	2,802
Shareholders' equity	108	108
Share capital	659	659
Share premium	299	38
Hedging reserve	1	
Translation reserve	2,192	1,996
Retained earnings	3,259	2,802

Source: Reproduced from easyJet (2018a, p. 120)

Cash flow statement

This document details the inflow and outflow position of cash over a defined time period and shows the overall (net) change in the airline's cash flow during the reporting period. Together with the income statement and balance sheet, the cash flow statement is a key financial document as it explains changes in the balance sheet over a specified period. Cash flow statements include details of dividend payments to shareholders, the purchase of property, plant and equipment (such as aircraft), repayment of bank loans and proceeds from the sale or leaseback of aircraft. An example of an easyJet cash flow statement appears in Table 12.6.

Cash flow from the business increased by £266 million to £1,215 million (prior year £949 million), which was used to pay a dividend to shareholders of £162 million (a reduction of £52 million on the 2017 dividend), which resulted in an increase in net funds generated of £298 million to £961 million. Asset investment (tangible and intangible) increased to £1,012 million (prior year £630 million), which was offset by a reduction in **money market deposits** of £269 million and sale and leasebacks of £106 million (prior year £363 million was added to money market deposits). Financing activities were minimal in the year at £27 million (activities the prior year were a net inflow of £214 million driven by the proceeds of a £451 million Bond issuance less share buy-backs of £220 million). Overall, cash and cash equivalents held at the year-end increased from £711 million to £1,025 million.

Money market deposits: a money market account is an interest-bearing bank account that will pay a higher interest rate than regular savings accounts some carry additional features but not so as to make them more attractive than a regular current account.

12.3 Financial ratios

One way to interpret financial statements and commercial performance is through the use of ratios. Financial ratios provide an indicator of a company's financial performance, and they can be used to compare the financial performance of different companies. Many different

Table 12.6 easyJet's consolidated statement of cash flows, 2018

Year ending	30-Sep-18	30-Sep-17
	£ million	*£ million*
Cashflows from operating activities		
Cash generated from operations	1,215	949
Ordinary dividends paid	(162)	(214)
Interest and other financing charges paid	(29)	(30)
Interest and other financing income received	11	9
Net tax paid	(74)	(51)
Net cash generated from operating activities	**961**	**663**
Cashflows from investing activities		
Purchase of property, plant and equipment	(931)	(586)
Purchase of intangible assets	(81)	(44)
Net decrease/(increase) in money market deposits	269	(363)
Proceeds from sale and operating leaseback of aircraft	106	115
Net cash used by investing activities	**(637)**	**(878)**
Cash flows from financing activities		
Purchase of own shares for employee share schemes	(17)	(10)
Proceeds from Eurobond issue	–	451
Repayment of bank loans and other borrowings	–	(220)
Repayment of capital element of finance leases	(6)	(7)
Net increase in restricted cash	(4)	–
Net cash (used by)/generated from financing activities	**(27)**	**214**
Effect of exchange rate changes	17	(2)
Net increase/(decrease) in cash and cash equivalents	**314**	**(3)**
Cash and cash equivalents at beginning of year	711	714
Cash and cash equivalents at end of year	**1,025**	**711**

Source: Reproduced from easyJet (2018a, p. 122)

types of ratios can be used to analyse performance. Financial performance ratios include the following:

- *Performance ratios*. These include operating ratios, net profit margin, return on investment and return on capital employed (ROCE), operating profit margin and asset utilisation ratio.

- *Stock market ratios*. These include dividend cover, dividend yield, market capitalisation, earnings per share, price/earnings ratio and net asset value per share.

- *Risk or gearing ratios.* These provide an indication of the financial strength (or weakness) of a company through debt/equity ratios.

- *Liquidity ratios.* These measure the availability of cash within a company to pay debts. They include the current ratio, the acid test or quick ratio and cash ratio.

Several financial ratios appear in easyJet's accounts (see Table 12.7) – including ROCE and gearing.

'Gearing' is a general term describing a financial ratio that compares the level of a company's debt against its equity capital. It is usually expressed as a percentage. Gearing ratios provide a measure of a company's financial risk and show the extent to which its operations are funded by equity (its owners or shareholders) versus its debts. A gearing ratio can be calculated by dividing a company's debt by its shareholders' equity and multiplying it by 100.

Usually a company with high gearing (high leverage) is considered to be riskier than one with a lower gearing. As with other financial ratios, an acceptable gearing is determined by the performance of other companies in the same industry. A company with high gearing is more vulnerable to economic downturns as it has to continue to pay its debts even when its revenue falls. Higher levels of equity provide more protection from the effects of a downturn (as unlike interest dividends, they do not have to be paid to shareholders) and can be seen as a measure of financial strength. Lenders examine an airline's gearing, as a high gearing may indicate that their loans are at risk of not being repaid. A company can reduce its gearing ratio by retaining profits, repaying loans, issuing new shares, reducing working capital and, where applicable, converting loans into shares.

! Stop and think

What are the benefits to investors and airlines of using financial ratios?

12.4 Financial KPIs

Key performance indicators (KPIs) measure factors that are critical to the financial and commercial success of a company. KPIs differ by sector and by company, but typical airline KPIs involve the use of financial ratios and measures of ROCE, and cost and revenue per seat. Evaluating changes in KPIs over time and comparing them to the performance of direct competitors can help a company measure its progress towards achieving its corporate targets. Table 12.7 shows the KPIs used by easyJet. This list is not exhaustive, and there are other KPIs that airlines may consider using, including seat occupancy, turnaround time and route kilometres flown, as well as safety and environmental measures.

easyJet's ROCE in 2018 at 14.4 per cent was an improvement over 2017 (11.9 per cent) but shows a declining trend from 20.5 per cent in 2014 and 22.2 per cent in 2015. The revenue and cost per seat measures are components of the profit per seat measure and show that revenues, after decreases in 2015–2017, grew again in 2018. Overall profit per seat has declined from a peak in 2015 of £9.15 to £4.68 per seat in 2018, indicating the business is under pressure on prices and costs in the marketplace.

Table 12.7 easyJet's financial KPIs

Key performance indicators	2018	2017	2016	2015	2014
Headline return on capital employed (ROCE)	14.4%	11.9%	15.0%	22.2%	20.5%
Net cash (total cash – borrowings) (£ million)	396	357	213	435	422
Total profit before tax per seat (£)	4.68	4.45	6.35	9.15	8.12
Headline profit before tax per seat (£)	6.07	4.71	6.18		
Revenue per seat (£)	61.94	58.23	58.46	62.48	63.31
Total cost per seat (£)	57.26	53.78	52.11	53.33	55.19
Headline cost per seat (£)	55.87	53.52	52.28		
Total cost per seat excluding fuel (£)	44.82	41.53	38.16	37.55	37.70
Headline cost per seat excluding fuel (£)	43.43	41.27	38.33		
Seats flown (millions)	95.2	86.7	79.9	75.0	71.5

Source: easyJet (2018a, p. 155)

Stop and think

Why do airlines report different KPIs in their financial statements?

12.5 Financial risk management

Airlines are subject to a range of financial risks that are outside of their control, including changes in foreign exchange rates, variations in interest rates on loans and fluctuations in fuel prices. Such variability can result in either gains or losses. Favourable exchange rates, for example, may enable an airline to make more money, whereas a sudden rise in the price of fuel can add millions of US dollars to an airline's costs and may ultimately result in the airline ceasing operations. To maximise financial gains and minimise losses, financial risk management strategies are required.

Exposure to movements in foreign exchange rates affect airlines' operating, financing and investing activities as large fluctuations in exchange rates may result in an airline paying more or receiving less than had been anticipated. The aim of foreign currency risk management is to reduce the negative effects of exchange rate changes on the business. Foreign exchange exposure can be reduced by matching the payments and receipts that are made in each individual currency and holding deposits in different currencies to provide protection from sudden changes in exchange rates. The effect of exchange rate changes can be considerable, running into millions of pounds per annum. There is also a risk that

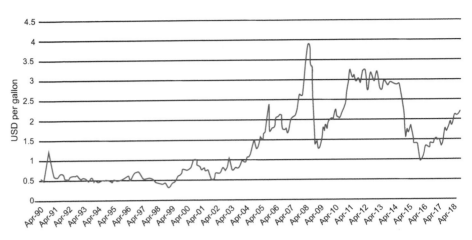

Figure 12.1 Jet fuel prices, 1990–2018

movement in a currency may impact on demand as it may make a tourist destination more or less attractive.

A further risk comes from the changing price of oil and jet fuel, as fuel comprises one of an airline's biggest cost items. By way of an indication, easyJet spent £1,184 million on fuel in the year ended 30 September 2018 (23.4 per cent of its total operating costs), and any increases in fuel price can have a considerable effect on airlines. During 2008, jet fuel prices spiked, but as shown in Figure 12.1, they remain highly volatile in terms of price. In 2008, a number of airlines worldwide ceased operating as a result of being unable to absorb the increased costs.

Hedging: a financial management strategy used to offset financial risks.

Many airlines engage in fuel price risk management (or **hedging**) to provide protection against sudden and significant increases in fuel prices and mitigate any volatility in the income statement in the short term. In order to manage their risk exposure, airlines may hedge a proportion of their short-term future fuel requirements at a set price, although there is a cost in undertaking this. If the fuel price increases during the hedging period, the airline enjoys a position of relative financial advantage over carriers that did not hedge. If, however, the fuel price falls during the hedging period, the airline pays more for its fuel than the market rate and is placed at a financial disadvantage against carriers that did not hedge (see Case Study 12.3).

FUEL HEDGING

An airline believes oil prices will rise to US$100 a barrel and so establishes a hedge to guarantee it at the equivalent of US$80 a barrel for the next year's supply. If the oil price rises above US$80 a barrel that year, the carrier has benefited from the hedge and its fuel costs are kept down. If the oil price drops to US$60 a barrel, the airline loses money compared with the price it could have paid and may be at a competitive disadvantage compared to airlines who did not hedge. In 2018, easyJet used forward contracts to hedge fuel purchases so although the average market jet fuel price increased by 32.5 per cent to $664 per tonne, the operation of easyJet's fuel hedging policy meant that the average effective fuel price decreased by 1.0 per cent to $590 per tonne, from $596 per tonne in the previous year (easyJet, 2018a, p. 34).

easyJet hedge fuel (US$ and € exposures) between 65 per cent and 85 per cent of their estimated exposures up to 12 months in advance and between 45 per cent and 65 per cent of estimated exposure from 13 to 24 months in advance (easyJet, 2019).

Hedging cannot eliminate the underlying risk of long-term fuel price variability, but it can reduce its short-term effects on the business. Hedging can be considered to be a form of insurance and, like any insurance policy, there is a premium to pay and a variety of cover available. This means that airlines have to decide not only how much of their future fuel requirement to hedge but also at what price so that they can achieve the corporate targets they have set.

Stop and think

What are the risks associated with fuel hedging?

Airlines can also be vulnerable to changes in interest rates that are payable on variable interest rate loans. This is especially an issue for carriers with high gearing that rely on large loans to continue their operations. Companies can use **interest rate swaps** to manage their exposure to changes in market rates of interest, although this is complicated to administer.

Interest rate swap: an agreement between two parties to exchange interest rate payments over a set time period.

12.6 Financial failure

Airlines operate in a very competitive and volatile market, not least because of the derived demand nature of the product they provide, which can very quickly reduce or disappear, for example, during times of recession, war or terrorist attack (➤Chapter 2). When demand does fall away, sharply significant short-term losses arise. Post-9/11, IATA reported airlines lost US$22 billion in revenues over the following three years.

In such a potentially hostile trading environment, airlines must employ robust financial, revenue and cost management strategies. If an airline is unable to service its debts, it becomes insolvent. Creditors may give the company time to pay its debts, but eventually they may require the airline to sell some of its assets. If a debt remains outstanding, the airline may be declared bankrupt. Bankruptcy describes a situation in which a company cannot meet its obligations to its creditors and seeks court protection to continue operating while it restructures or reorganises its business. In the US, this is called entering Chapter 11 bankruptcy protection (see Case Study 12.4), while in the UK it is described as entering administration. The key difference between the two is that under Chapter 11 protection the airline is run by its existing management team, whereas under UK administration an external team of administrators is brought in to reorganise the business to secure the best outcome for the airline's creditors and shareholders.

CASE STUDY 12.4

CHAPTER 11 BANKRUPTCY PROTECTION

If a US-registered airline is unable to service its debts or pay its creditors, it can enter Chapter 11 bankruptcy protection. Chapter 11 refers to the chapter of the US Bankruptcy Code, which permits US-registered businesses (including airlines) to reorganise their activities and continue to operate while being protected from their creditors. Over 60 US airlines have entered Chapter 11 bankruptcy protection since 1980, and a number have subsequently emerged as profitable businesses. Some non-US airlines have alleged that Chapter 11 offers US carriers an unfair advantage as it protects them from creditors while they restructure.

Restructuring and reorganisation seek to reduce costs in all areas of the business and may involve:

- returning aircraft to lessors;
- renegotiating aircraft leases on more favourable terms;
- deferring the delivery of new aircraft;
- renegotiating pension arrangements with their employees;
- renegotiating airport charges with airport operators;
- renegotiating employment and ground handling contracts;
- converting loans into shares to reduce interest payments;
- rationalising the route network and withdrawing unprofitable services;
- increasing ancillary revenue generation or introducing additional fuel levies;
- making staff redundant;
- selling assets.

If these measures are unsuccessful, the airline may be forced to sell off any remaining assets to generate cash (a process known as **liquidation**), and leased aircraft may be repossessed. With no aircraft and no assets, an airline cannot function and may be forced to leave the market. Market exit is particularly common among new entrant airlines that lack the capital and financial security of more established operators. In Europe, over 75 per cent of the LCCs that started flying between 1992 and 2012 left the market (Budd, Francis, Humphreys & Ison, 2014). More-established operators may also leave the market for reasons of new competition, a safety or security incident which damages customer and investor confidence in the business (➤Chapter 13), being merged with or taken over by a rival, or because of external events such as fuel price rises, geopolitical unrest in core markets, terrorist events, global recession or outbreaks of infectious disease.

Liquidation: the process of turning assets into cash in order to raise as many funds as possible towards an airline's debts.

Stop and think

Identify the reasons why an airline might fail.

!

Key points

- Finance forms a critical and integral part of airline operations and management.

- There are different sources of aircraft finance, namely the use of retained profits, bank borrowings, export credit finance, equity finance, manufacturer support, leasing and Islamic finance.

- Operating leases can be classified as dry, damp or wet leases, and they potentially offer flexibility to an airline's fleet and/or operations.

- Whether to purchase or lease aircraft is a key decision for airline operators.

- Airline operators, as with any company, have to present financial statements on an annual basis, namely an income statement, a statement of financial position and a statement of cash flows.

- Financial ratios, including performance ratios, stock market ratios, risk or solvency ratios and liquidity ratios, can be used as key performance indicators to assess an airline's performance.

- Hedging is a strategy used to protect against the risk of adverse changes in exchange rates, fuel prices or interest rates.

- Airlines operate in a volatile environment, and effective financial management is vital if they are not to be faced with insolvency and, ultimately, potential market exit.

- Insolvency, bankruptcy, liquidation and ultimately market exit are all situations that can face an airline, or indeed any company.

References and further reading

Airbus. (2018). *Airbus aircraft 2018 average list prices (USD m)*. Available at: www.airbus.com/content/dam/corporate-topics/publications/backgrounders/Airbus-Commercial-Aircraft-list-prices-2018.pdf

Boeing. (2019a). *Commercial market outlook 2019–2038*. Available at: www.boeing.com/commercial/market/commercial-market-outlook/

Boeing. (2019b). *Boeing current aircraft finance market outlook*. Available at: www.boeing.com/company/key-orgs/boeing-capital/current-aircraft-financing-market.page

Budd, L., Francis, G., Humphreys, I. and Ison, S. (2014). Grounded: Characterising the market exit of European low cost airlines. *Journal of Air Transport Management*, 34, pp. 78–85.

easyJet. (2018a). *The warmest welcome in sky: easyJet plc annual report and accounts, 2018*, Luton: easyJet plc.

easyJet. (2018b). *Full year results presentation*. Available at: http://corporate.easyjet.com/investors/reports-and-presentations/2018

easyJet. (2019). *H1 2019 results presentation*. Available at: http://corporate.easyjet.com/investors/reports-and-presentations/2019

Tozer-Pennington, V. ed. (2019). *The aviation industry leaders report 2019 tackling headwinds*, Stoke-on-Trent: Aviation News Ltd.

Vasigh, B. and Rowe, Z. C. (2019). *Foundations of airline finance methodology and practice*. 3rd edn. Abingdon: Routledge.

CHAPTER 13

Aviation safety
Mohammed Quddus

LEARNING OBJECTIVES

- To understand what is meant by aviation safety.
- To appreciate the difference between accidents, incidents and precursors.
- To recognise the principal causes of aircraft accidents.
- To appreciate how accident causation models and safety management systems can be used to improve safety.
- To understand emerging safety threats in aviation.

13.0 Introduction

Over 115 years, aviation has evolved from a new and hazardous mode of transport into one of the safest forms of long-distance mobility. Progressive developments in aeronautical design and propulsion have enabled the construction of faster, stronger, lighter and more reliable aircraft, which have not only improved safety standards but also reduced the financial cost of flying and stimulated unprecedented consumer demand for flight (➤Chapter 2). Aviation has developed into a highly complex system consisting of over 1,300 airlines operating almost 32,000 aircraft between 3,800 airports worldwide (ATAG, 2019). However, despite continued innovation, the physical environment 35,000 ft (10,650 m) above the earth remains unforgiving of mechanical failure or poor decision-making. Perhaps more than any other transport mode, the consequences of an aircraft accident are usually severe and may result in large numbers of people in the air and/or on the ground being fatally or seriously injured. Any aviation accident can change the public's perception of an airline and could contribute to the failure of the entire organisation.

In recognition of the importance of protecting life and property, this chapter focuses on the fundamental concepts of aviation safety. It introduces the definition of safety, details the principal causes of aircraft accidents and describes the causation models that can be used to reduce the likelihood of future occurrences.

In addition, this chapter also highlights an emerging safety concern to aviation, namely the growing use of drones.

13.1 Fundamentals of aviation safety

Boarding a scheduled commercial flight operated by a major airline in the developed world using a Western-built commercial aircraft is statistically one of the safest forms of transport. Indeed, passengers are far more likely to be fatally or seriously injured driving to the airport than they are once on board a flight. However, in the unlikely event that something does go wrong, the consequences can be severe, and the infrequency of such events means that safety incidents involving commercial aircraft attract considerable media attention. It is imperative, therefore, that the air transport industry continues to enhance its safety performance through: the development and rigorous testing of new technologies; the training of personnel; thorough accident investigation and learning from past events; cultivating an open and transparent reporting culture in which staff feel supported to raise concerns; and the implementation of safety management systems (Section 13.10). Proactive and predictive safety are also essential to enhancing safety performance.

Aviation safety: the theory of accident causation, investigation, categorisation and the analysis of aviation accidents or incidents and their prevention through the introduction of appropriate interventions related to enforcement (regulation), engineering (technologies) and education (training).

Safety is the state in which the risk of harm to people or property is reduced to or maintained at or below an acceptable level through a continuous process of hazard identification and risk analysis. However, **aviation safety** is more than simply the absence of an accident or the avoidance of harm. It is a culture, a way of approaching business and a way of performing daily operations that ensure that human life and property are protected. This involves identifying, analysing and eliminating, as far as possible, the development of circumstances that could lead to an accident and protecting against the consequences by, for example, installing fire retardant material in the cabin to reduce the effects of an on-board fire. Modern aircraft are highly complex and designed to routinely endure extremes of temperature, pressure and humidity. They encounter turbulence, hail and sandstorms and have to be capable of withstanding lightning bolts, bird strikes, in-flight fires and engine failure. In order to continually improve global aviation safety, international agencies, national regulators, aircraft manufacturers, airlines, airports and special interest groups routinely collate and analyse aviation safety statistics to understand the current situation, identify new trends in the data, make recommendations and issue safety directives to prevent potentially dangerous situations from (re)occurring. Examples of the agencies involved in the collation and analysis of safety statistics are presented in Table 13.1. Understanding how these different groups define and classify accidents is important, as it can lead to significant variations in published statistics. Appreciating who compiled the data, when, for whom and for what purpose is therefore crucial.

At the global level, the International Civil Aviation Organization (ICAO) sets the Standards and Recommended Practices (SARPs) that concern aviation safety, security, efficiency and environmental protection worldwide. The SARPs not only define best practice in these areas but also seek to balance the assessed risk against the risk mitigation strategies that can be imposed.

Table 13.1 Selected examples of agencies involved in the regulation of civil aviation safety and/or the collection and analysis of aviation safety statistics

International agencies	National regulators	Manufacturers	Special interest groups	Others
ICAO	FAA, US	Airbus	Flight Safety Foundation	Airport operators
IATA	NTSB (National Transportation Safety Board), US	Boeing	Aviation Safety Network	Air navigation service providers
EASA (European Aviation Safety Agency)	CAA, UK	Rolls-Royce		Ground handling agents
PASO (Pacific Aviation Safety Office)	CASA (Civil Aviation Safety Authority), Australia	GE		Airlines

Accidents, incidents and precursors

Air transport is a safety critical mode in which accidents are rare but their consequences can be severe. Accidents in aviation are therefore known as low-frequency, high-consequence hazardous events. Although the use of accident data is adequate in estimating the underlying risk and safety of road transport operations, they are not sufficient in estimating risk in the air transport industry. Consequently, the air transport industry records safety occurrence events known as incidents. This section introduces the concepts of **accidents**, **incidents** and **precursors**.

Aviation accidents

Since aviation is a global industry, a common definition of accident has been adopted. ICAO Annex 13 – *Aircraft Accident and Incident Investigation* – of the Chicago Convention 1944 (►Chapter 1) defines an accident as an occurrence associated with the operation of an aircraft which occurs at any point in a journey between a person boarding and disembarking an aircraft in which an individual is fatally or seriously injured as a result of:

- being on board the aircraft (excluding death by natural causes);

- coming into direct contact with the aircraft (or parts that have fallen from an aircraft) or being directly exposed to jet blast (unless self-inflicted); or

- the aircraft sustains damage or experiences structural failure which adversely affects its strength, performance or flight characteristics and which would require major component repair or replacement (contained engine failure and damage to wingtips, antennae, tyres or small dents are excluded); or

Accident: an occurrence associated with the operation of an aircraft in which a person is seriously or fatally injured and/or the aircraft is significantly damaged or destroyed.

Incident: a safety event in which an accident is about to happen but does not actually occur due to an intervention.

Precursor: a condition or event that could result in an accident or incident.

- the aircraft is missing or totally inaccessible.

In this context, serious injuries are defined as those which directly result from an occurrence and which require hospitalisation lasting over 48 hours, result in major broken bones, deep lacerations, second- or third-degree burns or burns which cover over 5 per cent of the body, result in exposure to radiation or infectious disease or in which injury to any internal organ is sustained. One accident that resulted in substantial damage to an aircraft but no loss of life occurred in New York in January 2009 (see Case Study 13.1). A similar event occurred outside Moscow in August 2019 when an A321 hit a flock of birds shortly after take-off and the aircraft was forced to make an emergency landing in a cornfield without any loss of life to the occupants or people on the ground.

CASE STUDY 13.1

US AIRWAYS FLIGHT 1549

US Airways Airbus A320 operating Flight 1549 from New York LaGuardia Airport to Charlotte, North Carolina, on 15 January 2009 struck a flock of geese two minutes after take-off at an altitude of 2,800 ft (850 m). The impact of the bird strikes damaged both engines and resulted in an almost complete loss of thrust. With power severely degraded and unable either to return to LaGuardia or make the alternative landing site at Teterboro Airport, the captain ditched the aircraft in the Hudson River. All 155 passengers and crew were able to evacuate onto the wings where they were rescued by river boats and ferries. One flight attendant and four passengers suffered serious injuries, and the aircraft was substantially damaged (NTSB, 2010). The captain's airmanship and decision-making combined with the flight management system on the A320 and the proximity of emergency responders meant that all the occupants survived.

Aviation incident

An incident is an occurrence, other than an accident, associated with the operation of an aircraft which affects, or could affect, the safety of operation (see Case Study 13.2).

CASE STUDY 13.2

AN AVIATION INCIDENT

On 8 October 2018, while landing at London City Airport, an unsecured bar trolley came out of its stowage area in the rear galley and rolled forward through the passenger cabin. The trolley was stopped by a cabin crew member who was sitting by the front forward door and the trolley was secured in the front forward galley. This meant that the front right door could not have been used in the event of a cabin evacuation. Four passengers were injured as the trolley moved through the cabin and they were treated on board and in the terminal on arrival. One of the passengers went to a local hospital for treatment. The armrest bumpers on two seats were also damaged. As a result of the incident, the aircraft operator reviewed and revised some of its Safety and Emergency Procedures to ensure that the cabin and galley areas are declared secure before landing (AAIB, 2019).

Precursors

These are conditions or events that precede and could result in an accident or incident. The terms 'near miss', 'close call' or 'partial failure' are often used when referring to precursors. A precursor can be considered as the first deviation from a normal operation or circumstance. This is known as a root event in a causal-effect sequence of the accident or incident development (see Figure 13.1). In a complex causation sequence such as an aviation accident or incident, multiple precursors may be responsible for a top event.

It is important to capture precursor data, and this data is primarily recorded in three ways:

1 *Industry's formal system.* Major aircraft and engine manufacturers developed their formal reporting systems to capture hazards, unsafe conditions and human factors contributing to safety occurrences. Airbus developed the Aircrew Incident Reporting System (AIRS) to help its customers establish their own confidential reporting systems.

2 *Regulator's formal system.* The provisions in Chapter 8 of ICAO Annex 13 require a country's aviation regulator to establish formal safety occurrence reporting systems to facilitate the collection of information on actual or potential safety deficiencies. In the EU, Regulation EU376/2014 details the practices for reporting, analysis and follow-up of occurrences in civil aviation. The Mandatory Occurrence Reporting (MOR) schemes in each EASA Member State reports to ECCAIRS (the European Co-ordination Centre for Accident and Incident Reporting Systems). It enables the collection, sharing and analysis of safety information amongst different countries and operators in order to identify potential industry-wide trends.

3 *Confidential reporting system.* These systems aim to protect the identity of the reporting person to ensure that voluntary reporting systems are non-punitive. Examples include the Aviation Safety Reporting System (ASRS) in the US and the Aviation and Maritime Confidential Incident Reporting (CHIRP) in the UK.

The inherent problems of reporting precursors include deciding which precursor events should be recorded and by whom, integrating confidential reporting systems with those of industry and ensuring the data is accurate. It is important to be aware of these issues when employing precursor data to:

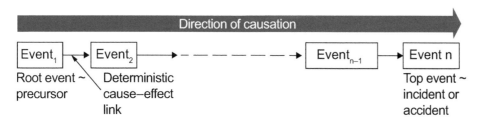

Figure 13.1 Relationship between a precursor, an incident and an accident

- understand the structure of a system's risk;

- monitor how risks are changing in one system;

- identify hazards;

- mitigate them before they lead to accidents;

- provide more safety performance data than is available from accident or incident data alone.

Stop and think

Explain the differences between accidents, incidents and precursors in the context of aviation safety occurrences.

13.2 Safety statistics and trends

Between 1942 and 2018, over 4,000 civil aviation accidents and 81,300 fatalities were recorded worldwide (Aviation Safety Network, 2019). This equates to an average of 52 accidents (one per week) and 1,070 deaths on commercial aircraft per year. However, as Figure 13.2 shows, these figures hide considerable annual variation. The worst year for accidents was 1948, when 99 accidents were recorded. This compares with only 14 in 2015 and 2017, despite the fact that passenger numbers grew from approximately 24 million in 1948 to over 4.3 billion in 2018.

Given the substantial growth in passenger numbers since 1948, it is necessary to calculate the global accident rate (expressed as the number of accidents per million departures) to

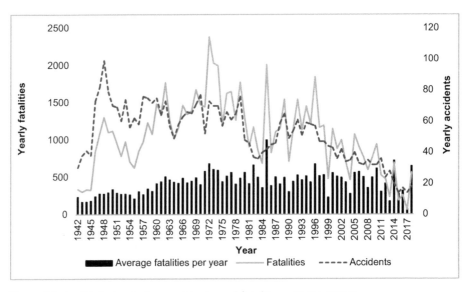

Figure 13.2 Global aviation accidents and fatalities, 1942–2018

enable year-on-year comparisons to be made. In 2016, the global accident rate was 2.1 accidents per million departures (the lowest ever recorded), an improvement on the 4.1 accidents per million departures recorded in 2009 (ICAO, 2018a). While the accident rate appears to be improving, the average number of fatalities per accident has increased over time in line with the introduction and utilisation of larger aircraft (see Figure 13.2).

As well as analysing annual trends, it is necessary to consider where these accidents occur, as there are considerable variations between world regions (see Table 13.2).

Table 13.2 Accident statistics by world region, 2017

Region	Estimated departures (in millions)	Number of accidents	Accident rate (per million departures)	% share of world traffic	% share of accidents
Africa	1.3	7	5.3	3.6	8
Asia-Pacific	11.8	20	1.7	32.2	22.7
Europe	8.7	12	1.4	23.8	13.6
Middle East	1.3	2	1.6	3.6	2.3
Americas	13.5	47	3.5	36.8	53.4
World total	**36.6**	**88**	**2.8**	**100**	**100**

Source: Data derived from ICAO (2018a)

The accident rate in Africa, at 5.3, is much higher than the world average, and the continent accounts for eight per cent of accidents worldwide despite only accounting for 3.6 per cent of global air traffic. Other regions, including the Middle East and Asia-Pacific, appear to perform much better. There are a number of possible explanations for the observed differences in regional accident rates. These include:

- the age and maintenance record of the airframes and engines being flown;
- levels of staff training, auditing and safety compliance;
- ability of national regulators to oversee and enforce safety standards;
- relative sophistication, availability and serviceability of navigation aids, ATC services, instrument landing systems (ILSs), radar, radio communications and runway lighting;
- frequency of severe weather conditions;
- proximity of hostile terrain to airports;
- provision, capacity and capability of airport rescue and fire-fighting services and local emergency services to respond to safety occurrences;

- capacity and capability of local hospitals to treat survivors;

- geopolitical and/or civil unrest and associated security threats.

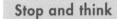

Stop and think

Why are there regional variations in global accident rates, and what could be done to improve aviation safety in Africa?

13.3 Accident categories

In addition to accurately recording the geographic location and consequence of every aviation safety occurrence, it is vital that there is a universal system for reporting and investigating accidents so that the principal cause(s) can be determined and action taken to reduce the likelihood of a reoccurrence. However, until relatively recently, there was no consensus as to which of the multiple different accident classification systems should be used. At the international level, IATA developed a Threat and Error Management (TEM) approach that focused on operations and human performance, while ICAO developed a taxonomy of 36 occurrence categories grouped into seven functional areas (Table 13.3). Individual countries also had their own classification schemes which made cross-border safety collaboration and data sharing problematic as comparisons could not easily be performed.

In order to improve international reporting and data sharing and enhance safety, ICAO, IATA, the US Department of Transportation and the European Commission established a

Table 13.3 Examples of ICAO occurrence categories

Airborne	Aircraft	Ground operations
Abrupt manoeuvre	Airframe system or component failure	Evacuation
Loss of separation	Engine failure	Ground collision
Loss of control in flight	In-flight fire or smoke	Runway excursion
Miscellaneous	**Non-aircraft related**	**Take-off and landing**
Bird strike	Aerodrome factors	Abnormal runway contact
Security related	Air traffic management	Obstacle collision
Medical		Undershoot/overshoot
Weather		
Icing		
Windshear/thunderstorm		
Turbulence		

Source: Derived from ICAO (2018a)

Global Safety Information Exchange (GSIE) in 2010. The GSIE resulted in a harmonised accident rate being introduced in 2011 which comprises eight accident categories (Table 13.4).

Of these, loss of control in flight (LOC-I) accounted for the largest number of fatal accidents between 1999–2018, followed by controlled flight into terrain (CFIT – pronounced 'see fit'), runway safety occurrences, other and unknown (Airbus, 2019). Of these, LOC-I and CFIT events accounted for over 31 per cent and 19 per cent respectively of all fatal accidents involving Western-built commercial jet aircraft between 1999 and 2018, while runway excursions (defined as a veer-off or overrun) accounted for the greatest proportion of **hull losses** (Airbus 2019).

Hull loss: when an aircraft is damaged beyond economic repair.

Table 13.4 GSIE harmonised accident categories

Category	Description
Controlled flight into terrain (CFIT)	Aircraft is airworthy but flown into terrain in a controlled manner.
Loss of control in-flight (LOC-I)	Unrecoverable loss of control during the airborne phase of flight.
Runway safety	Includes overshoots and undershoots, runway incursions and excursions, tail strikes and hard landings.
Ground safety	Includes ramp safety, ground collisions, taxi and towing events, ground servicing.
Operational damage	Damage sustained to the aircraft while operating under its own power. Includes in-flight damage and system or component failures.
Injuries to and/or incapacitation of persons	Includes turbulence related injuries, injuries to ground staff and passengers not related to acts of unlawful interference.
Other	Any event that does not fit into one of the other categories.
Unknown	Any event where the exact cause cannot be reasonably determined or there are insufficient facts available to make a conclusive decision regarding classification.

Source: Derived from IATA (2019, p. 143)

Loss of control in-flight (LOC-I)

LOC-I events can result from adverse weather conditions in the **cruise** (including icing, poor visibility and windshear/gusty winds), pilot error, aircraft system malfunction (such as an engine failure) or maintenance events. While LOC-I occurrences are rare, they are usually catastrophic. Between 2009 and 2013, 95 per cent of LOC-I events resulted in fatalities to passengers or crew, and Airbus (2019) states that LOC-I has been the single biggest cause of

Cruise: the level portion of a flight between take-off and landing that occurs at a constant airspeed and altitude.

fatal aviation accidents over the last 20 years. However, the introduction of more sophisticated flight management systems and flight protection envelopes (that prevent pilots from performing manoeuvres that would exceed an aircraft's structural and aerodynamic limits) combined with enhanced flightcrew training have resulted in fewer LOC-I events.

Controlled flight into terrain (CFIT)

The category which resulted in the second highest proportion of fatal accidents was CFIT. Most CFIT accidents occur during the approach and landing phases of a flight and may result from aircraft system malfunction or flightcrew errors. The introduction of glass flightdecks, more accurate global positioning systems (GPS) and enhanced ground proximity warning systems (EGPWS), which alert pilots to nearby terrain, have resulted in a significant reduction in CFIT accidents.

Runway excursions

Runway excursions (in which an aircraft veers off to the side of the runway or runs off the end) are usually caused by poor aircraft energy management during landing (in which the aircraft is too high or too fast or both) and/or contaminated runway surfaces which reduce friction and braking action. Unlike LOC-I and CFIT events, survivability is high, although the aircraft is invariably damaged beyond repair.

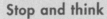

Stop and think

Why is it important to classify aircraft accidents?

13.4 Accidents by flight phase

In addition to descriptive categorisation, accidents can also be classified according to the phase of flight in which they occurred. Between 1999 and 2018, nearly 68 per cent of fatal accidents involving Western-built commercial jet aircraft occurred during descent/approach/landing and 22 per cent during the take-off/climb phases of flight. Only 7 per cent occurred during the cruise, and the remaining 3 per cent happened on the ground (Airbus, 2019).

During the descent/approach/landing, aircraft are operating closer to the ground and flightcrew have less time and space in which to react if something unexpected occurs. Aircraft are also aerodynamically more vulnerable as flaps and other high-lift devices on the wing will be deployed to slow the aircraft while generating sufficient lift to keep it airborne, and these create additional drag. The engines will generally be on a low or idle power setting and may take a couple of seconds to respond to sudden increases in power requirements. The proximity of terrain and the increased likelihood of encountering adverse weather (such as low visibility, icing, turbulence and windshear) combined with human factors, such as flightcrew fatigue, jet lag, perceived or actual pressure to ensure an on-time arrival and unfamiliarity with the approach procedure, compound the risk. On take-off and climb-out, aircraft are heavy and laden with highly flammable fuel, and the engines are working hard to deliver the required thrust. Any occurrence, such as tyre burst, bird strike or mechanical failure, has the potential to cause an accident.

> **Stop and think**
>
> Why is descent, approach and landing statistically the most dangerous phase of a flight?

13.5 Accident rates by aircraft type

A further factor that needs to be considered when analysing accident statistics is the type of aircraft that is operating the service, the airframe's manufacturer and the aircraft's age, service history and maintenance record. Statistically, **jet** aircraft have a lower accident rate than **turboprops**.

Turboprops are typically smaller and lighter than jet aircraft and are therefore more vulnerable to adverse weather conditions such as windshear and gusty conditions. They typically operate into smaller regional airports (some of which are situated in challenging terrain with associated weather conditions), which may not be equipped with precision navigation aids or ILSs, and serve some of the world's most remote airfields (➤Chapter 21).

13.6 Type of service

A further factor that can be considered is the type of service. Although the majority of accidents (79 per cent) occur to passenger aircraft, 16 per cent affect cargo aircraft and 5 per cent occur to repositioning flights that are not carrying revenue generating passengers or cargo. The unique operating characteristics of these services mean they are disproportionately affected by one particular type of safety occurrence. For example, incorrectly loaded or dangerous cargo that catches fire (or, in the case of animals, escapes) can lead to LOC-I events on cargo flights (➤Chapter 17), while incorrect loading on the ground can lead to tail stands (where the aircraft tips back onto its tail) or in-flight instability.

13.7 Accident costs

In addition to preventing accidents, safety has important implications for financial performance and business continuity. The financial cost of aircraft accidents includes not only direct costs such as damage sustained to aircraft and compensation but also indirect costs relating to the reputational damage and loss of consumer and investor confidence that may follow an accident, particularly one in which the airline is found to be negligent or liable. Although the Warsaw Convention (➤Chapter 1) established monetary limits for compensation due to passengers and consigners of air cargo in the event of an accident, even relatively minor incidents can result in substantial repair and compensation bills. In 2018, the estimated financial cost of accidents involving commercial jet aircraft was just under

Jet: an aircraft propelled by engines that produce forward motion as a consequence of the expulsion of exhaust gases from the rear of the engine.

Turboprop: (abbreviation of 'turbo-propeller') an aircraft propelled by an engine (or engines) that drives an external propeller which produces the principal thrust.

US$1,300 million, while the comparable cost for Western-built turboprops was over US$50 million (IATA, 2019).

13.8 Accident causation models

Aviation is a complex system with many different components, and aviation safety can be considered as a system property. The essential questions therefore are: why do accidents occur, and what causes them? If the source can be identified, then the potential accident can be avoided through the introduction of new interventions and measures related to technology, policy and education/training. Accident causation models have been developed to identify, represent, classify and organise causal factors associated with accidents and incidents. Accident causation models are used by the air transport industry to illustrate how accidents occur and to show the relationship between cause and effect. The models propose that most aircraft accidents are not caused by one single factor or mistake but occur as a consequence of lots of individual problems or errors coming together at the same time that make an accident an unavoidable outcome. While the cause of accidents may be attributed to 'pilot error', individual flightcrew are rarely solely responsible for accidents (unless they are self-inflicted). Air accident investigations have shown that accidents usually result from a combination of active failures, unsafe acts and latent and local triggering conditions, such as aircraft/airport design factors that may or may not have been reasonably foreseen or predicted.

Accident causation models and the techniques for accident analysis have been developed and refined by researchers from different disciplines, including engineering, psychology, sociology and medicine. HaSPA (2012) provides a historical perspective of accident causation models developed since the 1920s. Their research reveals that the evaluation of accident models exhibits common underpinning principles that can be classified into three distinct phases:

1 *Simple linear models.* These commonly used models contend that accidents result from a series of events or circumstances that occur sequentially. The primary objective is to identify problems and prevent accidents from (re)occurring. An example of a linear model is Heinrich's Domino Theory (1931), which conceptualises accidents occurring as a result of an adverse event, which causes a cascade effect, in much the same way as a line of dominos collapse one after the other when the first one is knocked over.

2 *Complex linear models.* These are the second-generation accident models which contend that accidents result from interactions between real-time unsafe acts by frontline staff and latent organisational conditions (e.g. top-level decision makers, line management) that exist within a complex system coming together in a linear sequence. One of the most famous complex linear accident causation models is the Swiss Cheese Model (SCM). First published by Reason in 1990, it proposes that accidents occur when multiple factors come together at the same time and in the same place. The SCM uses the analogy of individual layers of Swiss cheese piled up on top of each other. In most situations, a hole (i.e. hazard) in one layer that might contrib-

ute to an accident is blocked by the layer beneath it but, in certain situations, all the holes line up simultaneously and an unwanted outcome results. From a management perspective, the success of the SCM in preventing accidents relies on effective identification of the factors that create the holes in the cheese and then devising interventions to prevent their occurrence. However, complex linear models still adhere to the principles of sequential models as the direction of causality follows a linear path. Moreover, the SCM is insufficiently specific regarding the nature of the holes in the cheese and their interrelationships.

3 *Complex non-linear models.* These are the third-generation accident models in which accidents result from complex non-linear interactions of unfamiliar, unanticipated and/or unexpected sequences that may occur concurrently and which may interact with each other in complicated and unexpected ways which designers could not predict and operators cannot comprehend or control without exhaustive modelling or testing. An example of complex non-linear models is the Systems-Theoretic Accident Model and Processes (STAMP) model proposed by Leveson, who postulated that systems theory is a useful way to analyse accidents. In STAMP, accidents are treated as the result of flawed processes, in which the controls that were in place failed to detect or prevent changes, involving non-linear interactions among people, social and organisational structures, infrastructures and software system components. Table 13.5 analyses the relative merits of the three accident causation models.

Table 13.5 Relative merits of the three accident models

Model	Concepts/pros	Cons	Example
Simple linear	Models based on a temporal sequence of events, one of which prompts the next until an undesired outcome occurs; simplistic; identifies and eliminates broken links.	Only identifies one cause; too simplistic, especially as the complexity of aviation has increased over time.	Heinrich Domino Theory
Complex linear	Models based on unsafe acts, active failures and latent factors; defences/barriers against undesired outcomes; defences are dynamic in nature; suitable for complex systems; widely used.	Based on a sequential model, so can only consider one initial event; latent factors are not necessarily identifiable within the model.	SCM
Complex non-linear	Models based on tight coupling and complex non-linear interactions among the system components; capable of handling mutually interacting variables; monitor and control performance variability.	Interactions are not predictable unless data from normal flight operations is gathered.	STAMP

So successful has the air transport industry been in utilising causation models and improving safety standards that aspects have been transferred to other safety critical sectors. Recognising the success of pre-flight checklists in reducing incidents of aircraft being incorrectly configured for take-off, pre-operative checklists are now used by medical surgeons to ensure that they are preparing to perform the right procedure on the correct patient and that they have access to all the equipment they need (or may need in the event of a complication) to complete the operation.

!

Stop and think

Detail the principal differences between the three basic types of accident causation model and assess their relative merits.

13.9 Safety management systems (SMSs)

Safety management system (SMS): a clear, systematic and comprehensive approach to managing risks and improving safety.

Historically, aviation safety management was predominately based on the analysis of past events, but now a more proactive approach has been developed to help air transport service providers identify safety risks and implement strategies to minimise them. **Safety management systems (SMSs)** acknowledge the presence of hazards and provide a clear and comprehensive process for identifying, communicating and managing these risks to improve the overall level of safety.

The requirements for SMSs for air transport are contained within ICAO Annex 19 – *Safety Management*, which defines the organisational structures, processes of accountability, safety policies and procedures that must be established by air service providers, including aircraft manufacturers, aircraft operators, airports, **maintenance, repair and overhaul (MRO)** companies, air traffic control (ATC) and flight training schools. SMSs are designed to be an essential and intrinsic part of everyday operations that promote an active safety culture at all levels of the business. At a minimum, SMSs must:

Maintenance, repair and overhaul (MRO): a company that provides third-party maintenance for airlines.

- identify safety hazards at all levels of the business and develop a safety policy;

- manage risks by ensuring that action is taken to maintain an acceptable level of safety;

- monitor and continuously assess safety performance through regular audits;

- promote a proactive safety culture and aim for continuous improvements in safety performance;

- be capable of being overseen by the state.

There are four interrelated components to an SMS (CAA, 2015):

1 *Safety policy and objectives.* This section describes the aim and objectives that an aviation organisation will utilise to achieve specific safety outcomes.

2   *Safety risk management.* If risk cannot be eliminated, it must be minimised. An SMS must identify the hazards, assess and report hazards, evaluate potential consequential risks, monitor mitigation strategies and support the formation of hazard mitigation strategies.

3   *Safety assurance.* This uses auditing and performance surveillance to monitor an organisation's safety performance against its safety policy to drive continuous improvements in safety performance.

4   *Safety promotion.* This ensures all staff within the organisations who have a responsibility for safety receive training appropriate to their roles and are competent to perform their roles. The SMS should be communicated to employees to enhance the safety culture.

In addition to ICAO requirements, SMSs have also been incorporated into IATA's Operational Safety Audit (IOSA), an international evaluation programme which assesses airlines' operational management and control systems to improve safety.

13.10 Safety culture

A safety culture is essential to ensuring the safety of the air transport industry. Weaknesses in a safety culture often take the form of ineffective organisational structures, unclear communication strategies, insufficient or inappropriate equipment or inadequate technical and operational procedures. These are often fundamental accident triggers. A safety culture has five different elements, which were originally identified by Reason (2000):

1   *An informed culture.* Safety culture is managed through a top-down approach, and senior managers are responsible for emphasising safety and sharing information with front-line staff to develop their understanding of risks and hazards. Clear, concise communication of information is required, as otherwise an effective SMS cannot operate.

2   *A reporting culture.* The creation of a reporting culture is essential to ensure that incident and accident data is collected and analysed to assess risks and to enable mitigation to prevent future unsafe acts. Usually, reporting systems are non-punitive to eliminate fear of blame; otherwise incidents and accidents may not be recorded. Industrial, regulatory and confidential reporting systems are used to collect occurrence data.

3   *A learning culture.* Any organisation should learn from previous unsafe acts, to ensure safe operations (ICAO, 2018b). This is often sustained through monitoring, reviewing and evaluating mitigations through data collection, also known as a continuous improvement cycle.

4   *A just culture.* While honest errors will not be penalised, intentional and risky behaviour will be subject to disciplinary actions to discourage high risk-taking behaviour (ICAO, 2018b). To help facilitate the formation of a just reporting culture, all reporting systems share common non-punitive characteristics.

5 *A flexible culture.* Due to the dynamic nature of the aviation industry, a flexible culture is essential to allow for new circumstances and to enable operational procedures to be adapted to ensure safe operations.

In some instances, aviation safety incidents are not caused by individual, mechanical or institutional failure but by intentional illegal acts of terrorism or sabotage (➤Chapter 14).

!

Stop and think

What is a safety culture, and how can airlines develop one?

13.11 Emerging threats to aviation safety

Airprox: is a situation in which the distance between aircraft and their relative positions and speed have been such that the safety of the aircraft involved may have been compromised.

Drone: a small pilotless aircraft, also described as an unmanned aerial vehicle, an unmanned aerial system or a remotely piloted aerial system.

The introduction of new technology and procedures for improving airspace capacity and safety has been a continuous process since the beginning of flight. For example, the introduction of reduced vertical separation minima (RVSM) increases airspace capacity, while the traffic alert and collision avoidance system (TCAS) acts as another safety net to prevent air proximity (**airprox**) events and mid-air collisions. Modern aircraft are equipped with technologies including fly-by-wire, glass flightdecks, flight management systems, and improved navigation systems that reduce both CFIT and LOC-I accidents. Having said that, aviation accidents and incidents still occur. A new type of small unmanned air system (SUAS), which includes **drones**, are increasingly being used and coming into conflict with commercial aircraft.

Drones are primarily used for a range of commercial activities such as high-quality photography, capturing shots of inaccessible sites such as an active volcanic eruption, delivering medical supplies, mapping and GIS to make large composite images and 3D models from collected images and GPS data. However, drones are also considered to be a disruptive technology, as they can be used for delivery of explosives, weapons or drugs or for the purpose of disrupting flights at airports (see ➤Chapter 14). As such they create a high potential and emerging threat to government and military areas, nuclear facilities, commercial facilities and the aviation sector. A study conducted by the FAA concluded that a drone colliding mid-air with commercial and business aircraft inflict more physical damage than that of a bird strike due to the rigidity, shape and dynamics of impact. Commercial aircraft are designed to withstand strikes from birds weighing up to 8 lbs (3.6 kg) for the stabiliser and 4 lbs (1.8 kg) for the windscreen, but they are not designed to withstand metal drones (ASSURE, 2017). Examples of drone impacts on aircraft include the following:

- In October 2017, a SkyJet flight was struck by a drone while inbound to Jean Lesage International Airport in Quebec City. The Minister of Transport stated that the incident was the first time a drone had hit a commercial aircraft in Canada and that although the aircraft sustained minor damage, it was able to land safely (Transport Canada, 2017).

- In December 2018, an Aero Mexico B737–800 landing at Tijuana Airport was struck by a drone that shattered the nosecone and damaged the radome that protects the aircraft's radar.

CASE STUDY 13.3

AIRPROXES INVOLVING DRONES WITHIN UK AIRSPACE

Airproxes involving drones and aircraft within the UK airspace have increased from no incidents before 2009 to an average of 5 incidents a year over the period 2010–2014 and 138 incidents in 2018 (UKAB, 2019). Airproxes involving drones accounted for 125 out of the 138 small unmanned air system (USAS) incidents. According to reports compiled by the UKAB, the drone and its operator could not be traced for most incidents, resulting in a lack of information on the drone type and characteristics, and that most of the airproxes involving drones could not be assessed by the UKAB safety team during the investigation. Figure 13.3 illustrates the number of airproxes reported by drone operators in UK airspace.

In order to minimise the impact from drone activities, countries have developed policies, regulations and interventions. The UK government has introduced rules (also known as the 'Dronecode') for drone operations (UKAB, 2019):

1 All drones must remain below 400 feet unless approved by the Civil Aviation Authority (CAA);

2 For drones weighing over 250 g, the operator must pass the drone test and register with the CAA before commencing any operations;

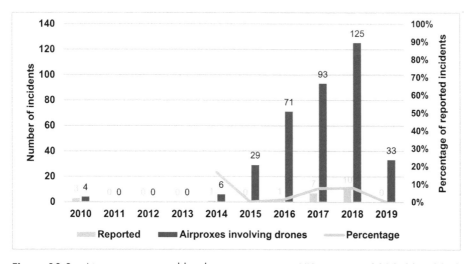

Figure 13.3 Airproxes reported by drone operators in UK airspace, 2010–May 2019

3   It is illegal to fly any drone at any time within airfield restricted zones (i.e. the airfield's aerodrome traffic zone) unless an operator has permission from air traffic control at the airport or, if air traffic control is not operational, from the airport itself.

!

Stop and think

Why do drones pose a safety risk to aviation, and what can be done to manage the risk they pose?

Key points

- Providing a safe air transport system in the face of diverse and newly emerging risks is vitally important for the continued operation of the air transport industry.

- It is important to identify and differentiate between accidents, incidents and precursors so that the industry can learn from past events and develop safety recommendations to prevent reoccurrence.

- The three most common accident categories are LOC-I, CFIT and runway safety occurrences.

- Increasingly sophisticated models have been developed to help understand causal factors.

- Drones are of increasing safety concern to aviation. With carefully formulated regulations, standards and awareness, drones and aircraft can co-exist.

References and further reading

AAIB [Air Accidents Investigation Branch]. (2019). *AAIB Bulletin 4/2019 EW/G2018/10/19.*

Air Transport Action Group. (2019). *Facts and figures, air transport action group.* [online]. Available at: www.atag.org/facts-figures.html

Airbus. (2019). *Commercial aviation accidents 1958–2019: A statistical analysis.* Blagnac Cedex: Airbus SAS.

ASSURE. (2017). *FAA and ASSURE Announce results of air-to-air collision study.* ASSURE.

Aviation Safety Network. (2019). *Accident statistics.* Available at: https://aviation-safety.net

CAA. (2015). *Safety management systems SMS guidance for organisations.* London: CAP795 Safety and Airspace Regulation Group London, CAA.

HaSPA [Health and Safety Professionals Alliance]. (2012). *The core body of knowledge for generalist OHS professionals.* Tullamarine: VIC, Safety Institute of Australia.

Heinrich, H. W. (1931). *Industrial accident prevention: A scientific approach.* New York, NY: McGraw-Hill.

IATA. (2019). *IATA safety report 2018.* 55th edn., Issued April 2019, Montreal: International Air Transport Association.

ICAO. (2018a). *Safety report 2018 edition*. Montreal: International Civil Aviation Organization.

ICAO. (2018b). *Safety management manual*. 4th edn., ICAO Doc 9859, Montreal: International Civil Aviation Organization.

NTSB [National Transportation Safety Board]. (2010). *Loss of thrust in both engines after encountering a flock of birds and subsequent ditching on the Hudson River US Airways Flight 1549 Airbus a320–214, n106us, Weehawken, New Jersey, January 15, 2009*. Accident Report NTSB/AAR-10–03 PB2010–910403, Washington, DC, NTSB.

Reason, J. (2000). Safety Paradoxes and Safety Culture. *Injury Control & Safety Promotion*, 7(1), pp. 3–14.

Transport Canada. (2017). *Statement by Minister of transport about a drone incident with a passenger aircraft in Quebec city*. Transport Canada [online]. Available at: www.canada.ca/en/transport-canada/news/2017/10/statement_by_ministeroftransportaboutadroneincidentwithapassenge. html

UKAB. (2019). *Current drone airprox count and information*. UK airprox board [online]. Available at: www.airproxboard.org.uk/Topical-issues-and-themes/Drones/

CHAPTER 14

Aviation security

David Trembaczowski-Ryder

LEARNING OBJECTIVES

- To understand what is meant by aviation security.
- To understand the threats to civil aviation.
- To examine the evolution of aviation security practices worldwide.
- To be aware of the role of international regulations and agencies in the formation and development of aviation security regimes worldwide.
- To understand the implications of cross functional threats from cyberattacks and drones.

14.0 Introduction

As aircraft and airports represent a target for terrorist attack, this chapter focuses on aviation security, an important aspect of air transport management. It examines why air transport is targeted, identifies the threats to civil aviation, discusses the evolution of aviation security regimes and details the mitigation measures that have been put in place to try to protect the industry from acts of unlawful interference.

14.1 Aviation security

Aviation security involves a combination of legal and regulatory measures and human and material resources that are collectively designed to protect civil aviation from acts of unlawful interference that jeopardise the security of civil aviation. The security threats facing aviation are diverse and take many forms, from the hijacking of aircraft, suicide bombers, landside and cyberattacks to the use of drones for malicious activity (Table 14.1).

In light of the long history of aviation security threats posed by terrorist activity, much of the contemporary international air transport security regime is focused on seeking to identify terrorists and prevent them from gaining access to aircraft and airports and disrupting the mobility of passengers and freight. In order to try to prevent attacks, it is necessary to understand the development of international rules.

The International Civil Aviation Organization (ICAO)

With the benefit of hindsight, it may seem hard to imagine how the need to address acts of sabotage, unlawful seizure of aircraft and the use of civil aircraft in terrorist attacks (as was the case on 11 September 2001) could have been overlooked by the drafters of the Chicago Convention (➤Chapter 1), the founding charter and cornerpiece for international technical legislation in the field of civil aviation of the International Civil Aviation Organization (ICAO). In 1944, however, no one foresaw such security threats and the need for security measures, even though there had been acts of hijacking civil aircraft since as early as 1919 (see Section 14.2).

When aviation security arose as a serious issue in the late 1960s, there was a need to adopt an international framework for addressing acts of unlawful interference. ICAO assumed a leadership role in developing aviation security policies and measures at the international level, and today the enhancement of global aviation security is a key objective of ICAO. Provisions for international aviation security were first disseminated as Annex 17 to the Chicago Convention in 1974 and since then have been improved and updated 16 times.

In addressing the evolving threat to civil aviation, ICAO relies on the advice of experts who sit on the Aviation Security (AVSEC) Panel. Established in the late 1980s, the Panel is currently comprised of 31 members nominated by States, as well as five observers from industry. Together with the ICAO Secretariat, the Panel actively develops ICAO security policy and responses to emerging threats as well as strategies aimed at preventing future acts of unlawful interference.

Aviation security and the European Union

Before the 9/11 terrorist attacks in the United States, the responsibility for aviation security in the European Union rested with the Member States as a national prerogative. Given the patchwork history of terrorist activity in Europe, the mitigation measures implemented by different states was dependent upon the perceived threat in each state. However, since 2002 the European Commission was tasked by the Member States to establish common rules in the field of civil aviation security aimed at protecting persons and goods from unlawful interference with civil aircraft. The rules establish a common framework for all EU Member States to adhere to the provisions in ICAO Annex 17.

Regulation (EC) 300/2008 of the European Parliament and of the Council lays down common rules and basic standards on aviation security and procedures to monitor the implementation of the common rules and standards.

The common standards comprise the following measures:

- Airport security (including: airport planning requirements; access control; screening of persons other than passengers [staff] and items carried; examination of vehicles; surveillance, patrols and other physical controls and prohibited articles);

- Aircraft security searches and protection of aircraft;

- Passengers and cabin baggage (including: screening of passengers and cabin baggage; protection of passengers and cabin baggage; potentially disruptive passengers; and prohibited articles);

- Hold baggage (including: screening of hold baggage; protection of hold baggage; baggage reconciliation and prohibited articles);

- Cargo and mail (including: screening and security controls for cargo and mail; regulated agents and known consignors; protection of cargo and mail; and security procedures for cargo and mail being carried into the union from third countries);

- Controls and screening of in-flight supplies and airport supplies;

- Staff recruitment and training;

- Security equipment standards and performance requirements.

Why aviation is targeted

Aviation remains a prime target for terrorists. Terrorists generally seek to achieve the following by conducting attacks to:

- inflict mass casualties to bring attention to an ideology or cause;

- cause economic disruption to society;

- make a symbolic statement against a particular target state, political establishment or religious belief;

- generate public fear so as to reduce consumer demand and impact on long-term economic growth.

Attacks on aviation targets may contribute to all of these objectives, hence the continued attractiveness of aviation as a focus of terrorist activity. The threat to aviation continues to broaden with new attack methodologies and potential methodologies continuing to be explored by terrorists, faster than old methodologies can be mitigated. New and evolving threats such as Remotely Piloted Aircraft Systems (RPAS) (drones) continue to broaden the range of risks that need to be managed.

Also, the threat to aviation is becoming ever more geographically dispersed. In the past, sophisticated plots could be developed only in places where there was a certain degree of terrorist "infrastructure" – the presence of a number of like-minded people, the ability to source, for instance, bomb making materials locally and the ability to operate without disruption by the authorities. The aviation plot uncovered in Australia in July 2017 showed that, by using the aviation system to distribute materials and the internet to distribute knowledge, sophisticated attacks could be launched from locations where just a small number of motivated individuals are present. 'Lone wolf' attacks (by individuals acting alone) are also becoming more common, and while they are mostly currently at a low level of sophistication, this phenomenon of what can be called 'mail order terrorism' may increase the sophistication and impact of such attacks in the future.

Table 14.1 Examples of aviation security threats

Person-delivered Improvised Explosives Devices (IED) on the body or in cabin baggage

IED in hold baggage

Landside attacks on the public areas of airports

Vehicle-borne IED

IED in cargo

Attack using Remotely Piloted Aerial Systems (drones) on aviation targets

Aircraft used as a weapon

Conventional hijack

Chemical, biological, and radiological threats

IED in services (catering, in-flight supplies)

Cyberattacks

Man Portable Air Defence Systems (MANPADS), particularly in conflict or proliferation zones

Insider threats posed by radicalised or disgruntled personnel

!

Stop and think

Why are aircraft and airports targeted by terrorist activity?
What type of threats does civil aviation face?

14.2 Terrorist attacks against aircraft

The first hijacking of a passenger-carrying aircraft is said to have taken place in 1919 and perpetrated by Baron Franz von Nopcsa Felső-Szilvás, a pretender to the Albanian throne, as he was seeking a new life in Vienna. He arrived at Budapest Airport with a gun and false documents and ordered the pilot to fly him to Vienna. However, it was not until the 1930s that a number of recorded incidents occurred, but with civil aviation in its infancy, there was no international body to establish civil aviation security rules. It was in the late 1960s, however, that international flights became terrorist targets. While there had been previous hijackings with political intent, 23 July 1968 was a significant date in the history of aviation terrorism. On that day, three members of the Popular Front for the Liberation of Palestine (PFLP) hijacked an El Al Israel Airlines flight while en route from Rome to Tel Aviv. In September 1970, the PFLP also hijacked four aircraft on the same day. The Dawson's Field **hijack** fundamentally changed the international aviation security regime, with the result that X-ray machines were introduced to screen cabin baggage and walk-through metal detectors were introduced to screen passengers for weapons and metal objects.

Hijack: the act of illegally seizing an aircraft in-flight and forcing it to divert to another destination or illegally taking control of it, usually involving (threat of) violence.

Following the introduction of screening of cabin baggage, terrorists began to look for other vulnerabilities. By 1985 terrorists had turned their attention to checked (hold) luggage. In that year there was no requirement for airports to screen checked luggage, and in June 1985, a bomb was placed in an unaccompanied suitcase loaded onto Air India Flight 182. The bomb exploded over the eastern Atlantic, killing all 329 persons on board. As a result, ICAO established the Aviation Security Panel to formulate rules and guidelines on aviation security. This did not prevent the attempted bombing of El Al B747 in 1986 or the bombing in December 1988 of Pan Am 103 over Lockerbie, Scotland. These incidents led to the introduction of passenger-baggage reconciliation to prevent unaccompanied suitcases from being loaded onto aircraft and the introduction of hold baggage X-ray screening equipment. Up until the turn of the millennium it was extremely rare for terrorists to act as suicide bombers and die for their cause, and as such, the global aviation security system was not geared to this type of attack. Additional screening measures were implemented for passengers, their cabin bags and hold bags, but the measures were insufficient to prevent determined suicide bombers. The events of 11 September 2001 (9/11) changed the perception (Case Study 14.1).

CASE STUDY 14.1

9/11

On 11 September 2001, 19 terrorists simultaneously hijacked four commercial airliners. Two aircraft were flown into the Twin Towers World Trade Centre in New York, causing their collapse, and a third was flown into the Pentagon in Washington, DC. The fourth aircraft crashed into a field in Pennsylvania when passengers overpowered the hijackers. The perpetrators enrolled in pilot training courses in the US after receiving visas to enter the US (one individual was refused a visa). The organisers carried out trial runs carrying box cutters on US domestic flights and were not intercepted, and all evidence points to the terrorists using knives and box cutters to overpower the crew. The 9/11 attacks caused over 3,000 deaths and resulted in new measures that required the installation of reinforced flightdeck doors and special procedures for opening flightdeck doors. Also, knives and cutting tools with blades of 6 cm or longer were prohibited. In addition, armed skymarshals were deployed on selected 'high-risk' flights.

Only three months after the attacks of 9/11 a male passenger, later described as the Shoe Bomber, attempted to detonate an explosive device packed into his shoes while on American Airlines Flight 63 from Paris to Miami on 22 December 2001. His attempt failed when he was subdued by other passengers and the attempt led to a closer examination of passenger shoes at security checkpoints.

In August 2006, a plot to destroy aircraft leaving London Heathrow for the US using liquid explosives hidden inside bottles of soft drinks was prevented (Case Study 14.2).

CASE STUDY 14.2

LIQUIDS PLOT

In August 2006 there was a terrorist plot to detonate liquid explosives, carried on board airliners travelling from the UK to the US and Canada, hidden in soft drink containers. The plot intended to destroy as many as ten aircraft in mid-flight using explosives brought on board in the suspects' hand luggage; the plot was intercepted by the UK security services. In the immediate aftermath of the attempt, passengers were forbidden from carrying any liquids, apart from baby milk, onto flights between the UK and the US. Immediately following the plots, no hand luggage was allowed except for essentials such as travel documents and wallets. Some restrictions were relaxed in November 2006 to allow limited volumes of liquids to be carried into the cabin. The failed attempt led to the introduction of restrictions on the carriage, in cabin baggage, of liquids, aerosols and gels (LAGs). The only items allowed were toiletries in small containers (up to 100 ml) onto the aircraft in a re-sealable plastic bag along with medicines and baby food (sufficient for the journey). See Figure 14.1. The restrictions still apply across a number of countries because the complete lifting of restrictions would cause significant operational difficulties at airport security checkpoints; each container would need to be screened for explosives.

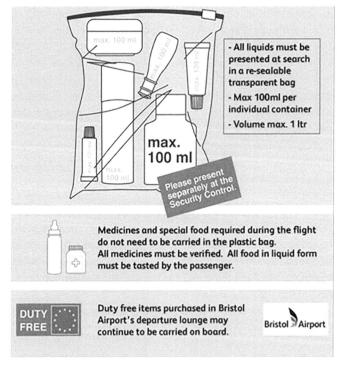

Figure 14.1 Passenger information poster detailing permissible volumes of liquids, aerosols and gels (LAGS) in hand luggage

Source: Bristol Airport

On Christmas Day 2009, a male passenger attempted to detonate plastic explosives hidden in his underwear on board a Northwest Airlines flight from Amsterdam to Detroit. During the attempted detonation he sustained first- and second-degree burns to various parts of his body. He was restrained by other passengers and arrested on arrival in Detroit. As a result, security scanners (known as Body Scanners or Advanced Imaging Technology – AIT) were introduced at many major airports. The security scanners use millimetre waves that can penetrate clothing and detect items hidden in or under clothing; the passenger image appears as an avatar or stick person.

While additional resources were being invested in enhancing passenger security, the security regime for air cargo remained vulnerable (➤Chapter 17). In October 2010, two explosive devices were discovered hidden inside US-bound printer cartridges. One device was discovered at East Midlands Airport in the UK, while the other was intercepted in Dubai. It was reported that both devices were timed to explode when the aircraft was in-flight above the US. As a result, the air cargo security regime has been tightened so that all cargo is routinely screened and shippers are monitored to ensure that all cargo can be traced and accounted for.

One common feature is that aviation security regulations have generally lagged behind threats, with mitigation measures covering the last known threat. However, industry and regulators are working hard to be more proactive in order to prevent security checkpoints from becoming technology shops with different detection equipment for each threat type.

Stop and think !

What security measures have been introduced as the result of attempted or successful attacks against civil aviation?

14.3 Airport security

There are a significant number of potential threats to the landside area (or public spaces) of an airport, not least because it can stretch a long distance from the airport perimeter. It is important to carefully distinguish which threats are applicable to which areas of the landside, and it is necessary to split the landside area into different zones depending upon the nature of the threat. The landside of an airport is anywhere that is before security control (➤Chapter 4). Given that the greatest threat to life is an attack where a large number of people gather, the terminal and immediate area are considered the most critical. Thus, in principle, the public areas of airports may be considered similar to other crowded places in terms of their potential as targets for terrorist attacks. Since 2000 there have been 18 attacks against civil airports in 15 different countries (Table 14.2).

Table 14.2 demonstrates that aviation continues to be a target. It also indicates that, while attacking aircraft in flight remains the desired objective for many terrorist groups, airports are becoming an alternative target because such attacks are easier to conduct. Just as for attacks on aircraft, airports have symbolic importance, their disruption brings potentially

Table 14.2 Examples of attacks on the landside area of airports

On 30 December 2006 at Madrid Barajas Airport, a van bomb exploded in the Terminal 4 parking area, killing two and injuring 52.

In June 2007, a Jeep loaded with propane canisters was driven into the front of the terminal at Glasgow Airport in Scotland.

In January 2011, a Chechen suicide bomber killed 35 people and injured over 100 more in the international arrivals hall at Domodedovo International Airport in Moscow.

On 2 March 2011 at Frankfurt Airport in Germany, a terrorist shot and killed two US airmen and seriously wounded two others.

On 18 July 2012 at Burgas Airport in Bulgaria, a terrorist attack was carried out by a suicide bomber on a passenger bus transporting Israeli tourists. The explosion killed the Bulgarian bus driver and five Israelis, and 32 were injured.

On 22 March 2016 Brussels Airport was attacked by three suicide bombers (one bomber escaped and his bomb was dealt with by bomb disposal experts). At the same time another suicide bomber attacked Maalbeek metro station in central Brussels. Thirty-two civilians and three perpetrators were killed, and more than 300 people were injured.

On 30 June 2016 Istanbul Atatürk Airport was attacked by three gunman who staged a simultaneous attack at the international terminal. Forty-five civilians and the three terrorists were killed.

On 6 January 2017 at Fort Lauderdale-Hollywood International Airport in Florida, a lone gunman shot and killed five people and wounded six others.

high economic impacts and they offer an opportunity to target specific national groups at predictable times and locations.

Best practice guidance looks at effective countermeasures, such as stand-off zones (i.e. closure of the forecourt area in front of the terminal building), hostile vehicle barriers, traffic and speed management measures, moving car parks away from the terminal, and improved building design standards (including glazing).

Recent attacks have demonstrated the vulnerability of public areas to attacks using person-borne IEDs and firearms. Defending against these forms of attack by restricting access is difficult, both because of their lay-out and because landside areas are legitimately used by a wide range of people other than passengers.

! Stop and think

What type of mitigation measures are used to prevent attacks in the public spaces of airports?

14.4 Passenger security screening

Data collection

Passenger profiling and Advance Passenger Information (API) systems oblige passengers to submit personal details, such as full name, nationality, passport number and date of birth, to their airline in advance of travel. Additional details (such as payment method) may also be attached to the personal information. The airline is then required to share this data with the security services and border officials in the country of destination (and, in some cases, with the security services in countries the aircraft is flying over) before departure. The data is then analysed using sophisticated algorithms and checked against databases of known or suspected criminals and terrorists. The security services then authorise or deny travel to the individuals who are booked on each flight.

Passenger Name Records (PNR) information is the generic name given to records created by the aircraft operators for each flight a passenger books. PNR records contain information provided by the passenger and information used by the aircraft operator for their operational purposes. PNR information may include elements of API. PNR information, along with API, is used by governments to conduct analysis that helps to identify possible high-risk individuals that may have been otherwise unknown to government authorities and to make, where appropriate, the necessary interventions.

Passenger security process

Once passengers arrive at an airport, they and their bags are subject to a number of pre-departure security **screening** protocols. Typically, these occur in two locations – at check-in and at the security search area. Passengers are required to confirm that they packed their bags themselves and that no one has interfered with them. They are also asked to confirm that there are no **prohibited items**, such as sharp implements or liquids over 100 ml in their hand luggage. It is an offence to knowingly give incorrect information or make security threats. Passports or other official identification documents are checked to confirm the identity of the individual who is intending to travel and the validity of any visas.

In order to access the departure lounge, passengers must pass through **security control**. This has the purpose of verifying the validity of a passenger's boarding pass and ensuring that passengers do not carry any unauthorised items on their person or in their hand luggage into secure airside areas. Security staff are trained to spot suspicious behaviour or dress, and a number of different technologies and techniques are used to identify prohibited items. These include walk-through metal detectors (WTMD), millimetre wave security scanners, explosive trace-detection systems, explosive detection dogs (EDD), X-ray machines and liquid explosive detection systems (LEDS)/bottle scanners, which are used to identify metallic or prohibited objects on the person, chemical compounds indicative of explosives, prohibited items in hand luggage and prohibited items concealed under the clothing of individual travellers. Despite having only a couple of seconds to view and assess the contents of individual bags, security personnel must be able to identify and remove not only obvious items such as guns or knives but also potentially malicious combinations of individually

Screening: the application of technical or other means which are intended to identify and/or detect prohibited items.

Prohibited items: weapons, explosives or other dangerous devices, articles or substances that may be used to commit an act of unlawful interference that jeopardises the security of civil aviation.

security control: the application of means by which the introduction of prohibited items may be prevented.

benign objects. Given the importance of this task, simulated threat items are frequently superimposed on the images to ensure that individual screeners remain alert to potential threats.

Other airport terminal security measures include the use of CCTV, behaviour detection officers and regular patrols by armed police, undercover security personnel and airport security staff.

In-flight security

In recognition that pre-flight and airport security regimes are not infallible, in-flight security continues in the air. Armed skymarshals are carried on some high-risk flights, and cabin crew are trained in restraint techniques. However, concern has been expressed about the safety implications of carrying armed security staff, not least because a mid-air exchange of fire between skymarshals and terrorists contributed to the destruction of an Iraqi Airways flight in December 1986 with the loss of 71 lives.

Other in-flight security interventions include cabin CCTV and reinforced flightdeck doors that can be opened only from the flightdeck. However, any intervention can have unintended consequences for flight safety. In March 2015, a Germanwings pilot deliberately crashed his aircraft into a French mountain after locking the captain out of the flightdeck, although this incident was not terrorist related.

!

Stop and think

To what extent could it be argued that aviation security has historically been reactive rather than proactive?

The security threats facing aviation are diverse and sophisticated. As a result, an international security regime that balances the threat of attack with the cost and inconvenience of mitigation strategies has been established. Based on the Standards and Recommended Practices (SARPs) contained within the ICAO Security Manual, Annex 17, airport and airline security programmes have to be clear, comprehensive, robust and flexible enough to identify and prevent emerging threats.

The ICAO Security Manual provides guidance to reduce the probability of an act of unlawful interference being directed towards airport facilities and users, and to minimise the effects of such an act, the following key elements should be integrated in the design of an airport:

a   perimeter protection;

b   physical security of buildings, including passenger terminal buildings;

c   access control;

d   screening and security control regime for non-passengers and the items they carry entering and/or within security restricted areas

In addition, airlines and airports have developed dedicated security management systems (SeMS) or appropriate quality control/audit systems to identify and nullify emerging threats. SeMS are an integral component of aviation business operations and are designed to create a security culture. SeMS define a company's security policy and its security management strategy and standards. SeMS must comply with national regulatory requirements and protect people and assets from acts of unlawful interference. Given the diverse nature of air service operations worldwide, SeMS must be appropriate to individual operating conditions and security environments.

Stop and think

To what extent can aviation be made totally secure, and where should the balance lie between safeguarding national security and protecting personal privacy?

14.5 Cyber security

Cyberspace is a non-physical environment made up of digital networks that communicate, exchange and store data and support various services and businesses that has broken down barriers between countries, communities and citizens, allowing interaction and sharing of information and ideas across the globe. For example, at airports, passengers are already used to online check-in, self-baggage drop services and automated border controls that speed up the process and improve customer service. In order to operate the global aviation system, the industry relies on information and communications technology (ICT) for critical parts, but many new technologies have no proper imbedded cyber security measures.

There are four types of cyber security incidents:

- *Natural disaster*: includes hurricanes, earthquakes, floods and storms. There are many cases when significant amounts of data and businesses were destroyed.

- *Malicious attack*: every minute, all around the world, cyberattacks are taking place: some are just disrupting the operations for the sake of it, others are for political reasons, but the majority are trying to gather secret information by tracking systems to steal money or leak confidential data.

- *Internal attack*: employees who have legitimate access may harm the IT systems or destroy/leak the data and disrupt the operations intentionally.

- *Malfunction and unintentional human error*: this includes power loss, equipment shutdown, internet cut-off, phone line, overwriting data, equipment damage by an accident or BYOD (Bring Your Own Device) virus.

All these types of incidents may result in long-term implications for civil aviation, such as:

- loss of business continuity (loss of operations for some period of time);

- loss of operational productivity (reduced throughput, e.g. HBS [Hold Baggage Systems]);

- destruction/leakage of data (operational processes);

- financial loss (delays/cancellation of flights);

- reputational loss/attention from media (impact on passenger experience);

- high recovery costs;

- reduction in competitiveness;

- potential bankruptcy.

In 2013, Istanbul Ataturk and Sabiha Gokcen International Airports experienced a cyberattack that shut down the passport control systems at the departure terminals. Due to the attack, passengers were forced to stand in lines at the passport control points for hours and the majority of flights were delayed. In 2014, hackers of the Islamic Cyber Resistance Group claimed to have breached the computer systems of the Israel Airport Authority; Japan Airlines reported a hacker attack in September 2014 with evidence confirming unauthorised access to its Customer Information Management System; American and United Airlines also reported that hackers managed to get hold of user names and passwords of frequent flyers in December 2014, and both British Airways' and Lufthansa's air-miles accounts were subject to cyberattacks in early 2015. On 12 April 2015, in Tasmania, Hobart International Airport's website was shut down for over 24 hours due to a cyberattack. Threats, the same as technology, rapidly evolve, so it is crucial that the industry maintains the highest levels of confidence and become increasingly connected in the most secure possible way.

!

Stop and think

What type of cyber incidents are there? What are the implications of cyberattacks for the civil aviation industry?

14.6 Drones

Incidents at airports involving drones have been increasing in prominence in the past few years, with attention having been focused on the matter since London Gatwick was closed for two days in December 2018 due to a series of drone sightings that presented a clear threat to aviation safety. The Gatwick incident disrupted 1,000 flights and over 140,000 passengers, at one of the busiest travelling periods of the year, highlighting the importance for airports of preventing such occurrences.

Many national regulators and airports have designated no-drone zones in the vicinity of airports and other national critical infrastructure. But as yet there is little clarity on which technologies may be safely deployed at airports to detect unauthorised drones and prevent them from interfering with airport operations. However, airports have developed Concepts of Operations and contingency plans for dealing with drone incidents, detailing procedures, lines of communication and responsibilities for drone-related incidents.

Clarity is required, however, as to which entities – airport operators, **Air Navigation Service Providers (ANSPs)** or state authorities – are responsible for surveillance, detection and enforcement, as private airport operators do not have a legal basis for disabling or destroying a drone that encroaches the airport boundary. Some national authorities, including the UK and France, have established a registration scheme for drones and their owners/operators, with any subsequent drone operations at the airport then subject to authorisation based on clear rules and procedures for the chosen drone scenario (including risk assessment, the drone type being used and the pilot operating the drone).

Regulators in the UK, France and Italy have already drafted laws that would set out requirements for a range of drone operations, depending on the drone being used, its purpose, the relevant airspace and the outcome of the mandatory risk assessment. Regulations will be able to reduce unintentional drone incursions (significantly reducing nuisance activity); however, it will not stop intentionally malicious drone activity.

Air Navigation Service Provider (ANSP): a body that provides air traffic services within a country.

Key points

- Providing a safe and secure air transport system in the face of diverse and newly emerging security threats is vitally important for the continued operation and financial sustainability of the air transport industry.

- The nature of the threats facing the industry constantly evolves, and new security protocols and screening technologies have been introduced to counter a diverse range of threats.

- The air transport industry needs to find a balance between providing optimal levels of security and not unduly inconveniencing passengers, imposing additional financial costs on operators or hindering air transport's continued development.

- If consumers perceive that aviation is unsafe, they may switch to other modes of transport and/or reduce the frequency with which they fly.

References and further reading

ACI EUROPE. (various dates) Policy papers on aviation security Available at: www.aci-europe.org/policy/position-papers Airport Security

Baum, P. (2016). *Violence in the skies: A history of aircraft hijacking and bombing*, London: Octopus Books.

Department for Transport. (2018). *Aviation cyber security strategy*. London: UK Department for Transport. Available at: https://assets.publishing.service.gov.uk/government/uploads/system/uploads/attachment_data/file/726561/aviation-cyber-security-strategy.pdf

European Commission. (2008). *Regulation (EC) 300/2008 of the European Parliament and of the council of 11 March 2008 on common rules in the field of civil aviation security and repealing regulation (EC) no 2320/2002.* Brussels: European Commission.

European Commission. (2015). *Commission implementing regulation (EU) 2015/1998 of 5 November 2015 laying down detailed measures for the implementation of the common basic standards on aviation security.* Brussels: European Commission.

ICAO WCO IATA. (2017). *Management summary on passenger-related information* ['Umbrella Document' version 2.0] – Montreal, International Civil Aviation Organization (ICAO), the World Customs Organisation (WCO) and the International Air Transport Association (IATA).

CHAPTER 15

Airspace and air traffic management

Lucy Budd

LEARNING OBJECTIVES

- To introduce the notion of airspace and recognise the importance of airspace sovereignty to air transport operations.

- To appreciate the structure and classification of airspace.

- To understand the function of air traffic control (ATC) and air traffic management (ATM).

- To comprehend the role of different technologies in the formation and maintenance of airspace.

- To assess current challenges and future innovations in ATM.

15.0 Introduction

Airspace is the medium through which aircraft fly. It is the aerial equivalent of roads, railway lines, canals or shipping lanes, except airspace is largely invisible, covers the entire surface of the earth and occurs in three dimensions (longitude, latitude, altitude and also, occasionally, time). The configuration of, and control over, airspace affects the safety, efficiency and cost effectiveness of air transport operations as well as the defence, national security and international relations of individual nation states.

The design and day-to-day management of airspace results in operational restrictions and safety implications for different users, financial implications for airline operators and nation states (see Section 15.7), geopolitical tensions between countries, social and environmental impacts for people on the ground (►Chapter 18) and degradation to the global climate (►Chapter 18). The purpose of this chapter is to introduce the concept of airspace, to explain how it is configured and controlled and to detail how processes of air traffic management (ATM) ensure the safe and efficient utilisation of airspace in accordance with strict international regulations under a situation of growing consumer demand for flights and increased capacity constraints. The chapter begins by introducing the concept of airspace and explaining how the sky is structured and classified. The role of air traffic control (ATC), specific technologies and ATM are then discussed. The chapter concludes by assessing current and future challenges in the provision of airspace and air traffic services. Although international standards govern many aspects of airspace and ATM, each country also has its own particular rules and regulations. Examples in this chapter are based on the situation in Europe. Different units of imperial and metric measurement are used to describe distance and/or vertical height above the earth. In each case, the unit used reflects the terminology that is used in the industry.

15.1 Airspace

Airspace: a three-dimensional volume of the earth's atmosphere in which aircraft and aerial objects fly.

Airspace is a defined volume of sky in which commercial, military and general aviation aircraft, UAVs (unmanned aerial vehicles, including drones) and other airborne objects, including hot-air balloons, gliders, kites and birds fly. Far from being 'free' and 'open', airspace is subject to international and national regulations that govern its use to ensure public safety and national security and bring order to the air. International airspace administration and governance derives much from maritime law, but only within the last 100 years, following the development of heavier-than-air powered aircraft, has systematic international agreement regarding the use of airspace been required.

The commencement of regular powered flights at the beginning of the 20th century caused countries to quickly recognise that the aerial territory above them represented an important strategic, military and commercial asset that needed to be strictly delimited and, if necessary, defended from incursions by unauthorised or otherwise unwanted or unwelcome users.

The formation of early airspace legislation

As long as a pilot took off, flew within a country's navigable airspace and landed within its national borders, there was no problem. International services, however, challenged the territorial integrity of individual states and produced one of the longest and most contentious debates in aeronautical politics, with each state seeking to seize control of as much airspace as possible while maintaining control of their borders for defence, national security and commercial reasons.

Although national claims to land, lakes, rivers and adjoining seas had been common for centuries, claims to aerial territory (or airspace) were entirely new concepts, and it was agreed that some form of international regulation was required (➤Chapter 1). The first attempt at airspace regulation occurred in 1909 when the French government suggested that a code governing international air navigation should be formulated to prevent unauthorised flights by foreign aircraft over French territory. This was followed, in Paris in 1910, by an attempt to bring international air services under unified control. However, the mutually incompatible positions of different states meant that agreement was not forthcoming. The most pressing issue concerned the right of access to airspace, and while some countries argued for complete freedom of the air above all territories, others maintained that the air was capable of being owned, just like land. The resulting debate was similar to challenges that had been encountered in the formation of international maritime law, which sought to reconcile the sea as a site of international transport, recreation and resource harvesting with the territorial aspirations and defence of individual nation states.

In 1911, the British government passed the British Aerial Navigation Act, which declared that Britain's airspace (including that of overseas colonies, dominions and mandates) was sovereign territory. The right of individual countries to claim sovereignty over their aerial territory (their airspace) was formally enshrined in Chapter One of the Paris Convention 1919, which stated, 'The high contracting parties recognise that every power has complete and exclusive sovereignty over the air space above its territory . . . and the territorial waters adjacent thereto.' However, in recognition of the need to facilitate the orderly development of international air services, Chapter Two stipulated that, during peacetime, states would grant 'freedom of innocent passage' through their airspace to aircraft registered in another country. However, international services still required the consent of the states that were overflown.

The lateral extent of airspace

Although the Paris Convention 1919 acknowledged the right to sovereignty of airspace, it failed to define its lateral or vertical extent, and it was not until 1944 that the Chicago Convention (►Chapter 1) defined the physical boundaries of national airspace. This definition was reaffirmed and strengthened in the 1982 United Nations Convention on the Law of the Sea (UNCLOS), which stipulated that a country's sovereign airspace corresponds to the definition of territorial waters and so extends for 12 nautical miles (22 km) from a coastline. Airspace that is not within a country's territorial limit is classed as international airspace and is not sovereign territory. However, an individual country may, subject to international agreement, assume responsibility for controlling sections of international airspace where it is considered in the global interest for them to do so. For instance, international airspace above the northern Atlantic Ocean is variously controlled by Canada, the US, Iceland, the United Kingdom and Portugal.

The vertical extent of airspace

Although the UNCLOS defined the lateral extent of territorial waters and airspace, it did not define the vertical extent of airspace. The international non-governmental organisation Fédération Aéronautique Internationale uses the Karman line, an invisible boundary 100 km above the earth, to define the upper limit of airspace or the division between the earth's atmosphere and space. This limit is politically and strategically important as it concerns boundary security. Beyond the Karman line, the atmosphere is so thin that aerodynamic lift becomes impossible and aeronautical activities (that require an atmosphere to generate aerodynamic lift) give way to astronautics (which relies on rocket propulsion, orbits and gravity). Consequently, a country can defend and enforce its aerial territory only up to this line. However, 100 km far exceeds the maximum altitude that can be attained by commercial aircraft (which only regularly cruise up to around 39,000 ft (11,887 m) above the earth) and military aircraft (which typically fly at 55,000–60,000 ft (16,764–18,288 m). For this reason, most states define the upper limit of airspace as being at an altitude of between 60,000 and 66,000 ft (18,288–20,117 m).

In addition to discussing the maximum vertical extent of airspace, consideration was also given to the lower extent of airspace. While most states agree that airspace starts at the surface of the ground (or sea level) and extends up to a defined level, they have found it necessary to strictly regulate air traffic operations that occur close to densely populated urban areas to protect public safety and personal privacy and minimise noise disturbance on the ground. These two issues are becoming increasingly important owing to the rise of privately operated drones and other UAVs.

15.2 The structure and classification of airspace

To fulfil its role as a medium of flight, airspace has to be:

- safe (both for airspace users and people on the ground);

- capable of being monitored, controlled and defended for reasons of safety, efficiency, national security and defence;

- flexible and able to accommodate the diverse (and often conflicting) operational requirements of different user groups (including commercial air traffic, military aircraft, recreational flyers and wildlife);

- designed and managed in such a way as to minimise the adverse environmental impacts of air traffic movements (including noise and pollution) on people and wildlife.

Like any transport network, airspace encounters constant fluctuations in demand depending on the season, the day of the week and the time of day. In order to separate and safely manage this traffic, airspace is divided into a number of discrete yet interfacing sectors which are subject to different degrees of monitoring and surveillance. All airspace is divided into Flight Information Regions (FIRs). Each FIR is managed by a controlling authority on behalf of a sovereign state. FIRs vary in size, and while some smaller countries may have only one FIR, larger states may have several. Some FIRs are also divided vertically into upper and lower sections. The upper section is called an Upper Flight Information Region (UIR), while the lower section is a FIR.

Within each FIR/UIR, airspace is further subdivided and classified as being controlled, uncontrolled or special-use, depending on the volume, density and type of air traffic that uses it. Areas with high traffic volumes (such as those near major airports) require strict monitoring and control, while peripheral areas with lower levels of air traffic require less intensive surveillance and pilots have more flexibility to operate as they wish, providing they adhere to basic aeronautical regulations. Special-use airspace describes areas in which certain types of air traffic are temporarily or permanently restricted. These include:

- prohibited areas within which all aircraft are banned (examples include airspace adjacent to nuclear power stations and around certain military installations);

- danger areas that present a significant hazard to aircraft (examples include airspace immediately above oil rig flare stacks and those around wildlife sanctuaries owing to the higher risk of bird strike);

- restricted areas within which aircraft operations are permitted only under certain conditions (examples include military air traffic zones and one-off events such as air shows).

All other airspace within a UIR/FIR is categorised into one of seven ICAO classes. These classes are identified by the letters A to G inclusive, where Class A airspace is subject to the most control and Class G the least. These classes determine the type of air traffic that can access the airspace, the conditions under which flights can operate and the level of air traffic service (if any) that is provided. Different criteria apply to each class.

Controlled airspace (CAS)

There are five classes of controlled airspace: A, B, C, D and E (see Table 15.1).

Controlled airspace can further be categorised as being a control zone, a control area, a terminal control area or an airway depending on its location and the function it performs:

- Control zones (CTZs) are located around certain aerodromes, and ATC is provided to all flights. CTZs extend upwards from the ground surface to a specific upper limit which varies according to location.

- A control area (CTA) is usually located above a control zone between defined flight levels.

Table 15.1 Controlled airspace classes

Class A: Used where air traffic flows are at their densest and most complicated in terms of the trajectory and vertical movement of aircraft. Aircraft cannot enter Class A airspace unless they are equipped with certain identification and navigational features, have filed a **flight plan** with air traffic control indicating their intended route and are piloted by individuals holding a valid IFR rating (see Section 15.3). Pilots are provided with an ATC service and are separated from each other.

Class B: Also subject to a high degree of control, but both IFR and VFR flights are permitted. Aircraft are provided with an ATC service and are separated from each other.

Class C: Both IFR and VFR flights are permitted (see Section 15.3). All flights are provided with an ATC service, and IFR flights are separated from both IFR flights and VFR traffic. VFR flights are separated from IFR flights and receive traffic information in respect of other VFR flights. There is no speed limit for IFR aircraft, but VFR aircraft are limited to 250 kt (NM/hr) below 10,000 ft. Two-way radio communication is mandatory. Clearances from ATC must be issued.

Flight plan: a written account of a proposed flight detailing the intended route, speed and altitude. Special international codes are used to concisely convey the information.

Continued

Table 15.1 Continued

Class D: Less busy areas of controlled airspace. Both IFR and VFR flights are permitted, and all flights are provided with an ATC service. In the UK, Class D airspace surrounds many regional airports and may extend from the ground surface to a specified altitude (often the base of Class A airspace).

Class E: Both IFR and VFR flights are permitted, and IFR flights are provided with an ATC service and separated from other IFR flights. All flights receive traffic information as far as is practical.

- Terminal control areas (also known as Terminal Manoeuvring Areas) may be established in the vicinity of one or more major airports.

- Airways are controlled areas of airspace between major airports that are used by en-route aircraft. They are the equivalent of aerial highways in the sky.

Uncontrolled airspace

There are two classes of uncontrolled airspace: F and G (see Table 15.2).

Table 15.2 Uncontrolled airspace classes

Class F: Advisory routes. IFR and VFR flights are permitted. All IFR flights receive an air traffic advisory service, and all flights receive flight information service if requested.

Class G: Falls under none of the aforementioned categories. Pilots using Class G airspace still have to adhere to basic aeronautical regulations, but they are otherwise able to fly in accordance with their licence restrictions. Both IFR and VFR flights are permitted and a flight information service is often available, if requested.

!

Stop and think

Why is it necessary to structure and classify airspace?

15.3 The rules of the air

All civilian air traffic is flown in accordance with one of two distinct rules of the air, Visual Flight Rules (VFR) or Instrument Flight Rules (IFR). These rules determine which sections of airspace can be accessed, by whom, when and the conditions under which that airspace can be used.

VFR

All qualified pilots can fly under VFR. Under VFR, the pilot in command is responsible for:

- the safety of the aircraft and its occupants;
- maintaining adequate separation from other aircraft (using the principle of see-and-avoid), both on the ground and in the air, to prevent collision;
- keeping clear of, and avoiding, terrain;
- navigation;
- ensuring adequate visibility and distance from cloud is maintained.

Under VFR, pilots must be able to remain clear of clouds by at least 5,000 ft (1,524 m) horizontally and 1,000 ft (305 m) vertically and maintain forward visibility of at least 8 km. For certain flights in some areas of airspace and at low altitudes, the requirements are less stringent. An aircraft cannot be flown at night or above 20,000 ft (6,096 m) without special permission. VFR flights can be performed only if strict visual meteorological conditions (VMC), which describe the distance from cloud, cloud ceiling (height) and visibility, are met. When the view from an aircraft is restricted and navigation cannot be performed visually with reference to the ground, instrument meteorological conditions (IMC) must be followed. IMC minima are below those specified for VMC, and pilots can fly under IMC only if they hold a valid instrument rating.

IFR

In adverse weather conditions or Class A airspace, flights must be operated in accordance with IFR. IFR training, qualification and equipment requirements are far more stringent than those for VFR. Aircraft must be equipped with suitable flight instruments and navigation equipment appropriate to the route being flown, and the pilot/s must hold a valid instrument rating. Unlike VFR, IFR flights can operate in all airspace classes.

15.4 Airspace charts

The boundary between different sectors of airspace, as well as information about the location of individual airways, waypoints and airports, is depicted on dedicated airspace charts. These are published in different scales and reflect the specific aerial navigation needs of VFR and IFR traffic. VFR charts are akin to regular terrestrial maps in that they use different colours and symbols to show the location of major roads, railways, rivers, estuaries and urban areas that are an aid to visual navigation, but they also have the boundaries of different airspace classes and the location of any restricted airspace or danger areas overlaid on top. VFR charts are larger in scale than IFR charts, which are used by aircraft that are flying faster and higher and not relying on ground-based features for navigation.

The scale used for IFR charts depends on the density of information that they have to convey. Unlike VFR charts, IFR charts feature little by way of terrestrial information (other than the location of coastlines, airports and information about minimum safe operating altitudes) as aircraft are navigating by instruments rather than by reference to the ground below. IFR charts are designed to be read easily in different lighting conditions, and so the most important information on airways, waypoints and very high frequency omnidirectional range (VOR) beacons is depicted in black. Specific cartographic symbols depict the location of airports, airspace boundaries and areas of restricted or dangerous airspace.

A further group of charts depict the arrival and departure procedures that must be followed at each individual airport, as well as the location of taxiways, aircraft manoeuvring areas and individual stands on the airfield. Like VFR and IFR charts, aerodrome charts are regularly updated to reflect changes in airspace structure, local operating procedures and new infrastructure. Originally, all airspace charts were printed on paper, but they now are being replaced by digital versions on tablet computers in each aircraft's **electronic flight bag (EFB)**. Replacing paper-based charts with EFBs confers considerable weight and cost savings for airlines. For example, a B777–200ER without an EFB would require almost 35 kg of paper to be carried in the flightdeck. An EFB typically weighs under 2 kg.

Electronic flight bag (EFB): an integrated electronic flightdeck information management system used by pilots to fulfil flight management functions including navigation and flight performance calculations.

! Stop and think

What is the difference between VFR and IFR, and to what extent do they influence the airspace pilots can access?

15.5 Air traffic services (ATS)

The global and safety-critical nature of airspace management means protocols and procedures have been standardised around the world to ensure that airspace is used safely and efficiently. The international Standards and Recommended Practices (SARPs) concerning the classification and maintenance of airspace and the provision of ATS, ATC and other related services are described in ICAO Annex 11 – *Air Traffic Services*. Globally standardised ATS provision is designed to:

- ensure the safety of aircraft, their occupants, and people and property on the ground by preventing collisions between aircraft that are in the air or on the ground;
- prevent collisions from occurring between aircraft and objects on an airfield;
- provide information and advice to pilots to aid the safe and efficient conduct of air services;

- maintain a safe and orderly flow of air traffic through the airspace;

- notify and liaise with national emergency services and military agencies in respect of search and rescue activities and unauthorised airspace incursion by foreign aircraft.

The level of ATS provided depends on: the airspace class; the volume, density and type of traffic it accommodates; and local weather conditions. ATS differs from Air Traffic Control (ATC) in that the former is a service that provides advice to pilots, whereas ATC is more active and interventionist as it issues clearances and ensures separation for aircraft operating in controlled airspace in addition to providing an advisory service to aircraft in uncontrolled airspace.

Stop and think

Why are ATC services provided, and what might happen if they were not?

15.6 ATC technologies

To safely and efficiently handle aircraft, a range of communication, navigation and surveillance (CNS) equipment has been developed. These typically use radio waves to identify and communicate with aircraft.

Radio

Radio is the medium of spoken communication through which instructions, requests and observations between pilots and controllers are passed and acknowledged. The introduction of two-way radios marked an important phase in the development of aviation as they enabled pilots to remain in contact with controllers while airborne. To ensure global comprehension and compliance, radiotelephony procedures have been standardised. English has been adopted as the universal language of aviation, and the English alphabet is spoken phonetically to ensure that phrases, words and numerals are clearly understood.

Numbers involving altitude, cloud height, visibility or runway visual range, which contain whole hundreds and whole thousands, are spoken individually (e.g. 'two thousand five hundred feet', not 'two and a half thousand feet'), while numbers in aircraft call signs, altimeter settings, flight levels (except FL100), headings, windspeeds and radio frequencies are all spoken separately. Thus, a controller addressing an American Airlines flight would say 'American nine one five, contact London on one one nine decimal seven two five' not 'American nine hundred and fifteen contact London on one hundred and nineteen point seven hundred and twenty-five'. Strict protocol determines which words can be used when and the order in which they must be spoken. Pilots, controllers and airfield operations staff have to be trained and tested to ensure that they deliver concise, clear and accurate information.

Radio messages are transmitted on dedicated airband frequencies, which are typically in the range of 110–140 MHz to avoid interference from public radio stations. Each sector of airspace is administered using a different frequency and, at major airports, different frequencies are used for arriving, departing and taxiing aircraft. All users are able to hear all the transmissions that are occurring on their frequency, enabling them to determine the relative position and intentions of other aircraft.

To ensure unambiguous communication, all commercial flights are allocated a call sign and a flight number. The call sign refers to the aircraft operator (e.g. the call sign of easyJet flights is 'easy', while international British Airways flights are prefixed by 'Speedbird'). This is followed by a numeric or alphanumeric designator. After establishing the identity of the flight being addressed, the controller then articulates his/her instructions or advice, including altitude and heading changes, speed restrictions, route clearances, taxiing or stand information, take-off or landing clearances and other information of relevance to the safe conduct of that flight. Air traffic controllers try to limit the number of instructions in any single transmission to three to prevent overloading the pilots with information. To ensure the message has been received and understood correctly, pilots read back the message in its entirety. Radio is therefore used to authorise clearances, decline requests and provide information, while flightcrew use it both to communicate with controllers to request new headings and/or altitudes and to (re)confirm instructions and to communicate with pilots of other aircraft in the vicinity.

Radio beacons

In the early years of flight, aircraft navigated with reference to the ground and followed key features such as roads and railway lines. As traffic volumes grew, this became increasingly dangerous, and separate arrival and departure routes were introduced to separate aircraft. The introduction of larger pressurised passenger aircraft in the 1950s meant that ground-based navigation was no longer practical as these aircraft flew at higher altitudes and often above cloud. In response, a network of ground-based radio beacons was established. Receivers on the flightdeck captured these signals and determined the aircraft's bearing from the beacon, allowing pilots to 'home in' on them from any direction and change direction at the intersection of two or more beams. Although radio beacons are now being replaced by satellite surveillance and multilateration radar systems (see later in this chapter), the use of beacon technology during the 20th century has locked the industry into a legacy way of working, particularly at the critical low and intermediate flightlevels.

Radar

Radar (radio detection and ranging) was developed in the 1930s to identify the presence and location of airborne aircraft using radio waves. Two complementary radar systems were developed – primary surveillance radar (PSR) and secondary surveillance radar (SSR). PSR sends electromagnetic radiation in the form of ultra-high frequency (UHF) radio waves from a rotating parabolic dish into the atmosphere at almost the speed of light (around 300,000 km per second). If the radio waves encounter an obstruction (an aircraft, high ground, storm

Table 15.3 Advantages and disadvantages of PSR

Advantages	Disadvantages
• Determines position of all objects within range of receiver.	• Only works on 'line of sight' and can suffer from blind spots.
• Determines range of all objects within range of receiver.	• Returned image can suffer from clutter from wind turbines, high ground or precipitation.
• Determines the relative speed and direction of travel of all objects within range of the receiver.	• Different aircraft can provide unevenly sized returns.
• Aircraft do not require any special equipment to be detected.	• Range is limited by curvature of the earth and high ground interrupting transmitted pulse.
	• No way of determining what has created the radar echo.

clouds), some of the original energy is reflected back to the dish in the form of an echo. Measuring the time that elapsed between the pulse being sent and the echo being returned determines the object's distance from the radar installation. The direction of the returned echo is also captured. This information is displayed as a blip on the radar screen. Each rotation of the radar dish updates the blips so controllers can determine whether an object is moving and, if it is moving, its relative direction of travel. PSR has a number of advantages but also limitations (see Table 15.3).

SSR helps to address some of the limitations of PSR. Unlike PSR, which relies on the strength of a reflected signal, SSR uses small radio transmitters in the aircraft. These transponders (transmitting responders) automatically respond to interrogation from ground-based radar pulses and send a unique coded four-digit identification 'squawk' signal back to the ground that uniquely identifies the aircraft. Squawks are transmitted on a different frequency from the ground station pulses, so SSR signals are stronger and more reliable. The word 'squawk' is believed to be a legacy of the forerunner of the SSR system that 'squawked' like a parrot when it was interrogated.

Modern squawk codes consist of four digits (such as 6425) which are produced and assigned to a particular flight before take-off. Some codes, including 7500 and 7600, are reserved for emergencies (and indicate radio failure and hijacking respectively) or for use by the military.

Ground-based decoders translate the transponder squawk back into flight data, providing controllers with information about the operator, altitude, call sign, origin/destination, aircraft type, air and ground speed, vertical speed (if it exceeds 500 ft (150 m) per minute) and the nature of any emergency. This additional information is then displayed alongside the relevant PSR blip. The principal advantages of SSR are the following:

- It enables controllers to positively identify individual aircraft on the radar screen.

- It is not subject to radar clutter or signal degradation in bad weather.

- It can be used to identify aircraft in distress.

- All aircraft appear as the same size on the radar screen.

Increasingly, multilateration radar (MLAT) systems, which use SSR signals received by multiple ground-based radar stations to precisely determine the location of individual aircraft, are being used. The principal advantages of MLAT are: increased accuracy, enhanced surveillance, greater oceanic range (using satellites) and relatively low implementation costs.

! Stop and think

What are the limitations of the present ATC technologies, and how might the system be improved in future?

15.7 Air traffic management (ATM)

While ATS describes the current and short-term tactical provision of airspace, ATM concerns the current and future dynamic management of airspace and the safe and efficient flows of aircraft within it. ATM consists of different functional elements including:

- air traffic services (including ATC);

- airspace management (ASM) – to maximise the utilisation of available airspace through sharing and segregation;

- air traffic flow management (ATFM) – to ensure the optimum flow of aircraft through an airspace, especially when demand exceeds capacity;

- aeronautical information service (AIS) – to provide information and advice to pilots.

ATM affects not only air traffic but also public safety and is a public service for which the state is responsible. Some countries have retained full public control and ownership of their ATM services, while others have partially transferred ATM services to the private sector. Providing ATM infrastructure and personnel is expensive and, in most countries, the cost of ATM provision is met by charging users (aircraft operators) for the services they receive.

! Stop and think

Explain the difference between ATC, ATS and ATM.

Airspace charging

The majority of countries charge commercial aircraft operators for the ATM services they receive while flying in their territory. In Europe, the cost of providing the infrastructure, staff, training, maintenance and other ATM services is funded through air navigation charges which are levied on aircraft operators who use European airspace. These are in addition to the landing fees levied by airports. Three different ATM charges are levied. These are en-route charges, terminal navigation fees and communication charges. The latter two are collected and administered locally, whereas the route charges are administered centrally by Eurocontrol on behalf of Member States. The route charges are non-discriminatory (the same charging rules and calculations are applied to all users), equitable (the user pays) and straightforward as a single currency unit is used to pay the charges. The en-route charges can be significant (see Tables 15.4 and 15.5). In cases where countries have denied airlines registered in (and/ or owned by) particular countries from accessing their airspace, the affected airlines have often had to fly much longer routes and pay higher fuel bills to avoid closed airspace, while neighbouring countries not engaged in the blockade have benefited from increased revenue from en-route charges from the re-routed flights.

Table 15.4 Top 10 airlines for route charges in Europe, January–June 2019

Airline	Route charges in € millions
Ryanair	333.8
easyJet	214.9
Lufthansa	187.2
British Airways	156.7
Turkish Airlines	139.8
Air France	129.8
Emirates	105.5
Norwegian Group	99.5
Qatar Airways	91.9
TUI Group	90.4

Source: Eurocontrol (2019)

Table 15.5 Top 10 countries for route charges received in Europe, January–June 2019

Country	Route charges in € millions
France	630.0
Germany	461.0
United Kingdom	352.8

Continued

Table 15.5 Continued

Country	Route charges in € millions
Italy	348.2
Continental Spain	330.5
Turkey	190.7
Austria	106.2
Sweden	95.1
Netherlands	92.2
Poland	91.8

Source: Eurocontrol (2019)

As the cost of delivering ATM services in individual countries differs, each country sets its own unit rate that enables them to recover their costs. The standard airspace charging formula that is applied and administered within Europe is based on this unit rate as well as the distance flown and the maximum take-off weight of the aircraft. Although all flights are technically liable for route charges, aircraft weighing less than two metric tonnes, flights operated on behalf of national governments, reigning monarchs or heads of state, and aerial search and rescue activities are generally exempt. Although the charging mechanism facilitates streamlined billing and payment collection, aircraft may fly further than the great circle route in a region, and so the charging formula does not penalise environmentally inefficient tracks.

Flight inefficiency

ATM inefficiencies can be assessed using a flight inefficiency metric. Commonly used ATM performance indicators quantify the difference between the theoretical minimum distance that could be flown and the actual distance that is flown between two points on the earth's surface to determine the average route extension over the great circle distance and hence the inefficiency of the flight.

$$\text{Horizontal flight inefficiency} = \frac{\text{Actual distance} - \text{optimal distance}}{\text{Optimal distance}} \times 100$$

Using this calculation, flying 200 miles (322 km) further than the great circle distance on an 800-mile (1,287 km) route would equal 25 per cent inefficiency. However, this metric is sensitive only to track extension over the ground, not to inefficiencies in the vertical dimension such as cruising at a sub-optimal flight level. The NATS 3Di metric addresses this limitation by determining both the horizontal and vertical inefficiency of a flight relative to its optimal trajectory in both the horizontal and vertical plane (see Case Study 15.1).

Possible sources of flight inefficiency vary by flight stage but all have the potential to increase fuel burn, increase airline costs, lengthen flight times and generate more pollution.

- *Departure phase.* Sources of inefficiency include long taxi routes, noise preferential routes (NPRs) that are effective from the runway end to a defined altitude, and which oblige aircraft to fly sub-optimal routes to avoid densely populated areas and lower the acoustic impact of aircraft operations on the ground, and sub-optimal climb trajectories that require aircraft to 'step up' to their cruising altitude through intermediate flight levels. Solutions include more widespread use of CCOs (continuous climb operations).

- *En route.* Sources of inefficiency include aircraft being assigned to sub-optimal cruise altitudes, convective weather (such as thunderstorms) that have to be avoided, routing round restricted airspace and avoiding expensive airspace.

- *Descent/landing.* Sources of inefficiency include stepped descents from cruising altitude (the solution is flying a continuous decent operations or CDO), holding and vectoring in a stack before landing, standard terminal arrival routes (STARs) and long taxi times to the terminal.

CASE STUDY 15.1

THE NATS 3DI INEFFICIENCY METRIC

NATS, the UK's Air Navigation Service Provider, developed a sophisticated three-dimension inefficiency score (known as 3Di) to measure the environmental performance of flights in UK airspace. Unlike other inefficiency metrics, 3Di measures both horizontal and vertical inefficiencies. The horizontal inefficiency is calculated in terms of track extension above the great circle distance, while the vertical inefficiency compares the actual vertical profile of a flight against the airlines' preferred trajectory. As aircraft performance varies during a flight, the 3Di metric applies different weightings to climb, cruise and descent. These factors are then used to give a combined 3Di score for each flight in UK airspace. Scores range from 0 (no inefficiency) to over 100. NATS's target is to reduce 3Di from 29.7 to 27.7 by the end of 2019 (NATS, 2015).

Owing to national ownership and control, the world's airspace comprises a number of discrete but interfacing zones of sovereign control. This fragmentation means that airspace is not optimised for efficiency or environmental performance. The situation is particularly acute in Europe owing to the close proximity of multiple relatively small sovereign states.

In February 2004, Eurocontrol, the European airspace network manager, received formal backing from European governments to develop a Single European Sky (SES) to increase capacity and make the continent's fragmented airspace structure more efficient. Initiatives have included the Advanced Flexible Use of Airspace (AFUA), in which airspace is segregated only on a temporary needs basis and contiguous volumes of airspace are no longer constrained by national boundaries, and Free Route Airspace, which enables aircraft operators to plan a route between defined entry and exit points. Currently 75 per cent of European airspace is covered. It is estimated that, once fully implemented, free route airspace will deliver savings of 500,000 nautical miles and 3,000 tonnes of fuel a day (Eurocontrol, 2019).

15.8 The future

The growth in the number of flights worldwide and the introduction of new technologies are creating both challenges and opportunities for airspace and ATM. The deregulation and liberalisation of airline markets in different countries and the development of new airline business models has stimulated new consumer demand and spread issues of air traffic congestion into new areas. The scheduling of flights (➤Chapter 11) has implications not only for airline operators but also for ATC, and the legacy of beacon technology was creating 'bottlenecks' in the sky as multiple flights all converged on a small number of beacons. Now, MLAT and satellite surveillance, combined with increasingly sophisticated flight management systems on aircraft, enable new possibilities for 4D (four dimensional) trajectory management. Other developments, such as remote air traffic control towers, which provide ATC services for regional and remote airports at a distance (➤Chapter 21), also offer potential efficiency enhancements. Such innovations will help to collectively drive new safety and efficiency improvements in ATM, but some obstacles remain.

1 *Change management.* Changes to airspace and the introduction of new ATM technologies impact not only pilots and controllers but also communities on the ground. Managing the expectations of local communities will be vital.

2 *Political.* Airspace remains sovereign territory, and few governments would consent to handing over control of their airspace to a foreign nation, even if it did improve efficiency.

3 *Prioritising users.* The needs of commercial users do not always align with those of military and general aviation users. Reconciling the diverse operational requirements without unduly hindering the activities of one user group is challenging and controversial.

4 *Social.* Communities living near an airport or airway often oppose airspace expansion, flight path changes and aircraft noise, making airspace use a socially contentious and political issue. The accuracy of modern navigation means that aircraft tracks (and hence noise) can be concentrated on a very localised area. Ways of managing and mitigating the social impact of noise will need to be devised.

5 *Safety minima and standards.* These are designed to ensure the safety of the ATM system. Although capacity could be increased by introducing new airways or flight levels, the safety case for doing so needs to be assured before changes are introduced.

6 *Existing runway infrastructure.* Runways are generally aligned into the direction of the prevailing wind and not the dominant direction of travel (➤Chapter 4). Given the operational necessity of landing into the wind whenever possible and the sunk costs of runway infrastructure, this represents an inefficiency which cannot be readily resolved.

Stop and think

!

Detail the main sources of flight inefficiency, and identify which inefficiencies may not be capable of being resolved.

Key points

- Airspace is a medium of flight that has been designed to ensure the safety and efficiency of airborne aircraft and to protect people on the ground.

- National airspace is sovereign territory that is subject to multiple regulations and jurisdictions governing its operation and use.

- Airspace is classified as being either controlled or uncontrolled, and different operating restrictions apply to each.

- Technologies, in particular radio waves that are used for spoken communications, navigation and radar, are vital for the construction and safe use of airspace.

- Global airspace is often highly fragmented and there are many sources of horizontal and vertical flight inefficiency, some of which are incapable of being resolved.

References and further reading

Eurocontrol. (2019). *Our data*. Available at: www.eurocontrol.int/our-data

NATS. (2015). *Environmental performance*. Available at: www.nats.aero/environment/3di/

CHAPTER 16

Aircraft manufacturing and technology

Andrew Timmis

LEARNING OBJECTIVES

- To identify the scale, scope and location of aircraft manufacturing and technological development.

- To understand the evolution of commercial aircraft manufacturers from component and sub-assembly producers to system integrators.

- To recognise the role of outsourcing in aircraft manufacturing and development.

- To appreciate the role of changing materials and manufacturing processes in new aircraft programmes.

- To examine the future challenges facing the aircraft manufacturing sector.

16.0 Introduction

The global aerospace technology and manufacturing industry is worth in excess of $800 billion per year and employs over 1.2 million people worldwide in high-skilled engineering, design and manufacturing roles (ATAG, 2018). The commercial aircraft industry is often seen as a symbol of a country's export leadership in product markets that require a high level of design and engineering innovation. Aircraft manufacturing has been a leading export sector in many countries for more than six decades. In 2017, China's aircraft manufacturing and repair industry generated $39 billion, while in 2018, the US aerospace sector exported $150 billion of products to international markets (US Department of Commerce, 2019).

An aircraft consists of two principal components: the airframe and the power plants (engines). Each of these in turn consists of multiple components and sub-systems; the airframe, for example, includes the fuselage, wings, tail assembly and landing gear. It is estimated that a Boeing 747–800 consists of over 6 million individual parts which are sourced from all over the world and then assembled on the final assembly line (FAL) outside Seattle in Everett, Washington.

Since the first heavier-than-air powered flight in 1903, aircraft designers and manufacturers have sought to develop ever safer, more reliable and efficient aircraft. The rate of technological change has been rapid. Within 30 years of the first flight, Boeing had launched the 247, a twin-engine, 10-seat passenger aircraft. The year 1949 saw the inaugural flight of the world's first jet powered commercial aircraft – the de Havilland Comet 1. In 1969, the twin-deck four-engine B747 'Jumbo Jet' performed its first flight, and by 1976, the supersonic Concorde was in revenue service (Table 16.1).

Table 16.1 Timeline of major developments in aircraft design and technology

1903	First flight of the Wright Brothers.
1914	Prototype of an 'automatic pilot' demonstrated.
1917	First all metal aircraft – the Junkers J4 – introduced.
1928	First electro-mechanical flight simulator introduced.
1935	DC-3 enters service. It is the first airframe that enables its operator to make a profit without postal subsidy.
1937	Jet engines are tested in Germany and the UK.
1949	The first flight of the Comet 1 – the world's first jet powered passenger aircraft.
1969	First flight of Boeing's B747 'Jumbo Jet'.
1976	Supersonic Concorde enters passenger service with Air France and British Airways.
1988	The A320 enters revenue service. It is the first commercial aircraft to feature fly-by-wire technology which replaces conventional flight controls.
1995	The Boeing 777, the first aircraft to be designed wholly on computers using CAD software, makes its first flight.
2005	First flight of the A380 'Super Jumbo'. It enters revenue airline service in 2007.
2011	B787 enters revenue service.
2018	The E190-E2 regional jet enters revenue service.

Original equipment manufacturer (OEM): a company that makes/assembles a final product.

As aircraft became larger and more expensive to design and develop, the number of manufacturers decreased. In the period between 1903 and the mid-1940s, small scale aircraft production via multiple companies dominated, but the capital investment required to design and manufacture aircraft meant that these firms often merged to consolidate their operations and reduce competitive pressure. The resulting few manufacturers of commercial aircraft (including Airbus, Boeing, Embraer and Bombardier) are described as being **original equipment manufacturers (OEMs)**. They have also evolved from being in-house producers of entire airframe assemblies to system integrators. Boeing and Airbus, in particular, have opted for a 'systems integration' mode of production in which key components and sub-assemblies are designed and manufactured by external risk-sharing partners and suppliers and then assembled at fixed FALs (Figure 16.1). While this represents a logical financial strategy, a potential drawback is that foreign subcontractors and/or risk-sharing partners must receive a transfer of technology or **tacit knowledge** from the systems integrator to make the business model work, and overseeing complex supply chains, often at a distance, can be challenging.

> **Tacit knowledge:** information gained through experience and which cannot be formally taught.

Due to the increasing demand for air travel, it is estimated that nearly 40,000 new aircraft will be required by 2040 (Airbus, 2019). Two-thirds of these new deliveries will be required to meet future growth in demand, particularly in the Asia-Pacific region, as opposed to replacing aircraft retired from service. An airframe typically has a service life of 25–30 years, and the development of new aircraft models is expensive and often characterised by delay and cost overruns. The chapter now examines the reasons for the historical location of OEMs and describes how the structure of the aircraft manufacturing sector has evolved into an increasingly global and decentralised industry. Particular attention is given to the manufacturing processes involved in the design and assembly of large commercial jet aircraft seating 100 or more passengers (Table 16.2). The chapter concludes by discussing future trends in aircraft manufacturing technologies.

Figure 16.1   An A350-900 on the assembly line in Toulouse, France.
Source: © Airbus - Master films – photo by Jean-Baptiste Accariez

Table 16.2 Development of new commercial aircraft programmes (over 100 seats)

Country/region	Commercial aircraft OEM	Entry into service	Model	Seat capacity
US	Boeing Commercial Airplanes	2011	787	280–360
		2017	737Max	130–190
		2020[a]	777X	350–400
Europe	Airbus Group	2014	A350XWB	300–350
		2015	A320neo	140–200
		2018	A330neo	250–310
Canada	Bombardier Aerospace/Airbus Canada Limited Partnership	2016	A220*	110–135
Brazil	Embraer	2018	E Jet E2 Series	90–140
China	Commercial Aircraft Corporation of China (COMAC)	2015	ARJ21 Regional Jet	100–120
		2021[b]	C919	170–200
Russia	United Aircraft Corporation	2011	Sukhoi Superjet	80–110
		2021[c]	Irkut MC-21	150–230

Notes: *Developed by Bombardier Aerospace and formerly branded Bombardier C-Series. Now marketed by Airbus and built in a joint venture partnership in both Canada and the US. [a] Johnson (2019), [b] Qiu (2019), [c] Trimble (2019).

16.1 Industrial location of aircraft OEMs

Aircraft manufacturing demands significant sources of capital for research and development and access to highly skilled labour, manufacturing materials, components and markets. Historically, the development of civil aviation technology can be traced to military applications, and many global aircraft OEMs have significant aerospace and defence activities (see Section 16.4). Three main factors affect the industrial location of commercial aircraft OEMs:

- *Demand*. Historically, major aircraft OEMs were located in countries that had a strong domestic demand for air transport and/or countries that had advanced military aircraft manufacturing capability. The last few decades have seen a progressive consolidation of the global aircraft manufacturing sector, as large OEMs have acquired or merged with their former competitors to gain increased market share and access to new markets (countries and product segments).

- *Funding and policy*. Government funding to support the development of their domestic aerospace industries provides incentives for OEMs and their suppliers to locate production in that country. For example, European governments provided Airbus with financial support via **launch aid** to develop the Airbus A380. Faced with an increasingly competitive market, commercial aircraft OEMs have responded through downscaling, joint ventures, mergers and various types of international **subcontracting** arrangements. A government may also use tariffs (a form of tax) on imported goods to protect domestic manufacturers from foreign competition.

- *Outsourcing*. OEMs seek to lower final assembly and development costs by outsourcing production to supply chain partners. This has had a major impact on developing the build and design capabilities of the global aerospace industry. This has resulted in an evolution from a simple 'build to print' subcontractor relationship to a full 'design and build' risk-sharing partnership. These contracts have allowed global partners of the OEMs to develop new capabilities for production capacity, tooling, design and final assembly. This is leading to a restructuring of the commercial aircraft industry and a change in industrial location, regional markets and, ultimately, jobs.

Launch aid: Financial support offered by a government to an OEM repaid through royalties on future sales of a product.

Subcontracting: a business practice in which one company hires the services of another to perform part of its activities.

Stop and think

Identify the factors that affect the industrial location of aircraft OEMs worldwide.

!

16.2 Structure of the aircraft manufacturing sector

Figures 16.2a and 16.2b illustrate the deliveries and orders of commercial aircraft since 1998. Since 2000, the production of large commercial jet aircraft has increased by over 50 per cent. Despite this significant increase in production, Boeing and Airbus have a combined order backlog in excess of 13,000 aircraft. Outstanding orders are dominated by demand for single-aisle narrow-body aircraft such as the Airbus A320 and Boeing B737. This backlog in orders is due in part to the annual variation of orders by airline operators and leasing companies which can be dependent on wider economic conditions and increasing demand for air travel outpacing OEM production capacity.

Commercial aircraft OEMs have sought to improve the efficiency of their operations by lowering costs, accessing specialised production and knowledge, increasing production rates and reducing development expenditure through the outsourcing and 'offshoring' of some of their production and development. Offshoring of manufacturing capacity to an overseas country that has lower labour costs and/or a more favourable tax regime may result in reduced production costs. For example, STELIA Aerospace Maroc, a subsidiary of Airbus

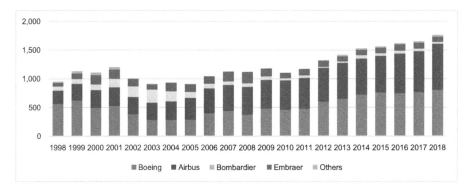

Figure 16.2a   Annual deliveries of large commercial jet aircraft
Source: Statista (2019)

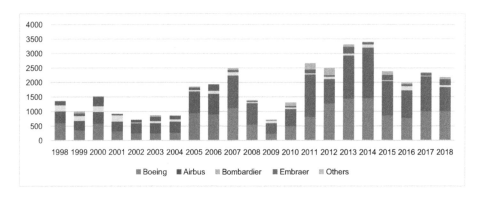

Figure 16.2b   Annual orders
Source: Statista (2019)

based in Morocco, produces relatively low-value components such as landing gear doors and the flightdeck lining for the A320.

Offshoring and outsourcing of components may result in financial benefits but may also create potential risks as the OEM must transfer technology and tacit knowledge to enable the supply chain to develop. As such, industrial offsets have been common throughout the history of aircraft manufacturing. The first major **industrial offsets** in aircraft manufacturing occurred in the 1960s when the US-based Douglas Aircraft Company subcontracted the fuselage assemblies for its DC-9 and DC-10 jetliners to Alenia in Italy. As a result of these transactions, Douglas secured substantial sales of aircraft to Alitalia, the flag carrier of Italy. One of Boeing's early offsets was with Japan in 1974, when Mitsubishi was given contracts to produce inboard flaps for the Boeing 747. Major sales of 747s to Japan followed.

Today, the foreign content of the Boeing 787 is around 30 per cent. This compares with 2 per cent on the B727 in the 1960s. Although the proportion of foreign content in Boeing products is likely to increase (as order backlogs for older Boeing models that have a higher

Industrial offset: a form of compensatory trade agreement whereby the exporter (in this case the aircraft OEM) grants concessions to the importer (a supplier in an overseas country). These concessions typically take the form of production-sharing agreements.

Table 16.3 Boeing 737 and 787 manufacturing strategy

Component	737 Programme	787 Programme
Sourcing strategy	Outsourced 30–50%	Outsourced 70%
Supplier relationship	Traditional supplier relationship	Strategic partners with tier 1 suppliers
Supplier responsibilities	Developed and produced parts for Boeing	Developed and produced sections for Boeing
Number of suppliers	Thousands	Approximately 50 tier 1 strategic partners
Supply contracts	Fixed-price contracts with delay penalty	Risk-sharing contracts
Assembly operations	30 days for Boeing to perform final assembly	3 days for assembly of complete sections

Source: Adapted from Tang and Zimmerman (2009)

domestic content are low and more production of later models has been outsourced to foreign suppliers/partners), this has more to do with cost savings than it does with securing sales to foreign airlines.

The B787 Dreamliner represents a significant change in airframe architecture compared to previous models. New materials and production techniques would have required significant capital investment by Boeing in production facilities if it were to be produced in a similar manner to earlier airframes. As such, Boeing developed a new manufacturing and supply chain strategy to expedite development and production times, mitigate the risks of launching a new innovative airframe, reduce the costs of development and increase production capacity (see Table 16.3).

In this new supply chain strategy Boeing has developed strategic partnerships with several key **tier 1 suppliers** who assemble large components and transport them to the final assembly line in the US. For example, Alenia in Italy constructs the central fuselage section. These sub-assemblies are then transferred as airfreight to the US for final assembly. The Japan Aircraft Development Corporation, formed by Fuji Heavy Industries, Kawasaki Heavy Industries and Mitsubishi Heavy Industries, funded the development and design of key airframe components including the centre wing box and main landing gear wheel well. It is the responsibility of these tier 1 strategic partners to manage the **tier 2 suppliers**. The primary role of Boeing is to coordinate this complex just-in-time international supply chain.

Tier 1 supplier: a company that directly supplies OEMs.

Tier 2 supplier: a company that supplies tier 1 companies.

!

Stop and think

Why is the foreign content of aircraft that are finally assembled in the US and Europe increasing, and is this trend likely to continue?
How might international supply chains change the risks faced by OEMs?

16.3 New technology

A characteristic of aircraft manufacturing has been constant innovation. In part, this innovation is driven by the need to lower costs at all stages of the supply chain and to construct aircraft with improved operational performance such as increased fuel efficiency and lower operating costs. Due to the significant cost of developing new airframes, OEMs have typically developed incremental improvements of existing models. The original Boeing 737 airframe first entered service in the 1960s, and the Airbus A320 in 1988, and both remain bestselling aircraft due to their continued refinement and development.

The recent development of new aircraft by major OEMs have been beset by significant delays and cost overruns. Manufacturing challenges as a result of the novel design of the Boeing 787 meant that the first aircraft was delivered to the launch customer (ANA) three years late, and the development cost overran the original $6 billion forecast. Independent analysis at the time of the delayed launch of the 787 programme estimates Boeing's development costs exceeded $15 billion (Gates, 2011). The C919 produced by Chinese OEM COMAC, as a domestically produced rival to the Boeing 737 MAX and Airbus A320, was originally scheduled to enter service in 2017 but has been delayed until 2021 (Qiu, 2019). Likewise, the first commercial delivery of the Sukhoi MC-21, developed by United Aircraft Corporation in Russia, has been delayed until 2021 due in part to economic sanctions imposed by the US limiting the transfer of technology from US-based aerospace partners (Trimble, 2019).

One of the most significant recent innovations has been to reduce the weight of the airframe by replacing heavier metal alloys with new lighter-weight composite materials. The adoption of new composites and manufacturing technologies (see Case Study 16.1) has changed manufacturing processes, lowered lead times and reduced the costs of manufacturing commercial aircraft.

CASE STUDY 16.1

FACTORY OF THE FUTURE

Airbus and Boeing recognise that digitisation, automation and the Internet of Things will revolutionise how aircraft are manufactured, monitored and maintained. The adoption of this new technology will increase the efficiency of production, reduce waste and lower costs. When the Boeing 777X enters production in 2020, the final assembly of forward and rear fuselage sections will utilise robotic technology. The robots will install up to 60,000 fasteners per aircraft more quickly, more cheaply and more accurately than the manual workers they will replace. Likewise, Airbus's Wing of Tomorrow programme is utilising new capabilities in advanced robotics and composite materials to develop highly automated production processes to develop new wings for future variants of the A320.

Composite materials are lighter and stronger than conventional aircraft-grade aluminium alloys and they confer significant efficiency improvements for aircraft operators. They also offer other design advantages such as higher cabin pressure, large windows and higher humidity. Disadvantages include the difficulty of reclaiming and recycling composite material at the end of an aircraft's useful life and the release of toxic particles and fibres in the event of a fire.

The use of composite materials on commercial aircraft dates back to the use of fibreglass on the Boeing 707 in the 1950s. In the 1980s, the Airbus A310–300 contained 5 per cent composite material, and by the 1990s, 12 per cent of the Boeing 777 was constructed from composite material. These proportions have continued to increase, and today over 50 per cent of the Boeing 787 and Airbus A350XWB are made from composite material.

Since 2005, the commercial aircraft industry has begun to use aluminium-lithium alloys in their new aircraft programmes as the cost and availability of this material has improved. However, there is a limit to the application of aluminium-lithium material on large commercial jet aircraft, and metal is still required on the leading edges of the wing and the engine pylons for reasons of bird strikes and fire protection. Although composites are now widely used on wide-body aircraft, composites do not yet downscale for single-aisle fuselage applications.

Stop and think

What are the advantages and disadvantages of manufacturing aircraft from composite materials?

16.4 Global shifts in commercial aircraft manufacturing

In terms of the global market share for large passenger aircraft, the US moved from an almost complete monopoly in the 1960s to a more competitive market position by 2019. Part of this shift can be explained by the emergence of the pan-European OEM Airbus, which moved from zero market share in 1970 to nearly 50 per cent by 2018. Airbus and Boeing control over 90 per cent of the market for large commercial jet aircraft, and they compete to see who can secure the highest number of aircraft orders. In saying this, the entry of Bombardier (Canada), Embraer (Brazil) and COMAC (China) into the single-aisle passenger aircraft market is challenging the Airbus-Boeing duopoly in the single-aisle market. While once it seemed inconceivable for the US to lose market share to a European competitor, the next major shift is from West to East, with more commercial aircraft deliveries going to the Asia-Pacific region and countries in South America and Asia developing their own aircraft manufacturing capabilities.

United States

The commercial aircraft industry is a crucial part of the US industrial base in terms of skilled production jobs, applied research, foreign exports and inter-industry multiplier effects. Boeing, the US's sole remaining producer of large passenger jets, has opted for a 'systems integration' mode of production to reduce unit costs, simplify final assembly procedures and speed up product development. Under a systems integration model, risk and costs are spread across a network of domestic and foreign partners. While the final product is assembled within the US, major parts of the airframe are subcontracted to foreign suppliers. The final assembly lines (FAL) for all wide-body Boeing models are located at Everett, near Seattle, in the world's largest manufacturing building (Boeing, 2019). There is a second FAL for the B787 in North Charleston, South Carolina. The FAL for the 737 family of aircraft is in Renton, Washington State.

In 2015 Airbus opened a FAL in Mobile, Alabama, to manufacture narrow-body A319/320/321 aircraft for airline customers in North America. It also began manufacturing the A220 (formerly the Bombardier C-Series) for the US domestic market in August 2019.

! Stop and think

Assess the relative merits of outsourcing as a manufacturing strategy.

Canada

Bombardier Aerospace, based in Montreal, designs and manufactures commercial, business, specialised and amphibious aircraft. Bombardier manufactures the A220 (formerly branded C-Series) of commercial jet aircraft (seating 100–149 seats) in partnership with Airbus, the CRJ regional jet series (seating 60–99) and the Q series of turboprops (seating up to 86). Like Boeing, Bombardier works with a range of overseas suppliers (currently around 3,000 suppliers) based in 17 countries worldwide, including Brazil, China, India, Japan and the UK.

In October 2017, Bombardier entered into a partnership with Airbus (who hold a 50.01 per cent interest in the joint venture) to manage, manufacture and market the C-Series, subsequently rebranded the A220. The aircraft features advanced new structural materials that confer significant weight savings. The A220's FAL and nose assembly is located in Montreal, with major sections of the aircraft being shipped from around the globe. A second FAL has been established in the US.

Although Canada has historically enjoyed a strong comparative advantage in regional jet aircraft production, this advantage is weakening in light of growing international competition from lower cost competitors in Brazil, Russia and China.

Europe

Historically, almost every major European country had its own domestic commercial aircraft manufacturing industry. Over time, consolidation has occurred and once-independent national OEMs have been taken over, merged or have ceased production. Consequently, once-familiar names in European aircraft manufacturing, including British Aerospace and de Havilland (UK), Fokker (Netherlands), Saab (Sweden), Junkers and Dornier (Germany) and Aérospatiale (France), have either ceased production of aircraft or left the market.

The main European OEM of large commercial passenger aircraft is Airbus. Airbus has developed into a multinational aerospace corporation with interests spanning commercial

Table 16.4 European production sites of Airbus commercial aircraft, 2019

Country	Site	Responsibilities
France	Toulouse	Engineering design, testing, flight tests
		Final assembly lines for A320, A350XWB, A330 and A380
	Saint-Nazaire	Structural assembly
		Forward fuselage assembly of A320
		Forward and central fuselage assembly of A330 and A380
		Nose fuselage of A350XWB
	Nantes	Central wing boxes
		Carbon fibre reinforced plastic structural parts
		Radomes, ailerons and air inlets for A350XWB, A380 and A320neo
		Painting and equipment installation
	Paris	Research and development at Suresnes
	Marignane	Civil helicopters
Germany	Hamburg	Structural assembly and outfitting of A320
		Major component assembly of A380
		Manufactures rear fuselage sections for A330 and A350XWB
	Bremen	Design and manufacturing of high-lift wing devices for all aircraft
	Stade	Vertical tail planes for all aircraft
		Carbon fibre reinforced plastic components
UK	Broughton	Wing production
	Filton	Engineering research and development
Spain	Madrid	Aeronautical component engineering, design, production and assembly
	Seville	Main component assemblies for A330 and A380 at Tablada pre-FAL

Source: Derived from Airbus (2019)

aviation, defence and space and helicopters. Airbus has a product line of aircraft ranging from 100 to 500 seats. Its single-aisle aircraft is the A320 series, which includes the A319, A320/A320neo (new engine option) and A321/A321neo models. The company's wide-body aircraft include the A330/A330neo, A340 family, A350XWB and A380 superjumbo. By October 2019, Airbus had delivered over 12,000 aircraft and attracted 19,500 orders (Airbus, 2019). Production is based at dedicated production sites in the UK, France, Germany and Spain (Table 16.4). Components are transported between the sites by air on dedicated modified A330–200 freighters (the Beluga XL) and by road and water on specially adapted roll on-roll off ferries and river barges.

A second important European OEM is Avions de Transport Régional (ATR). Based in Toulouse, ATR is a joint partnership between Airbus and Leonardo and has sold over 1,500 airframes (ATR, 2019). ATR manufactures regional turboprop aircraft, so it is not in competition with Airbus as their aircraft families serve different markets and customers.

Brazil

Embraer is the third largest manufacturer of commercial jet aircraft and one of Brazil's largest exporters of industrial products. Embraer has a global workforce of over 19,000 employees and has delivered over 1,000 E-Jets. In 2013, Embraer launched the E2 series, a new variant of the E-Jet family of commercial aircraft that seats 80–132 passengers. The E2 has new technologies, including full fly-by-wire, Pratt & Whitney geared turbofan (GTF) engines and high-aspect ratio wings with swept tips, giving it a 16 per cent improvement in fuel consumption. The first aircraft entered service with Norwegian airline Widerøe in 2018.

In February 2019 Boeing established a joint venture partnership with Embraer's commercial aircraft division creating Boeing Brasil-Commercial. The new joint venture will see Boeing take an 80 per cent stake in Embraer, with the aim of expanding the global market share of the E-Jet and E2 series. At the time of writing, this venture had been approved by the shareholders of both companies and was awaiting approval by international antitrust authorities. This deal may further reduce competition in the international market for large commercial jet aircraft following the Airbus-Bombardier partnership.

China

China is projected to be the largest market for commercial passenger aircraft in the next 20 years. This attempt to gain market access, combined with lower labour costs, has led major aircraft OEMs to outsource some of their assembly to Chinese aerospace companies and/or establish FALs and delivery centres in China. International companies investing in China must form joint venture (JV) partnerships with domestic companies. These JVs require the transfer of technology and tacit knowledge to China. Western aircraft manufacturing at Chinese factories includes:

- Airbus at Harbin (composite manufacturing), Beijing (engineering centre), and Tianjin (FAL for A320, A330/340 families, A330 completion and delivery centre);
- Boeing at Zhoushan – B737 family completion and delivery centre.

CASE STUDY 16.2

AIRBUS A320 FINAL ASSEMBLY LINE, TIANJIN, CHINA

In 2008, Airbus China, Aviation Industry Corporation of China (AVIC) and Tianjin Free Trade Zone Company (TFTZC) signed a joint venture agreement for an A320 final assembly line to be located in Tianjin. The factory is a 'copy' of Airbus's Hamburg plant (Airbus owns the tooling and TFTZC owns the building). The current production rate is six A320s a month, which are sold to Chinese airlines. In the first ten years of operation, 380 aircraft were produced (Airbus, 2018).

China is also actively developing its aircraft OEM capability by investing in research and development to support the production of Chinese mid-size passenger aircraft. In May 2008, a new state-owned company, Commercial Aircraft Corporation of China (COMAC), was created to develop, manufacture and commercialise Chinese passenger aircraft. COMAC is overseeing the development and production of the C919 aircraft (seats 168–190) and ARJ21 regional jet (seats 75–90). At the time of writing, the C919 development had been significantly delayed and was likely to enter service in 2021.

China has a long-term commitment to developing a family of aircraft that meet Western certification standards. With the aim of becoming more competitive as a low cost producer with high quality and better productivity, the Chinese commercial aircraft industry has decided to take advantage of its centres of competence from decades of industrial cooperation with the main global OEMs. These strategic alliances and joint ventures allow the Chinese to develop leaner cost structures.

Russia

In 2007, the Russian government consolidated the civil aircraft industry into one state-owned enterprise, the United Aircraft Corporation (UAC). UAC is majority owned by the Russian government (80 per cent). The UAC aims to:

- create sales and technical services for domestic and international aircraft markets;
- develop the Russian aircraft industry so that it can compete in international markets for aircraft products;
- work with international partners;
- create a modern research and development infrastructure.

The UAC has two civil aircraft programmes: the Sukhoi Superjet 100 (80–110 seats), now in production; and the Irkut MC 21 (150–230 seats), which first flew in 2019. The main challenge facing Russian civil aircraft has been out of country service support for foreign airline customers. The Sukhoi Superjet 100 programme is trying to address this by partnering internationally with the Italian–Russian Superjet International joint venture. At the time of writing, orders for the Irkut MC 21 had been dominated by domestic Russian airlines.

Both Boeing and Airbus have developed international collaborations with Russia. Boeing has a Russian technical research centre and a design centre that work on research projects and structural design, and, since 2001, Boeing has advised on management, marketing, certification and after-sales support for the Sukhoi Superjet 100 programme.

Mexico

The development of Mexico's aerospace clusters is due to a national strategic programme to trade market access for co-design and production of Western-certified aircraft sub-assemblies and components. The aim of the Mexican government has been to support the development of a cradle-to-grave aerospace industry. In 2019, the manufacturing industry included over 350 organisations, many of whom supply large international OEMs. The US–Mexico Bilateral Aviation Safety Agreement of 2007 means that aircraft sub-assemblies, components and parts produced in Mexico can be certified by the FAA and exported globally.

India

India is another country with a rapidly emerging commercial aircraft manufacturing industry. One of the major aerospace companies, Hindustan Aeronautics Limited in Bangalore, have secured a range of contracts to manufacture and supply airframe parts and components for OEMs including Airbus and Boeing. The company is also involved in development of a proposed new Indian Regional Jet.

Japan

Mitsubishi Aircraft Corporation is developing the new SpaceJet family of regional jet aircraft. The M90 and M100 will seat between 76 and 92 passengers with an entry into service date expected in 2024.

Stop and think

To what extent do you think Boeing and Airbus might lose market share to OEMs in other countries in the future, and how might they seek to protect their existing market shares?

16.5 Future trends

Future demand growth for air travel would suggest that demand for new aircraft should remain resilient. Deloitte (2018) forecast that by the mid-2030s aircraft deliveries should increase 25 per cent to nearly 2,000 aircraft per year. However, aircraft manufacturers and OEMs face many challenges, not least that other countries, including Japan and India, are actively developing their own domestic aircraft manufacturing capabilities. This globalisation of commercial aircraft OEMs will have profound implications for the future of aircraft manufacture. OEMs are likely to witness some erosion to their market share. The actions of Boeing and Airbus to acquire or partner with potential rivals is evidence of their adapting to this potential change in competition and the difficulties they would face to enter airframe market segments with newly developed aircraft. It is possible that supply chains will become more complex, automation will continue and costs will fall, but only up to a point. There is a risk that continually driving down costs will create sub-standard products that do not deliver the operational cost savings they promise, and reputational damage will result. It is thus critical that OEMs, if they pursue a strategy of global outsourcing, manage and monitor their suppliers and contractors to ensure that products are delivered to time, to budget and, crucially, to the specified manufacturing standard.

With aircraft expected to have a service life of 25–30 years, and given the environmental impact of air travel (➤Chapter 18), airframe and engine manufacturers must work to reduce greenhouse gas emissions and noise. In order to achieve this, radical redesigns of aircraft, both the airframe and engines, will be required, and work has already commenced. Airbus established a partnership with Siemens and Rolls Royce to develop a hybrid electric aircraft, the E-Fan X. Boeing, in partnership with NASA, established the SUGAR programme (Subsonic Ultra Green Aircraft Research) with the aim of developing hybrid electric aircraft, truss-braced wing aircraft and long-term hydrogen fuel-cell aircraft. These initiatives will require significant capital investment in research and development, new materials, supply chains and production processes.

Key points

- The commercial aircraft industry is a symbol of a nation's innovation and technology prowess.

- Aircraft design, development and manufacturing is capital intensive.

- Historically, countries with a large domestic air transport market and military requirement led commercial aircraft manufacture.

- Consolidation has been a feature of the global aircraft manufacturing industry.

- Large commercial aircraft manufacture is dominated by Boeing and Airbus.

- Bombardier and Embraer are key companies in the regional and specialist aircraft markets.

- New technologies and automated production processes are lowering lead times and production costs.

- Commercial aircraft OEMs have evolved into system integrators that outsource to foreign suppliers/risk-sharing partners.

- Outsourcing confers both benefits and risks to OEMs.

- China, Russia, India and Mexico are actively developing OEM capability through state subsidy and state-sponsored technological support.

- Rapid innovation in materials, production techniques and supply chains will continue.

References and further reading

Airbus. (2018). Airbus' China assembly facility marks 10 years of quality manufacturing for A320 family jetliners 28th September. Available at: www.airbus.com/newsroom/news/en/2018/09/airbus – china-assembly-facility-marks-10-years-of-quality-manufa.html

Airbus. (2019). *Our locations in Europe*. Available at: www.airbus.com/careers/our-locations/europe. html

ATR. (2019). *ATR aircraft*. Available at: www.atraircraft.com/

ATAG. (2018). *Aviation benefits beyond borders*, Geneva: Air Transport Action Group.

Boeing. (2019). *Current products and services*. Available at: www.boeing.com/commercial/

Deloitte. (2018). *2019 global aerospace and defense industry outlook* Available at: https://www2.deloitte. com/global/en/pages/manufacturing/articles/global-a-and-d-outlook.html

Gates, D. (2011). Boeing celebrates 787 delivery as program's costs top $32 billion. *The Seattle Times* [online]. Available at: http://old.seattletimes.com/html/businesstechnology/2016310102_ boeing25.html

Johnson, E. (2019). Boeing's 777x faces engine snags, questions rise over delivery goal. *Reuters*. [online] (Last updated 5th June 2019). Available at: www.reuters.com/article/us-boeing-777x/boeings-777x-faces-engine-snags-questions-rise-over-delivery-goal-idUSKCN1T62KU

Qiu, S. (2019). COMAC pushes back C919 jet's China certification target to 2021. *Reuters*. [online] (Last updated on 7th August 2019). Available at: www.reuters.com/article/us-china-comac-c919/comac-pushes-back-c919-jets-china-certification-target-to-2021-idUSKCN1UX0A7

Statista. (2019). Number of jets added to the global aircraft fleet from 1998 to 2018, by manufacturer (in units). Available at: https://www.statista.com/statistics/622779/number-of-jets-delivered-global-aircraft-fleet-by-manufacturer/

Tang, C. S. and Zimmerman, J. D. (2009). Managing new product development and supply chain risks: The Boeing 787 case. *Supply Chain Forum: An International Journal*, 10(2), pp. 74–86.

Trimble, S. (2019). U.S. sanctions trigger one-year MC-21 schedule delay. *Aviation daily*. [online]. Available at: https://aviationweek.com/commercial-aviation/us-sanctions-trigger-one-year-mc-21-schedule-delay

US Department of Commerce. (2019). *Aviation: China*. Available at: www.export.gov/article?id=China-Aviation

CHAPTER 17

Air cargo and logistics
Martin Dresner and Li Zou

LEARNING OBJECTIVES

- To identify the functions of companies involved in air cargo operations.

- To make the business case for air freight shipments and highlight past industry growth.

- To understand the type of products that are shipped by air and appreciate the role of product segmentation.

- To highlight the differences between air cargo and air passenger operations.

- To appreciate the innovations, challenges and future prospects in the air cargo industry.

17.0 Introduction

In 2018 approximately 55 million tonnes of freight, about 3 per cent of total trade (by weight), were transported by air on scheduled services, with freight tonne-km (FTK) totalling over 231 billion. Although 97 per cent of trade moved by other modes, air freight accounted for over one-third of world trade by product value, according to the estimates by the International Air Transport Association (IATA). The implication is that air cargo carriers move goods that have a high value-to-weight ratio, leaving the lower-value cargo to other modes.

During the period 1973 to 2018, world air freight traffic (measured by FTKs) expanded at an average of 6.24 per cent per year, higher than the 4.77 per cent annual growth in world merchandise exports, and the 2.94 per cent annual growth in world gross domestic product (GDP) (see Figure 17.1). Therefore, air cargo's share of world trade and its share of the world's economy has increased over the past 40 years. The global financial crisis in 2008 heavily impacted the cargo industry, resulting in two consecutive years of declining air cargo volume. Since then, however, the air cargo market has recovered, with growth based on global economic recovery, expanding international trade, and rapidly increasing e-commerce.

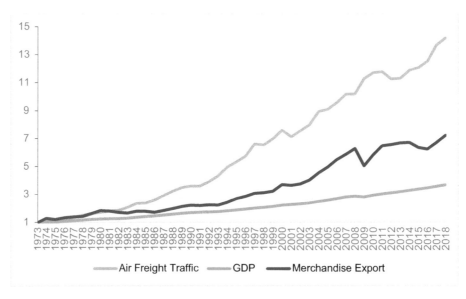

Figure 17.1   The historical growth of world air freight traffic, GDP, and merchandise export, 1973–2018

In 2017, the world air cargo industry had its strongest growth in terms of cargo volume, load factor, yield, and revenue since the global financial crises. Between 2016 and 2017, air cargo demand increased by 9 per cent (measured by FTKs), about twice the growth rate of global merchandise trade. Airlines worldwide experienced cargo demand growth, with the highest year-over-year growth rate experienced by African airlines and the lowest by Latin American carriers (Table 17.1). The strong air cargo demand growth slowed down in 2018, growing only 3.5 per cent worldwide over the previous year, and the African region experienced an air cargo decrease of 1.3 per cent.

Although the growth in world air cargo has outpaced the growth in global GDP, and international merchandise trade, this trend may not continue. There are indications of weaknesses in the market. Following the rapid growth in cargo traffic in 2017, the ratio of air cargo annual growth relative to the GDP growth declined from 2.93 in 2017 to 1.95 in 2018 (see Figure 17.2). The air cargo volume worldwide, as measured by total FTKs, had decreased by 4.8 per cent in June 2019 over the same month in the previous year. This negative year-over-year air freight decline has persisted for eight consecutive months. The weakness in air cargo shipments may be due to increased global economic uncertainties including the US-China trade war. The imposition of higher tariffs by the US and China increased the cost of traded goods, leading to a reduction in international cargo shipments, including air cargo. According to the US Census Bureau, goods exported from China to the US declined by more

Table 17.1 The air cargo performance of airlines by region in 2017 and 2018

Regional air cargo performance	Demand growth in 2017 over 2016 (FTK)	Demand growth in 2018 over 2017 (FTK)	Capacity growth in 2017 over 2016 (AFTK)	Capacity growth in 2018 over 2017 (AFTK)	Market share (FTK) in 2017 (2018)
Asia-Pacific airlines	7.8%	1.7%	1.3%	5.0%	37% (35.4%)
North American airlines	7.9%	6.8%	1.6%	6.8%	20.5% (23.7%)
European airlines	11.8%	3.2%	5.9%	4.3%	24.2% (23.3%)
Middle Eastern airlines	8.1%	3.9%	2.6%	6.2%	13.7% (13.3%)
Latin American airlines	5.7%	5.8%	3.1%	3.4%	2.7% (2.6%)
African airlines	24.8%	-1.3%	9.9%	1.0%	1.9% (1.7%)
Worldwide	9.0%	3.5%	2.9%*	5.4%	100% (100%)

* This is an estimated value based on regional capacity growth and market share by freight traffic. (A)FTK = (Available) Freight Tonne Kilometre

Sources: www.iata.org/pressroom/pr/Pages/2018-01-31-01.aspx; www.iata.org/pressroom/pr/PublishingImages/freight_infographic_-18.png

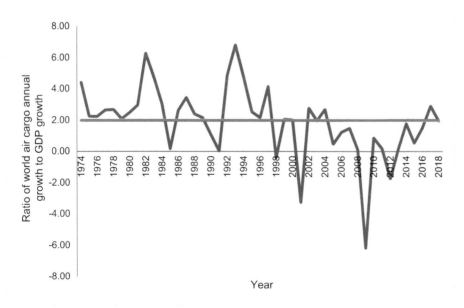

Figure 17.2 The ratio of world air cargo annual growth to GDP growth

than 12 per cent during the first six months of 2019, compared to the same period in 2018 while US merchandise exports to China fell by 18.9 per cent during the same period.

In general, air cargo demand is vulnerable to downturns in the business cycle, with its growth rate highly impacted by economic activity, as measured by indicators such as the global manufacturing Purchasing Managers Index (PMI), new export orders, and the consumer confidence index. During a downturn in economic activity, economic growth slows, as does world trade. Moreover, firms look to reduce transport costs and switch to lower-cost modes. Finally, during economic downturns, firms may increase inventory due to slower sales, so replenishing stock does not have to be as time critical which may favour surface transport over air cargo.

!

Stop and think

Outline the relationship between air cargo growth and global GDP.

17.1 The air cargo industry

Integrator: a company that offers an integrated door-to-door service, normally using their own fleet of aircraft and road vehicles.

Air freight forwarder: a company responsible for arranging the entire shipment of goods from A to B using a range of transport providers.

The air cargo industry consists of many different companies specialising in various aspects of the industry. The best known companies are the **integrators**, such as FedEx, UPS, and DHL Express, that provide customers with a door-to-door service for mainly time sensitive and relatively small shipments.

In addition to integrators, air cargo is carried by airlines that operate dedicated freighter aircraft and by airlines that carry both passengers and freight in the same aircraft. The first type of company is called a pure cargo carrier, while the second type is called a belly-hold cargo carrier, as cargo is carried in the hold (the belly) of passenger flights. Finally, there are companies that facilitate the carriage of air freight although they do not own any aircraft themselves. These companies, known as the 'travel agencies' for air cargo, are called **air freight forwarders**. Table 17.2 provides a listing of the world's largest air cargo operators, while Table 17.3 details the largest air freight forwarders.

Table 17.2 The world's top ten air cargo operators by scheduled FTK, 2018

Rank	Airline	FTK (million)	Number of freighter aircraft in service	Types of freighter aircraft
1	FedEx	17,499	364	B757–200F; B767–300ERF; B777–200F; A300–600F; A310F; DC-10F/MD-10F; MD-11F
2	Emirates	12,713	11	B777–200LRF
3	Qatar Airways	12,695	20	B747–8F; B777–200LRF; A330–243F
4	UPS Airlines	12,459	239	B747–400F; B747–8F; B757–200F; B767–300F; B777–200F; A300F; MD-11F

Rank	Airline	FTK (million)	Number of freighter aircraft in service	Types of freighter aircraft
5	Cathay Pacific Airways	11,284	21	B747–400BCF; B747–400ERF; B747–8F
6	Korean Air Lines	7,839	23	B747–400ERF; B747–8F; B777–200LRF
7	Lufthansa	7,394	17	B777–200LRF; MD-11F
8	Cargolux	7,322	26	B747–8F; B747–400ERF; B747–400F
9	Air China	7,051	15	B747–400F; B757–200F; B777–200LRF
10	China Southern Airlines	6,597	14	B747–400F; B777–200LRF

Source: Air Cargo News, 16 October 2019, fleet data retrieved from Airline Monitor, October 2018 with aircraft type data supplemented by FlightGlobal

Table 17.3 Top 25 air freight forwarders by air freight tonnage, 2018

Current rank (2014 rank)	Freight forwarder	Air freight tonnes	Gross revenue (US$) (in millions)
1 (1)	DHL Supply Chain & Global Forwarding	2,150,000	28,120
2 (2)	Kuehne + Nagel	1,743,000	25,320
3 (3)	DB Schenker Logistics	1,304,000	19,968
4 (6)	Panalpina	1,038,700	995,900
5 (7)	Expeditors	1,011,563	985,549
6 (4)	UPS Supply Chain Solutions	935,300	9,814
7 (8)	Nippon Express	899,116	18,781
8 (9)	Bolloré Logistics	690,000	5,415
9 (16)	DSV	689,045	12,411
10 (5)	Kintetsu World Express	600,849	4,752
11 (10)	Hellmann Worldwide Logistics	587,007	3,646
12 (12)	Sinotrans	530,100	10,174
13 (11)	CEVA Logistics	476,600	7,356
14	Apex Logistics International	430,000	1,350
15 (13)	Agility Logistics	415,000	4,400
16 (17)	Kerry Logistics	409,127	4,875
17 (15)	Yusen Logistics	380,000	4,820
18 (18)	GEODIS	363,451	6,645
19	Dachser Air & Sea Logistics	344,900	7,602
20	Crane Worldwide Logistics	337,300	916

Continued

Table 17.3 Continued

Current rank (2014 rank)	Freight forwarder	Air freight tonnes	Gross revenue (US$) (in millions)
21 (19)	NNR Global Logistics	315,011	1,146
22 (24)	Hitachi Transport System	300,000	6,283
23 (20)	FedEx Logistics	276,400	3,170
24 (22)	Pilot Freight Services	232,256	798
25	C.H. Robinson	225,000	16,631

Source: Air Cargo World, June 2019

Most of the world's largest air cargo operators and air freight forwarders are concentrated in Asia, Europe, and North America, reflecting air cargo demand on the major trade routes between these continents. However, the rise of Emirates from the 9th largest cargo operator in 2007 to the 2nd largest in 2018, and Qatar from 34th in 2007 to the 3rd in 2018, reflects the dramatic growth of Gulf carriers in recent years. These carriers, based in the Middle East, use their strategic location between Europe and East/South Asia to create powerful hub-and-spoke networks. They can then transship cargo through their hubs, serving the major trade routes around the world on a one-stop basis. Finally, these airlines have invested heavily in wide-body aircraft that provide the capacity necessary to carry belly cargo.

Integrators

Compared to air cargo operators that rely on freight forwarders to sell their cargo space, integrators have two advantages: fully integrated operations, and direct interface with clients. These enable integrators to differentiate their services from other air cargo operators by providing:

- increased efficiency in cargo handling and movement;

- reduced dwell time and in-transit time;

- enhanced security monitoring and control;

- elimination of duplicate paper documentation and handling processes;

- transparent information sharing, including provision of real-time tracking and tracing options to shippers;

- improved on-time performance and flexibility in handling customer requests.

The integrators own large fleets of aircraft devoted entirely to the carriage of air cargo. Integrators, in common with other cargo airlines, also use other carriers to provide additional capacity on a contractual basis; for example, the integrators contract with the major passenger carriers for belly cargo space. Additionally, the integrators own trucking fleets or work with contract trucking firms to pick up and drop off shipments, to transport cargo to its final destination, and to carry packages that are not particularly time sensitive. By offering fast, reliable, seamless, end-to-end and value-added services to customers, integrators have been successful in differentiating their services. In less than 20 years, integrators have grown into the largest air cargo operators worldwide (see Case Study 17.1).

THE RISE OF FedEx

Before the deregulation of the US domestic air travel market in 1978, UPS was the dominant firm in the express package industry, with a market share of between 80 per cent and 90 per cent. Its main competitor, the US Postal Service (USPS), was unable to compete on service quality and rates without government subsidies. The conventional wisdom is that firms in a regulated industry with high barriers to entry exploit a stable industry setting to achieve economies of scale and higher profits due to a lack of competition. Also, in a market with limited competition, firms may not be innovative. As organisations grow in size, they often become more hierarchical, rigid, and slower to adapt to the changing environment. In a regulated environment, both UPS and USPS provided standard services with little thought to a customer-driven value chain.

Deregulation lowered the entry barriers faced by new entrant airlines and stimulated competition. Benefiting from the rapid development of information technologies, the express package rival FedEx (then known as Federal Express), which was established in 1971, grew dramatically. FedEx entered a package delivery market sector underexploited by UPS and USPS – overnight air express. By 1983, FedEx was generating US\$1 billion in revenue, more than any other air delivery carrier in the US, and had overtaken UPS in the express market. Since then, UPS has worked hard to catch up with FedEx. In particular, UPS has been successful in focusing on international markets.

The integrators, however, face new challenges, as Amazon and other large online retailers, such as Alibaba in China, expand their own delivery capabilities to compete directly with the integrators. Faced with competition from one of its own customers, FedEx ended its contractual relationship with Amazon and will seek business elsewhere (see Case Study 17.2).

AMAZON AND FedEx

Within three years of its launch in 2016, Amazon Air expanded its freighter fleet from zero aircraft to a 50-freighter fleet. As of August 2019, Amazon Air's fleet consisted of a mix of B767 and B737 aircraft, with a fleet size projected to grow to 70 by 2021. With an estimated \$1.5 billion investment, the future hub of Amazon Air at Cincinnati/Northern Kentucky International Airport (CVG) will be Amazon's largest fulfilment centre in the world, with a 900-acre area and a 3 million square foot facility. Technology employed will include robotics, machine learning, and artificial intelligence, allowing Amazon Air to accommodate over 200 take-offs and landings per day, along with the shipments they will bring in and send out. This Amazon Air super hub is expected to be ready for use by 2021. Although Amazon Air intends to have major freighter operations at its CVG super hub, it will also conduct operations on a point-to-point basis, connecting more than 20 gateway locations in Amazon Air's network in the US and overseas.

As the big e-commerce companies rapidly build their own logistics networks, their business relationships with integrators have evolved from contract-based partnerships to increased competition. When Amazon established its own delivery network and shipping capacity, FedEx decided to end its contract business with this e-commerce giant. There was some thought that Amazon was choosing to undertake the lower-cost deliveries itself, leaving the higher-cost deliveries to integrators such as FedEx. Instead of delivering packages for Amazon, FedEx plans to increase its business with other e-commerce retailers, such as Target Corp. and Walmart Inc., as well as with small and medium-sized shippers. Reducing reliance on Amazon will allow FedEx greater leverage in price negotiations. The termination of the FedEx–Amazon relationship will create more business opportunities for other integrators, such as UPS and the US Postal Service. In the long term, however, these firms may encounter a similar situation to FedEx and will need to decide whether to continue making high cost deliveries for Amazon at low contract prices or to cut that business and seek revenue elsewhere.

Passenger-cargo combination carriers

Payload: the total weight of passenger, baggage, and cargo carried on an aircraft for which revenue is derived.

Unit load device (ULD): a container or pallet used to load freight and luggage on wide-body and some types of narrow-body freighter aircraft.

Most passenger airlines, including some low cost carriers, carry belly freight in their cargo holds. Wide-body aircraft are especially capable of carrying large **payloads**. Much of this freight travels in specially designed air cargo containers called **unit load devices (ULDs)** that allow a large number of individual items of freight to be packed in a single unit. The ULDs are manufactured in different sizes and will fit only specific types of aircraft. Some passenger airlines also operate dedicated cargo aircraft. The most popular types of freighter aircraft operated by passenger–cargo combination airlines are long-range, high-capacity aircraft, including the Boeing 777 and Airbus 350 families. Compared to integrators, passenger–cargo combination airlines (e.g. Cathay Pacific and Korean Air) or their fully owned cargo subsidiaries (e.g. Emirates SkyCargo, Lufthansa Cargo, and Singapore Airlines Cargo) have much smaller fleets of dedicated freighter aircraft. Nevertheless, the access to the belly holds on passenger flights and extensive route networks with frequent scheduling options provide a distinct advantage for passenger–cargo combination airlines. As a result, it is estimated that about 50 per cent of air cargo traffic worldwide (RTK) is carried by aircraft that are also carrying passengers, using their belly-hold capacity. The carriage of belly-hold cargo can complement passenger services and make an otherwise unprofitable route into a sustainable one (see Case Study 17.3).

CASE STUDY 17.3

LAN'S CARGO OPERATIONS

LAN Airlines of Chile carries air cargo traffic using the belly-hold capacity of wide-body passenger aircraft on international routes between Latin America and the US and Europe. The cargo carried from Latin America to the US and Europe consists mainly of perishable goods, such as seafood, flowers, and vegetables. The majority of return shipments are high-value, low-weight merchandise, including computer hardware, smartphones, and auto parts. The benefits from operating cargo and passenger services in tandem include: increased utilisation of aircraft assets; reduced break-even load factors; and more diversified revenue streams. By strengthening its operations, LAN has been able to preclude entry of potential competitors and obtain relatively high freight yields as a result.

In contrast to many Asian, European, and Middle Eastern airlines, US airlines derive, on average, only a small share of revenues from cargo. This is because many US airlines use narrow-body aircraft with limited belly-hold capacity on their domestic routes, as they are driven primarily by passenger demand and also have short turnaround times between flights which do not allow for loading and unloading of cargo. Even on international routes, US carriers generally prioritise passenger transport over cargo traffic, and their networks are not optimised for cargo traffic.

All-cargo operators

All-cargo operators, sometimes called traditional cargo carriers, provide point-to-point services on a scheduled or charter basis. Since these operators use freighter aircraft, they can often take shipments of large items, such as oil and gas drilling parts or power generators. Indeed, much of their traffic consists of heavy-lift shipments. In addition, the all-cargo carriers will often transport items to remote regions, far from the nearest airports served by the regularly scheduled carriers. Since all-cargo carriers often do not provide a door-to-door service or have extensive marketing departments, they rely on air freight forwarders to book loads and to arrange for freight pick-up and delivery.

One of the keys to success for the all-cargo operators is to provide an excellent dedicated service to shippers. According to the European Shipper's Council (ESC), there are four areas where shippers expect carriers to provide improved services:

1 more reliable, efficient, door-to-door seamless logistics services;

2 paperless chain of air cargo flow from consignor to consignee;

3 fair, transparent and reasonable surcharges;

4 industry-wide standards for carbon emissions and information on the emissions generated by each shipment so shippers can compare environmental performance among airlines.

To address the gap between shipper expectations and the service provision of air cargo operators, the ESC advocates strategic collaboration and cooperation between shippers, freight forwarders, cargo handlers, and cargo airlines.

Air freight forwarders

Air freight forwarders are central to air cargo operations as they connect suppliers and customers. Freight forwarders typically reserve space on cargo flights and consolidate deliveries from multiple shippers, arranging for the consolidated loads to arrive at the airport of departure ready for shipping. They then arrange for the delivery of shipments to end destinations from the arrival airports. Freight forwarders will also provide value-added services to shippers, such as documentation, customs processing, insurance, goods storage, packing, handling, and distribution.

Around 40,000 freight forwarders are members of the International Federation of Freight Forwarders Associations (FIATA). While the majority of them are small, non-asset-based businesses, the world's top air forwarding companies have extensive geographic coverage and large-scale operations, provide diversified logistics services, and enjoy dominant positions in the rail, sea, road, and air cargo markets.

! Stop and think

Detail the structure and organisation of the global air cargo industry.

17.2 The business case for air cargo

The time value of money

In evaluating the cost-time trade-off faced by shippers, they need to consider:

- the cost of having funds invested in products that are delayed in transit;

- the perishability or speed of obsolescence of the product.

For example, when Apple released the iPhone 6 and iPhone 6 Plus, it was expected that 60 million devices would be shipped from assembly sites in Chinese cities, including Shenzhen and Zhengzhou, to customers in 115 countries by the end of the year. Nearly all of these shipments would be by air. Tim Worstall, a contributor to Forbes.com, asked why these devices were being shipped by air when sea transport is much cheaper. Given fuel prices at the time, he calculated the cost to ship the iPhones to Europe by sea at about 1.2 cents per device. But the shipment time from Shanghai to Rotterdam would be 25 days, with another five days for customs clearance and transport to an inland city, such as Paris or Frankfurt. On the other hand, transit time by air via an all-cargo Boeing 777F, such as the type of aircraft employed by the express air freight company FedEx, would be approximately 15 hours. Apple was willing to pay 50 times the ocean freight price to ship the iPhones by air.

The value of centralised distribution

Fast fashion retailers rely on getting their product to market as quickly as possible in order to take advantage of the latest fashion trends. These companies recognise the time value of money. Missing shipment deadlines to retailers for the spring season can result in large inventory discounts, as retailers move onto the next season's fashions. So, delivery speed is very important. Companies may, therefore, decide to transport their products by air.

An alternative model would be to establish distribution centres in their major markets, ship their product by sea to these distribution centres, and then supply individual stores from the closest distribution centre. In theory, an order placed, for example, from a Houston, Texas, store could be delivered from a nearby distribution centre by road faster than it could be delivered by air from a centralised distribution centre in Europe or Asia.

There are other advantages to centralised distribution that outweigh the potential time savings from a decentralised distribution network. In particular, centralised distribution allows a company to operate with less inventory and lower warehousing expenses than a decentralised system does. Fashion retailers can save considerable capital and operating expenses by centralising their inventory and distribution capabilities and using air cargo to ship their merchandise worldwide. Centralised distribution systems allow companies to ship new fashions faster than a decentralised system would. The elapsed time from product design to in-store is reduced as is the possibility of inventory obsolescence. Although a decentralised system can result in shorter transit time once an order is received, considerable time will elapse in shipping products to geographically dispersed warehouses by sea. In a fast changing fashion industry, styles and demands could change while the product is in transit, rendering the inventory obsolete.

The value of certainty and the reduction in delivery time variability

Although increasing delivery speed is important in achieving high levels of customer service and minimising inventory costs, reducing the variance in delivery times can be as important. If a transit option via sea takes an average of 30 days with a standard deviation of three days, the shipper can be only 97.5 per cent sure that the shipment will arrive within 36 days. Therefore, the cautious shipper, or its customer, not only has to finance 30 days of inventory associated with normal sea transit time, but an additional 6 days of inventory, called safety stock. If the safety stock is not financed, the firm faces an increased risk that its business will be disrupted due to the late receipt of goods. On the other hand, if the mean transit time by air is 15 hours, and the standard deviation is 1.5 hours, the shipper can be 97.5 per cent certain that the shipment will arrive within 18 hours, providing the shipment doesn't miss its booked flight. Given the cost differential between an extra three hours of inventory versus an extra six days of inventory, it is even more beneficial to ship goods by air.

The value of security and the mitigation of loss

Shippers are increasingly concerned about the security of their goods. As business has become increasingly globalised, goods are being transported over much greater distances, thereby increasing the risk to the products in transit. Many natural or human interventions can damage goods in transit. Goods can be lost, stolen, dropped, or mishandled. The goods can be transported at too hot or too cold a temperature, causing deterioration in quality. The chances of goods being lost, stolen, or damaged are much higher when being transported by sea than by air. The transit time with sea transport is not only longer, but goods travelling via sea tend to sit in warehouses or on docks for much longer than goods travelling by air. Water damage and piracy are also much more prevalent for sea transportation.

Although the rates charged by ocean shipping carriers are generally much lower than the charges for air cargo, there is a business case to be made for shipping many products by air instead of sea. The next section discusses the characteristics of shipments that may be most cost-effective to ship by air.

! Stop and think

Discuss the relative merits of transporting goods by air.

17.3 Characteristics of air cargo

The previous sections of this chapter discussed two types of products that are often shipped to their destination by air – consumer electronics such as Apple iPhones and fast fashion clothing items. This section discusses more generally the characteristics of products that lend themselves to transit by air. Air cargo products have four key characteristics.

1 *High value-to-weight ratio*. Gold, for example, would more likely be shipped by air than coal. Insurance considerations aside, the shipping cost for 1,000 kg of gold should be approximately equivalent to the shipping cost of 1,000 kg of coal. However, the purchase cost of coal is about US$50 for 1,000 kg. If it is assumed that gold is trading for US$49,000/kg, the 1,000 kg shipment of gold would, therefore, be valued at US$49 million. Clearly, a shipper sending goods valued at US$49 million would be much less concerned about transport costs and much more concerned about inventory costs and the potential for loss or damage than would a shipper with products valued at US$50. Conversely, the coal shipper would be much more concerned about transit costs, given that these costs may surpass the actual value of the product being shipped. As a result, a 1,000 kg intercontinental gold shipment would most certainly travel by air, while a similar-sized shipment of coal would almost certainly travel by sea. Although a high value-to-weight ratio is a good judge of products that may be suitable for air cargo shipments, some products that fit this description but have low densities are expensive to transport by air. Household products are a good example. Assembled furniture may have considerable value, especially to its owner, and may not weigh that much compared to other similarly sized shipments. This is because assembled furniture consists of mainly empty space. Given space constraints on aircraft, airlines charge for shipments on the basis of **dimensional or volumetric weight**; that is, heavier products are assessed on the basis of their weight, while less dense products are assessed on

Dimensional or volumetric weight: a calculation of the volume of the shipment compared with the actual weight. The higher value is used to calculate the cost.

the basis of their dimensions or volume. Consequently, a product that consumes considerable space will be very expensive to ship by air, even if its weight is relatively low. In general terms, belly-hold freight, which travels in the hold of a passenger aircraft, will tend to be volume constrained, whilst pure freighter aircraft will be weight constrained.

2   *Time sensitive*. A time-sensitive product will quickly diminish in value as time elapses. Fresh flowers, seafood, and certain pharmaceutical products, for example, are perishable. They have a limited shelf-life and will deteriorate quickly as time elapses. Time-sensitive products need to be transported to their destination as soon as possible to maximise returns for their sellers and are often shipped by air.

3   *Uncertain demand*. For example, sophisticated production machinery, such as the kind necessary to operate electric power plants, may break down. Some parts that are prone to regular wear and tear will have readily available replacements on site or close by. Other parts, however, which rarely fail may not be kept close by. The demand for these parts may be difficult to predict, and replacements could be needed only infrequently. However, if these parts are crucial to the operation of the power plant, they need to be ordered and installed as soon as possible. If air freight is the fastest means for acquiring these parts, then they will be transported by air, even at a much higher cost than surface freight alternatives. Cargo airlines have critical rates that are available for products which must fly.

4   As with passengers (➤Chapter 10), product segmentation and differentiation is important. This can range from a distinction being made between premium and non-premium products to a range of different products and services being offered such as constant climate control, secure delivery, carriage of human remains, courier services, and time critical shipments. The transport of different types of products affects not only facilities in the aircraft but also those on the ground, such as warehousing, surface transportation vehicles, and the need for specialised ground handling vehicles.

Even if a product meets all three characteristics, it still may not be shipped to its destination by air. If the product is manufactured only a short distance away from the customer and there is a good highway or railway connecting the two, then the product is much more likely to be shipped by road or rail than by air. Distance between origin and destination and the relative speed and difficulty of using surface transport modes must also be considered. The faster and cheaper the surface alternative, the more likely it is to be chosen over air freight. In some cases, surface transportation can compete with air even over long distances (see Case Study 17.4).

Stop and think

Identify goods that would not normally be transported by air, and discuss the reasons why they might be unsuitable for aerial shipment.

RAIL TRANSPORT VIA CR EXPRESS BETWEEN EUROPE AND CHINA

Asia-Europe air cargo routes have experienced competition from rail express in recent years. Since its inception in 2011, China-Europe Freight Train Service (CR Express) has had rapid development and growth in terms of route coverage, market scope, operating capacity, service quality, and variety of goods carried. Although these trains are considerably slower than air cargo lines, they offer cost savings to shippers of goods that are not extremely time sensitive. Through its various operations, CR Express provides transcontinental rail services over 7,000 miles connecting China, Europe, and countries in Central Asia along the 'Belt and Road' routes. To date, over 28 cities in China have been connected through CR Express to 10 cities in Europe, including Duisburg and Hamburg in Germany, Moscow in Russia, Warsaw in Poland, Pardubice in the Czech Republic, Madrid in Spain, Rotterdam in the Netherlands, Minsk in Belarus, and Lyon in France. The cargo train provides shipping time from China to these destinations in Europe of 16 days, much faster than the sea routes originating from the Chinese sea ports on the Pacific.

A wide variety of goods are carried by CR Express, including laptops, shoes, clothes, furniture, and other non-perishable items on Europe-bound freight trains, while wine, milk, cosmetics, auto parts, and medical equipment are transported from Europe to China. Many of those goods are high-value products that would otherwise be shipped by air. With a planned $43 billion infrastructure investment, it is estimated that by 2020, 42 transport hubs will be developed on 43 CR Express lines, providing an annual capacity of 5,000 cargo trains per year between China and Europe. The lower-priced CR Express will offer air cargo strong competition for its safety, convenience, efficiency, and lower carbon footprint.

17.4 Global air cargo operations

Air cargo traffic differs from air passenger traffic in several important respects:

- Air passengers tend to travel in even flows from Point A to Point B and from Point B to Point A, balancing the flow of traffic on any given city pair. In contrast, most cargo goes in one direction only, giving rise to a unidirectional demand for cargo capacity. For example, the volume of air cargo traffic moving from Asia to North America far exceeds the cargo flow moving westbound. To deal with the problem of unbalanced traffic volume, it is important for cargo operators, especially those with dedicated freighter aircraft, to implement the fifth freedom of the air (➤Chapter 1) on international routes so that they can make multiple stops, picking up and dropping off cargo en route to their final destination. Examples of 'round-the-world' routes are those operated by AeroLogic Boeing 777Fs (Hong Kong–Cincinnati–Bahrain–Hong Kong and Hong Kong–Los Angeles–Leipzig–Hong Kong) and Cargolux (which operates a B747–8F on a three-day trip, Luxembourg–New York–Mexico City–Houston–New York–Lagos, Nigeria–Accra, Ghana–Nairobi, Kenya–Maastricht–Luxembourg).

- Air travellers prefer to fly non-stop, directly to their destinations whenever possible. If a transfer is needed, short connecting times are desired. In contrast, air cargo customers may be indifferent to where a transfer is made as long as the shipment arrives at its destination on time. Several factors may impact the total transit time or dwell time (waiting time for cargo on the ground) of a cargo shipment. These factors include connecting flight schedules, airport congestion, weather conditions, and the capability and efficiency of ground handlers in breaking down and rebuilding air cargo containers/pallets. Unlike air passengers, the transfer of cargo from one flight to another requires additional unloading and reloading but short transit times remain desirable.

- Air cargo revenue management may be more complex than passenger revenue management as a result of cargo's three-dimensional characteristics, greater demand uncertainty, and a larger variety of cargo rates and routing options.

Stop and think

Why do passenger and air cargo services differ?

17.5 Specialised air cargo

Even when the overall growth in air cargo is slow, there are often lucrative and growing markets for shipping specialised products, including live seafood, fresh produce, pharmaceutical goods, and temperature-sensitive goods. According to *Air Cargo World* (2014b), perishable and pharmaceutical goods account for about 12 per cent of the total revenue of the world's air cargo industry and have a much higher growth rate in traffic volume than general cargo. Moreover, the average cargo yield for pharmaceutical goods is about three times as high as that for general cargo. It is estimated by the IATA that the global spending on temperature-controlled biopharma logistics will grow to more than $16 billion by 2021. Case Studies 17.5 and 17.6 describe speciality shipments facilitated by two cargo operators.

Stop and think

Outline the benefits to airlines of specialising in particular types of cargo shipment.

EMIRATES SKYCARGO COOL CHAIN OPERATIONS

To ensure a cool chain environment for perishable goods during a shipment, Emirates SkyCargo has invested millions of dollars in special equipment and facilities both in the air and on the ground. Investments have been made in temperature-controlled containers and temperature- controlled handling and storage facilities. The airline provides three levels of integrated cool chain services for shipping fresh, perishable goods, including Emirates Fresh, Emirates Fresh Breathe, and Emirates Fresh Active, which enable customers to select a mix of different service components including whether the shipment has dedicated ramp operations, quick ramp transfers, thermal blankets, cool dolly services, and dedicated handling and storage at Emirates' Dubai Hub. Similar bundles of products and services are offered to shippers of pharmaceuticals, live animals, valuable/vulnerable products, and specialised cargo. In 2018, Emirates SkyCargo carried over 73,000 tonnes of pharmaceuticals and more than 400,000 tonnes of fresh produce, meat, and seafood products using its 11 dedicated Boeing 777–200 freighter aircraft and the belly cargo capacity on the airline's passenger flights to 143 destinations in 78 countries worldwide. Emirates SkyCargo was given the Diamond Award for first place in Air Cargo Excellence in the large carrier category (more than 1 million tonnes of cargo annually), based on the surveys conducted by *Air Cargo World*.

SHIPPING SEAFOOD BY KOREAN AIR

As a leading country in lobster fishing, processing, and shipping, Canada exports lobsters to 50 countries, and the lobster industry is estimated to have a US$1.7 billion annual impact on the country's economy. In July 2014, Korean Air started a cargo route between its hub at Seoul Incheon and Halifax, Canada, using Boeing 777F aircraft shipping a minimum of 40 tonnes of live lobster every Sunday during the lobster harvesting season. This new cargo service is offered to leverage the growing appetite for live lobsters among Asian consumers. Trade has been further enhanced given the signing of a free trade agreement between Canada and South Korea in September 2014.

17.6 Airports as air cargo logistics nodes

Airports provide a vital connection for the movement of air cargo. To accommodate the continued expansion of freighter aircraft and the growing adoption of cargo-friendly wide-body aircraft, several airports worldwide have expanded or modified their infrastructure. These investments are designed to attract more cargo. Case Study 17.7 details how Hong Kong Airport is facilitating growth in air cargo. Table 17.4 provides a list of the world's top 20 cargo airports.

CASE STUDY 17.7

AIRPORTS AND AIR CARGO

In 2010, Memphis International Airport (MEM), the primary hub airport for FedEx, was surpassed by Hong Kong International Airport (HKG) in terms of cargo volume. Since then, HKG has remained the world's largest airport by cargo throughput (measured in metric tonnes) in the world. HKG now handles about 37 per cent of Hong Kong's international trade by product value and has a total of 7.4 million tonnes of annual cargo handling capacity. The estimated cargo volume at HKG was over 5 million metric tonnes in 2018. The airport offers two tiers of cargo handling services. The first-tier services are provided through four cargo-dedicated terminals for cargo loading/unloading from aircraft, while the second-tier facilities include the Air Mail Centre, Airport Freight Forwarding Centre, Marine Cargo Terminal, and Tradeport Logistics Centre. These centres are designed to provide multimodal value-added logistics services. HKG and its primary cargo handler, Hong Kong Air Cargo Terminals Ltd., are the initial participants in the Center of Excellence for Independent Validators (CEIV) for fresh and perishable products, known as CEIV Fresh, an IATA programme to enhance industry-wide standards for shipping and handling seafood, fruit, meat, vegetables, and other perishable goods. The recent completion of the Hong Kong–Macau–Zhuhai bridge will help Hong Kong maintain its gateway status for e-commerce and cargo trans-shipments for high-value cargo to Mainland China, and in particular to the Pearl River Delta region, as the surface travel time from Hong Kong to Zhuhai is reduced from two hours to 40 minutes.

Table 17.4 The world's top 20 cargo airports ranked by metric tonnes in millions, 2018 (2008)

Rank	Airport	Cargo	Change	Rank	Airport	Cargo	Change
1 (2)	HKG, Hong Kong, HK	5.121 (3.661)	39.88%	11	DOH, Doha, QA	2.198 (0.441*)	398.41%
2 (1)	MEM, Memphis, US	4.470 (3.695)	20.97%	12 (10)	SIN, Singapore, SG	2.195 (1.884)	16.51%
3 (3)	PVG, Shanghai, CN	3.769 (2.603)	44.79%	13 (7)	FRA, Frankfurt, DE	2.176 (2.111)	3.08%
4 (4)	ICN, Incheon, KR	2.952 (2.424)	21.78%	14 (6)	CDG, Paris, FR	2.156 (2.280)	-5.44%
5 (5)	ANC, Anchorage, US	2.807 (2.340)	19.96%	15 (12)	MIA, Miami, US	2.130 (1.807)	17.87%
6 (11)	DXB, Dubai, AE	2.641 (1.825)	44.71%	16 (18)	PEK, Beijing, CN	2.074 (1.366)	51.83%
7 (9)	SDF, Louisville, US	2.623 (1.974)	32.88%	17 (26)	CAN, Guangzhou, CN	1.891 (0.686)	175.66%
8 (15)	TPE, Taipei, TW	2.323 (1.493)	55.59%	18 (19)	ORD, Chicago, US	1.807 (1.332)	35.66%

Continued

Table 17.4 Continued

Rank	Airport	Cargo	Change	Rank	Airport	Cargo	Change
9 (8)	NRT, Tokyo, JP	2.261 (2.100)	7.67%	19 (16)	LHR, London Heathrow, UK	1.771 (1.486)	19.18%
10	LAX, Los Angeles, US	2.210 (1.630)	35.58%	20 (14)	AMS, Amsterdam, NL	1.738 (1.603)	8.42%

Sources: ACI-NA World Airport Traffic Report, September 2019, retrieved from https://aci.aero/news/2019/09/17/worlds-top-five-fastest-growing-airports-for-passengers-and-cargo-revealed/ on 16 October 2019; ACI Cargo Traffic Summary 2008 Final, retrieved from https://aci.aero/data-centre/annual-traffic-data/cargo/2008-final-summary/ on 16 October 2019

* DOH is not in the top 30 cargo airports in 2008, and its cargo traffic volume in 2008 was retrieved from FlightGlobal.

17.7 Air cargo: challenges and opportunities

Safety and security

An important challenge is to provide optimum security for cargo shipments and the least possible disruption and delay to the flow of goods. In October 2010, explosives were found in printer cartridges carried on a freighter aircraft bound for the US. This incident raised concerns about the vulnerability of air cargo to security threats. In response, the EU imposed a new security rule that requires a carrier not to accept cargo unless it has been fully screened and secured and the carrier is aware of the contents, the origin, and the transit history of the shipment. Other countries have similar rules.

Airlines are also examining ways to enhance the safety of their shipping process. In some cases, regulatory agencies will forbid the shipment of products if they are deemed to be safety risks. This has been the case with some shipments of lithium batteries. Hong Kong-based Cathay Pacific has developed methods to ship these batteries safely (see Case Study 17.8).

CASE STUDY 17.8

LITHIUM BATTERIES

While lithium batteries are popular power sources for a wide variety of consumer products, they present a major safety concern due to fire risk when not packaged or shipped properly. The fatal crash of the UPS 747–400F flight from Dubai to Cologne on 3 September 2010 was due to an uncontained fire. The fire was found to start in the main-deck section of the aircraft, which was carrying a large number of lithium batteries and other combustible materials.

In February 2016, the ICAO Council adopted a recommendation that prohibits the carriage on passenger aircraft of lithium batteries when not contained in or packed with equipment. In addition, the rule sets 30% of the rated capacity as the maximum charging volume for batteries carried on board freighter aircraft. However, lithium batteries must still be transported from manufacturing facilities to customers. As a result, Cathay Pacific developed protocols for the safe transport of standalone lithium batteries. With a new fire-containment bag designed by Cathay Pacific, the airline has now been approved by the US Federal Aviation Administration (FAA) to carry lithium batteries on its freighter aircraft to the US.

The environment

The global aviation industry accounts for about 2 per cent of all human-generated carbon dioxide (CO_2) and 12 per cent of the carbon emissions produced by the transport sector. Given that air transport carries only 3 per cent of freight (by weight), its carbon footprint is much larger per tonne carried than the footprints of other modes. Emissions from airside vehicles and volumes of plastic waste and wrapping are other significant environmental challenges (➤Chapter 18). The air cargo industry is committed to decreasing its carbon emissions. Many cargo operators are replacing old, inefficient aircraft with more fuel-efficient ones. The greening of the air freight business has been embraced by air freight forwarders, and some shippers are offered the opportunity to participate in carbon-offset programmes.

Innovations

Air cargo operators have been innovative in the use of technology for assuring on-time, fast, and reliable cargo delivery. Modernising documentation (e.g. with electronic air waybills (eAWBs)) has been a goal of the air cargo industry for some time. Waybills provide key information on the cargo that is being shipped, such as shipper, destination, and contents. In January 2015, over 75 per cent of AWBs were still issued in paper format. By October 2018, penetration of electronic AWBs had reached over 58 per cent. E-commerce is also extending to online cargo bookings, and airlines are now developing their own dedicated cargo websites. Automation is also extending to the use of autonomous ground handling vehicles which can improve the precision and efficiency of aircraft turnarounds.

Other innovations are connected with new transportation routes. For example, DHL Express has launched helicopter services in Los Angeles, New York, and London to move important documents and small packages from major airports directly to a heliport in the city centre to avoid traffic congestion. Drones are also being developed and introduced by cargo shippers. Case Study 17.9 provides a description of potential uses for drones from flights inside warehouses to long-haul routes.

CASE STUDY 17.9

DRONE SHIPMENTS

Nemo is a small California-based drone company established in 2016. The company aims to develop a (relatively) large, long-distance commercial drone, similar in carrying capacity to a Boeing 777 freighter aircraft, to transport cargo globally at half of the current air freight cost. Flights by these cargo drones will enable shippers to transport 90-tonne cargo loads between Los Angeles and Shanghai in 30 hours at a cost of $130,000. This is much faster than the three-week shipment by sea, and 50 per cent less costly than shipping the goods by air freighter.

Many companies, such as Boeing, Amazon, DHL Express, and EHang, are at different stages of developing and testing propeller-based drones that can be used in both urban and rural areas for last-mile delivery as a substitute for surface transportation. In addition, GEOGIS, a logistics company, and DELTA Drone, a French UAV company, have started producing 'plug and play' drones designed for warehouse operations, designed to fly at low altitudes for short distances within a building.

In summary, a variety of drones are being developed to suit many purposes with carrying capacities ranging from a few kilos to many tonnes. Together, they will be able to reduce the cost of air freight and provide time-definite deliveries in both crowded urban and remote rural areas.

Hybrid transport such as sea-air, rail-air, and express truck-air may become more popular as their complementary services provide customers with increased speed, cost efficiency, and market reach. Strong trucking networks are already in evidence in parts of Europe, the US, and Asia. Many shippers can now choose multimodal shipping options that include air services. For example, Athens International Airport has collaborated with COSCO, a Chinese shipping company. Goods are shipped via COSCO to the seaports in Greece. On arrival in Greece, the goods are trucked to an airport and then shipped to Northern Europe or elsewhere by air.

17.8 The future of air cargo

Despite the challenges facing the air cargo industry, the outlook is optimistic. With the increasing growth of internet commerce, the continuing importance of global trade, the growth in sales of high-value consumer electronics, the shrinking size of many products (making them more amenable to air shipment, such as mobile devices that have supplanted laptops, which in turn replaced desktop computers), shorter product life cycles, and the emphasis on customer service by shippers, there will be a central place for air cargo in the future transport system.

!

Stop and think

What might the global air cargo industry look like in the future?

Key points

- Air cargo is an important but often overlooked aspect of the commercial air transport industry.

- Air cargo enables the routine and on demand movement of high value-to-weight, time-sensitive, and/or perishable products from their point of origin to their point of consumption.

- Air cargo services are provided by dedicated cargo airlines, integrators, and passenger–cargo combination carriers.

- Air cargo can be shipped as belly-hold freight on scheduled passenger flights or as pure freight on freighter aircraft. Outsized and hazardous cargo may be transported by specialist operators.

- Air cargo offers a number of advantages over surface modes, including faster delivery, lower transit times, and reduced likelihood of theft and mishandling.

- Airports that handle large volumes of air cargo need specialist warehousing and customs facilities, and a number of airports have developed to primarily serve the particular needs of air cargo operators.

- Air cargo faces a number of challenges, including the introduction of new security directives, volatile fuel prices, and global economic uncertainty, which have depressed demand and led to a rise in alternative forms of cheaper land and sea transport.

- Air cargo needs to be flexible in its approach and innovative in its business approach to take advantage of new opportunities that are presented by changing patterns of consumer demand and the introduction of new technology, for example drones and airships..

References and further reading

Air Cargo World. (2014a). *Halifax sends live lobsters to Asia.* Available at: http://aircargoworld.com/halifax-sends-live-lobster-to-asia-9800/

Air Cargo World. (2014b). *ESC proposes changes to airfreight*, 104(4), p. 19.

Air Cargo News. (2019). *Top 25 Cargo Airlines.* Available at: www.aircargonews.net/airlines/top-25-cargo-airlines-fedex-at-the-top-as-qatar-closes-in-on-emirates/

Casadesus-Masanell, R. and Tarziján, J. (2012). when one business model isn't enough. *Harvard Business Review*, January. Available at: https://hbr.org/2012/01/when-one-business-model-isnt-enough

IATA. (2019). *Press release no. 46 air freight in decline for eighth consecutive month.* Available at: www.iata.org/pressroom/pr/Pages/2019-08-07-01.aspx

Lee, J. (2019). Cathay freighters to carry lithium batteries again. *Cargo Facts*, August 1, Available at: https://cargofacts.com/cathay-freighters-to-carry-lithium-batteries-again/

Mazareanu, E. (2019). Number of e-air waybills worldwide from January 2016 to July 2019, by month, Available at: www.statista.com/statistics/531554/volume-of-e-air-waybills-by-month-worldwide/

Sandler, E. (2017). Cargo drone startup Natilus reveals first "nemo" design. *Air Cargo World*, 107(7), p. 8.

Woods, R. and Kauffman, C. (2019). DHL express, Boeing test refined smart drones. *Air Cargo World*, 109(5), p. 8.

Worstall, T. (2013). *It's cheaper to send Apple's iPhones by air than by sea.* Available at: www.forbes.com/sites/timworstall/2013/09/12/its-cheaper-to-send-apples-iphones-by-air-than-by-sea/

Xinhua News. (2017). Infographic: China-Europe freight train service. Available at: www.xinhuanet.com/english/2017-04/21/c_136225717.htm

Ziobro, P. and Mattioli, D. (2019). FedEx to end ground deliveries for Amazon. *The Wall Street Journal*, August 30. Available at: www.wsj.com/articles/fedex-to-end-u-s-ground-deliveries-for-amazon-11565182535

CHAPTER 18

Environmental impacts and mitigation

Thomas Budd

LEARNING OBJECTIVES

- To appreciate why the environment represents such a challenging issue for aviation.

- To identify the principal sources of pollutants and the resulting environmental impacts of air transport operations at global and local levels.

- To examine the measures that can be adopted to help mitigate air transport's environmental impacts.

- To explore the impact of a changing climate on air transport operations and the adaptation measures that may be adopted.

18.0 Introduction

In addition to promoting international connectivity and trade, air transport creates a range of adverse environmental impacts at global and local levels. Reducing the severity of these impacts is one of the key challenges facing the industry and society as a whole. The chapter begins by setting the context for contemporary debates about air transport and the environment. The key global and local environmental impacts of aviation, including climate change, local air quality, noise, and energy, water, and waste are then examined. This is followed by a discussion of how these issues can be addressed. This includes a discussion of policy interventions and the role of new technology. The chapter concludes by considering how a changing global climate will impact on air transport.

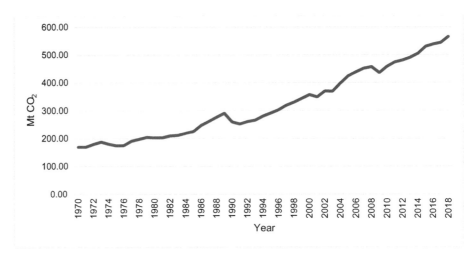

Figure 18.1 CO_2 emissions from international aviation, 1970–2018
Source: Crippa et al. (2019)

18.1 Air transport and the environment

In 2018, commercial aircraft consumed 95 billion US gallons (approximately 359 billion litres) of fuel and emitted 905 million tonnes of carbon dioxide (CO_2) into the atmosphere, accounting for approximately 2 per cent of total global CO_2 emissions (IATA, 2019a). Although technological advances in airframe design and materials used in airframe manufacturing mean that modern aircraft are up to 80 per cent more fuel-efficient and up to 75 per cent quieter than the first generation of jet aircraft (ATAG, 2018), these improvements are being negated by rapid air traffic growth. Between 1970 and 2018, CO_2 emissions from *international* aviation increased from 169.2 Mt (million tonnes) to 564.6 Mt (Figure 18.1).

By 2050 ICAO forecasts that they could increase by as much as 400 per cent (Fleming and de Lépinay, 2019). The long life-cycle of aircraft (typically 25–30 years in service) and technological lock-in mean that it can take many years for technological changes and design innovations to become commonplace in the global fleet.

As a consequence, it is anticipated that aviation's share of global carbon emissions will increase as traffic demand grows and other sectors of the global economy, including power generation (currently the largest source of GHGs worldwide) and road transport (currently the largest generator of GHG emissions from the transport sector worldwide), decarbonise and transition to cleaner renewable energy sources including wind and solar power. In Europe in 2016, aviation accounted for 3.6 per cent of all **greenhouse gas (GHG)** emissions, and 13.4 per cent of emissions from all transport modes (EEA et al., 2019). In the UK, air transport accounts for around 7 per cent of the country's GHGs, but this share may increase to as much as 25 per cent by 2050 as other sectors decarbonise (DfT, 2018).

In recognition of air transport's effects on the global climate, IATA adopted the following CO_2 emissions targets in 2009:

- an average fuel efficiency improvement of 1.5 per cent per year from 2009 to 2020;

Technological lock-in: the process by which prevailing social, economic and/or regulatory systems create barriers to technological change and innovation.

Greenhouse gases (GHGs): gases which contribute to global heating by absorbing infrared radiation and warming the earth's surface.

- a cap on net aviation CO_2 emissions from 2020 (carbon neutral growth);

- a reduction in net aviation emissions of 50 per cent by 2050, relative to 2005 levels.

Meeting, and preferably exceeding, these three targets demands innovation in technology (including more sustainable fuels), more efficient operations, improved infrastructure, and the creation of a single global market-based measure (see Section 18.4). Collectively, these measures form the four pillars of IATA's aviation climate change strategy (IATA, 2019b).

18.2 Sources of pollution

In order to create sufficient thrust and lift to overcome the effects of aircraft weight and aerodynamic drag, aircraft engines burn kerosene – an energy-rich fossil fuel. Modern high-bypass jet engines work by sucking in large volumes of air through the front fan. A high proportion of this cold air bypasses the central engine core and is mixed with hot exhaust gases at the rear to create additional thrust. The remaining air enters the central compressors and combustion chamber where fuel is injected and ignited. The resulting **exothermic** reaction produces a range of gaseous and solid pollutants – including carbon dioxide (CO_2), nitrous oxides (NO_x), water vapour (H_2O), and particulate matter (PM) – which change the chemical composition of the atmosphere and generate different effects depending on the location and altitude at which they are released. Aircraft also generate noise pollution from the engines and airframe. Aircraft-derived atmospheric and noise pollution are inherently mobile, whereas aviation's other environmental impacts occur at fixed locations at and around airports.

Exothermic: a reaction which generates heat

18.3 The environmental impacts of air transport

Air transport's environmental impacts occur at both a global and a local level, but they are inextricably linked (Table 18.1). Local impacts (particularly noise and odour) are concentrated at particular locations, are experienced immediately and are likely to demand instant attention and action. Global impacts, in contrast, are dispersed and take longer to manifest themselves.

Table 18.1 Air transport's principal environmental effects

Global impacts	Local impacts
Climate change, leading to:	Local air quality
• rising sea levels	Noise
• changing frequency of storms	Odour
• changing severity of storms	Pollution from energy generation
• changing precipitation patterns	Waste generation and disposal
• higher average and extreme temperatures	Water management (particularly for airports located in water stressed areas) and contaminated surface run-off from the airfield

Climate change

Kerosene is the primary fuel source for aircraft. When burnt in aircraft engines, carbon dioxide (CO_2), water vapour (H_2O), nitrogen oxides (NO_X), particulates (PMs), and other pollutants are produced (Box 18.1). The majority of these emissions are released at high altitudes (typically 30,000–39,000 ft), where their impact on the climate is greater than if they were released at sea level. CO_2, water vapour, and NO_X are greenhouse gases (GHG) which contribute to global heating and **climate change** by absorbing infrared radiation from the sun and warming the planet. Their impacts are assessed using a metric called **radiative forcing (RF)**, which is measured in Watts per square metre (W/m^2). RF measures the influence a particular pollutant has in altering the balance of incoming and outgoing energy in the atmosphere. RF can either be positive (leading to atmospheric warming) or negative (leading to atmospheric cooling). Given the growth in the air traffic and aviation's contribution to rising GHG emissions, managing, and reducing, aircraft emissions is a key goal that the air transport industry is seeking to address through the development of new technology (including more sustainable fuels), enhanced operating procedures and the introduction of a global market-based measure (Section 18.4).

Climate change: long-term alterations in the global climate.

Radiative forcing (RF): a measure of the influence a particular pollutant has in altering the balance of incoming and outgoing energy in the atmosphere.

Box 18.1

Key aircraft emissions and their effects on the global climate

Carbon dioxide (CO_2). Burning 1 kg of aviation fuel creates 3.16 kg of CO_2. CO_2 is a GHG which has a strong positive effect on RF regardless of the location or altitude at which it is emitted. CO_2 has the longest atmospheric life of any aviation-derived GHG, and it is the key pollutant the air transport industry is trying to reduce to lessen the severity of climate change.

Water vapour (H_2O). Water vapour can promote the formation of condensation trails (or contrails), the white linear clouds that form in cold humid conditions at high altitudes (low level contrail formation is rare) when water vapour condenses around soot particles in the engine plume. It is thought that water vapour leads to positive RF as it traps outgoing long-wave radiation.

Nitrous oxides (NO_x). Reactive nitrogen (NO_x) is emitted from aircraft engines in the form of nitrogen oxide (NO) and nitrogen dioxide (NO_2). These alter levels of ozone (O_3) and methane (CH_4) in the atmosphere. CH_4 is a potent GHG implicated in global heating.

Particulate matter. The two main classes are black carbon (soot) and sulphate particles from aircraft engines. Although their combined RF impact is relatively small, soot particles absorb radiation and have a warming effect, while sulphate particles reflect radiation and have a slight cooling effect. Particulates may also lead to increased cirrus cloud formation, as water vapour from the engine plume condenses around them to form contrails.

Stop and think

Why is aviation's contribution to climate change such an important issue?

!

Local air quality

In addition to high altitude emissions, aircraft also generate emissions when they are landing and taking off and taxiing on the ground. Owing to the changing thrust settings that are required during individual stages of the **landing and take-off (LTO) cycle**, different concentrations of pollutants are produced, but all of them degrade local air quality around airports. Local air quality is also affected by emissions from ground access transport vehicles (principally private cars and taxis), power generation, and vehicles on the airfield.

The most significant local emissions are nitrous oxides (NO_x), volatile organic compounds (VOC), and particulate matter (PM). When emitted at low altitudes, NO_x promotes the formation of petrochemical smog and ozone (O_3). Although ozone forms a vital natural shield from incoming ultraviolet radiation high in the atmosphere (the ozone layer), at ground level it is hazardous to human health and can lead to respiratory problems, including bronchitis, asthma, and emphysema. VOCs are a class of toxic chemicals, some of which are known to be carcinogenic (cancer causing) to humans. An indicator of poor local air quality around airports is often a strong chemical odour, and some airports have systems to monitor and record complaints made about odour by local communities. PM is classified according to the diameter of the particle in microns (μm). These range from PM_{25} to ultrafine PM_1 and $PM_{0.1}$. Generally, the smaller the particle size, the worse the health impact as they can be inhaled, deposited in the lungs and passed into the bloodstream.

Air quality is measured according to the atmospheric concentrations of individual pollutants at different locations around an airport. These concentrations are commonly expressed in terms of micrograms per cubic metre ($\mu g/m^3$) and are automatically recorded at air quality monitoring stations positioned around an airport. In the same way that noise contours can be developed to reflect levels of aircraft noise, air quality contours can be used to show spatial variations in air quality around an airport. For simplicity and to aid dissemination, air quality is usually reported on a specified scale or index. In the UK, air pollution is reported on a 1–10 scale from 'Low' to 'Very High' to indicate relative pollution levels and associated risks to human health. Many airports, including London Heathrow, have dedicated websites showing current air quality (Case Study 18.1).

Landing and take-off (LTO) cycle: covers four modes of engine operation – idle, approach, climb out and take-off – each of which is associated with a specific thrust setting.

HEATHROW AIRWATCH WEBSITE

The Heathrow Airport Airwatch website (www.heathrowairwatch.org.uk) and local air quality monitoring scheme provides publicly available reports on the current and previous air quality at 26 monitoring stations around the airport. The scheme is funded by a joint partnership between Heathrow Airport Ltd, neighbouring Local Councils and London Boroughs, and British Airways.

Concerns about local air quality have resulted in tighter regulatory standards being imposed. Although air quality regulation varies by country, airport operators increasingly record and report emissions in terms of Scope 1, 2, or 3 emissions (these are not to be confused with Scope 1, 2, and 3 emissions reduction for CO_2):

- **Scope 1.** Direct emissions from sources owned and/or controlled by the airport, including airside vehicles or airport buildings.

- **Scope 2.** Indirect emissions, predominantly from off-site electricity generation (for use at the airport).

- **Scope 3.** Indirect and/or optional emissions that result from the airport's operation. This includes aircraft and ground access transport emissions.

Scope 3 emissions represent the largest share of an airport's total emissions (as much as 95 per cent). They are also the ones over which the airport operator generally has the least control. This is due to the numerous different tenant companies, government agencies, public transport operators, ground handling agents, airlines, and passengers involved in airport operations. This can make the task of reducing Scope 3 emissions challenging.

!

Stop and think

Why should airport operators be concerned about local air quality, and what can they do to improve it?

Noise

Noise is another major local environmental issue at airports, the impact of which is most likely to provoke action and complaint from local airport communities. Aircraft noise (see Example 18.1) has been an emotive, politicised, and controversial issue since the introduction of the first jet aircraft in the 1950s. Although modern aircraft are individually up to 75 per cent quieter than earlier generations of aircraft, rising numbers of flights, combined with increased community awareness of, and sensitivity towards, noise as well as expanding residential communities near airports, means that it remains a major issue. Limiting or reducing the number of people exposed to aircraft noise is a key priority.

Example 18.1
What's in a name? Sound versus noise

Sound describes acoustic vibrations that can be detected by the human ear, whereas noise is unwanted or intrusive sound. Sound can be measured objectively, in terms of levels of acoustic energy, typically measured in terms of average sound exposure in decibels over a given period of time (dB(A)). Noise, on the other hand, is highly subjective, culturally specific and difficult to quantify. The way in which different sounds are perceived, and framed, depends on a host of factors relating to its source and the time and place in which it occurs. For example, while sound from road traffic and industrial operations are typically considered intrusive, in the past they were considered to represent the sounds of prosperity and economic development.

Exposure to aircraft noise causes a range of adverse physiological effects, including increased blood pressure and sleep disturbance, the severity of which is influenced by factors including an individual's age, their general health, social conditions, lifestyle characteristics, and the time of day or night when the noise event occurs. Since annoyance is subjective, there may be little direct correlation between objectively recorded noise and levels of community annoyance.

One of the challenges associated with aircraft noise is the number of different methodological techniques and metrics that can be used to quantify it (Example 18.2). The simplest measure of a noise event, such as an aircraft flying overhead, is the maximum sound level that occurred during the event. This is measured in dB(A) (the highest noise level), where the greater the value, the greater the risk of disturbance. The sound exposure level (SEL) of a noise event is the sound level in dB(A) that would be recorded if the energy of the entire event was compressed into a constant sound level lasting one second. However, measuring individual aircraft noise events in this way does not assess the full impact of the noise exposure, as it does not take account of the combined impact of many aircraft over longer periods of time or the effects of noise at anti-social times of the day.

Example 18.2
Measuring noise

In the UK, aircraft noise is measured as an 'equivalent continuous sound level', or Leq. This reflects the average sound levels at a specific location between 07:00 and

23:00 local time. Following a number of studies, the UK Government stated that an equivalent continuous sound level of 57 dB(A) represents the onset of 'significant noise annoyance.' In the EU, a similar measure is used, but the average Leq is taken over a full 24-hour period and additional weightings of 5 dB(A) are applied for noise events that occur during the evening (19:00–23:00) and 10 dB(A) for noise events during the night (23:00–07:00). The results of these noise monitoring exercises are typically presented as contours on a map relative to the airport and indicate both the extent of the noise and the local communities affected.

Although it is possible to objectively measure levels of acoustic energy, assessing their impact on people (in terms of stress, annoyance, and health impacts) is far more challenging. The difficulty of measuring the non-acoustic impacts of aircraft noise, combined with wider issues surrounding levels of trust between airports and local communities, has often contributed to the formation of airport protest groups who seek to limit current operations and/or prevent future expansion. Managing community expectations of noise represents a key management issue for airports.

!

Stop and think

Why is aircraft noise contentious, and what are the challenges associated with measuring it?

Energy, water, and waste

Air transport's environmental impacts also extend to energy, water, and waste.

Energy

Airports use electricity for lighting, heating, and air conditioning and to power computers and equipment in the terminal and on the airfield. In recognition of the CO_2 impacts of conventional power generation, airports are increasingly investing in on-site renewable power sources (including solar, wind, and biomass), reconfiguring terminal buildings to provide more natural light and ventilation, and replacing incandescent and fluorescent lightbulbs in the terminal and on the airfield with more energy-efficient LEDs (Case Study 18.2).

CASE STUDY 18.2

CHP AT HEATHROW AIRPORT

At London's Heathrow Airport, a 10 MW wood chip fuelled Combined Heating and Power System (CHP) supplies around 20 per cent of the energy to Terminals 2 and 5, and saves an estimated 13,000 tonnes of CO_2 per year compared with the same electricity being sourced from the national grid. Other technological innovations for reducing energy consumption include 'smart' energy monitoring and control systems for heating and lighting. Such systems also allow escalators and walkways to operate on demand, rather than continually running 24 hours a day.

Water

Airports consume large volumes of water for drinking, catering, cleaning, and sanitary purposes. Reduced flush lavatories and rain and grey water collection and recycling systems can be employed to reduce demand for fresh water.

Airports must also be able to quickly drain rain water away from runways and taxiways. The run-off must be intercepted and managed, as it may be contaminated with aircraft fuel, hydraulic fluid, de- and anti-icing compounds, and fire-retardant foam from emergency simulation exercises. **Balancing ponds** are required to hold and treat contaminated water before it can be discharged. Any failure in the contaminated water management system can be lengthy, expensive to resolve, and potentially hazardous to local ecosystems and human health. Table 18.2 provides a summary of other water management issues at airports.

Balancing pond: an element of a drainage system that is used to control flooding by temporarily storing water before it can be treated and discharged.

Table 18.2 Summary of water management issues at airports

Main sources of water consumption	Areas requiring monitoring and management
Potable (drinking) water	Airfield and terminal hard surfaces and run-off
Catering and retail operations	Aircraft maintenance and ground handling
Cleaning (both buildings and aircraft)	Washing parked aircraft on stand
Toilets and hand basins	Airfield maintenance activities
Airfield and terminal maintenance	Winter operations and de- and anti-icing
	Fire service training

Waste

Air transport operations also generate large quantities of solid waste from airside, landside, infrastructure development and in-flight operations (Table 18.3). Airport operators will typically have most direct control over waste originating from offices, engineering, security, and airfield maintenance (such as grass clippings and rubber from aircraft tyres that is removed from runways). Airlines, or their designated service providers, are responsible for the waste that is generated from in-flight catering, cleaning, and retail activities. Worldwide, it is estimated that 5.7 million tonnes of cabin waste is produced every year, which costs $500 million US to process (IATA, 2019c). Although recycling rates are improving, the international sanitary regulations which are imposed by individual countries mean that it is often illegal to recycle or compost certain types of waste in case it contains (or has been contaminated with) food products (Case Study 18.3). Consumer awareness of the need to reduce the use of single use plastics (SUPs) and the introduction of SUP bans are encouraging service providers to review their waste management practices and, where possible, introduce more sustainable material.

Table 18.3 Summary of waste sources at airports

Source	Example types of waste
Airside	Foreign object debris (FOD) such as baggage straps and litter
	Green waste (grass clippings) from airfield maintenance
	Hazardous waste such as oil/hydraulic fluid
Terminal	Office waste
	Security and check-in operations (including discarded personal items)
	Packaging, food/catering, retail
	Human waste
Landside	Food/catering, retail
	Ground transport
	Human waste
Infrastructure development	Construction waste
	Inert/hardcore waste
In-flight operations	Food/catering/cabin waste
	Newspapers and magazines
	Human waste
	Retail

Large quantities of refuse from retail operations, hardcore and specialist waste from construction and maintenance activities, and toxic waste in the form of obsolete electrical and security screening equipment and chemicals are also generated. Some of this can be reclaimed and recycled, whilst other materials will require special treatment and processing. Airports will seek to prevent the creation of waste at source but, where it is unavoidable, the reuse, recycling, or recovery of energy from waste is preferable to disposal, which has both environmental and cost implications.

INTERNATIONAL CATERING WASTE (ICW) AT UK AIRPORTS

ICW from international flights arriving in the UK is classified as a high-risk category 1 animal by-product and, by law, it must be incinerated or sent to deep landfill and not be composted or used in biogas plants in case it is contaminated with foreign pathogens. This regulation extends to plastic drink containers which would otherwise be recyclable but which may have come into contact with milk products that have been produced outside the EU. The airline or owner of the aircraft is responsible for ensuring compliance with the regulation.

18.4 Environmental mitigation policy

Given the scale and diversity of air transport's environmental impact and the absence of quick and easy methods of decarbonisation, it is imperative that mitigation measures are developed and implemented. New forms of electric or hybrid aircraft will not be able to transport commercial volumes of passengers in the near future. Refining capacity for Sustainable Aviation Fuel (SAF) is insufficient to meet present needs (Case Study 18.4), **electrofuels** are at the early stages of development, and other proposed technological and aircraft design innovations (Table 18.4) do not, at present, offer the step change in environmental performance that is required. The opportunities for mitigating aviation's environmental impact in the short term fall into three categories – enhancing operations through regulation and more efficient use of existing technology and infrastructure (this includes retrofitting existing aircraft with more modern technology); introducing market-based measures to modify consumer and operators' behaviour; and encouraging changes in consumer behaviour through public information campaigns.

Electrofuel: a liquid hydrocarbon that is produced from carbon dioxide and water using electricity. If renewable electricity is used and CO_2 is captured from the air, in theory a zero carbon fuel can be produced.

SUSTAINABLE AVIATION FUEL (SAF)

SAF offers an opportunity to reduce air transport's reliance on kerosene. There are different types of SAF.

Biofuel

Biofuels are derived from renewable biological feedstocks. These feedstocks absorb CO_2 as they grow, which acts to offset the carbon released from aircraft operations when the fuel is burnt and also the processing and transportation of the fuel. While not carbon neutral, biofuels have a lower carbon footprint. Advanced biofuels can be generated from waste and residues.

The first flight using blended biofuel occurred in 2008 and, since then, over 150,000 flights have been powered wholly or partly by biofuel. However, in 2019 only Bergen, Brisbane, Los Angeles, Oslo, and Stockholm airports had regular biofuel distribution, and the 15 million litres of biofuel that was produced in 2018 accounted for a very small proportion of aviation's total fuel consumption (IEA, 2019; ICAO, 2019). To increase biofuel use, it will be necessary both to increase production and to make biofuels cost competitive with kerosene.

Synthetic e-fuels (electrofuel) or Power to Liquid (PtL) fuels

Fuels are created by using (ideally renewable) electricity to extract hydrogen from water through electrolysis. The resulting hydrogen is then combined with CO_2 captured from the atmosphere, to produce a liquid hydrocarbon fuel that can be 'dropped in' to existing fuel lines and engines without any modification.

In order to act as a drop-in replacement for kerosene, SAFs must:

1 Be capable of being mixed with conventional jet fuel, use the same infrastructure and supply lines, and require no adaptation to current aircraft or engines.
2 Match or exceed the chemical specifications and performance of aviation kerosene, particularly in respect to energy content (minimum of 42.8 MJ/kg) and cold resistance (–47°C).
3 Deliver lifecycle carbon reductions (and, in the case of biofuel, require little use of fresh water resources and do not compete with food crops or necessitate land use changes).

Table 18.4 Selected technological and design innovations that could improve environmental performance

Airframe design	• Raked wingtip or wingtip fences to reduce drag • Laminar flow technology to reduce drag • Blended/hybrid wing body aircraft
Engine design	• Ultra High Bypass Ratio Engines • Open Rotor or Propfan Engines
Alternative power sources	• Electric and hybrid propulsion • All electric aircraft • Hydrogen aircraft
Sustainable aviation fuels	• Biofuels or electrofuels made from sustainable biomass
Airport operations	• Electric aircraft tugs • 'Smart' technologies in terminals

Enhanced operations, infrastructure management, and regulation

Enhanced air traffic management and operational procedures can improve the efficiency of flight routings and trajectories to reduce fuel burn (▶Chapter 15). On the ground, single-engine taxi operations and sophisticated pre-departure sequencing and Airport Collaborative Decision Making (A-CDM) can improve ground movement coordination to minimise taxiway holds and reduce fuel burn and emissions. On stand, fixed electrical ground power (FEGP) units and electric aircraft tugs can replace noisy and polluting **auxiliary power units (APUs)** and diesel-powered tugs.

Airports are also actively seeking to reduce the share of passengers, employees, and other airport users who access the airport by private vehicles and increase mode share by public transport (➤Chapter 6). For passengers, particular focus is on reducing the number of drop-off/pick-up journeys (as these generate additional vehicle trips and exacerbate problems of congestion) and increasing public transport ridership. However, car parking is an important revenue stream for airport operators which can lead to an incompatibility between commercial and environmental objectives. Airport operators also target employee surface access travel to reduce single occupancy vehicle use through financial and practical incentives including interest-free loans for purchasing public transport season tickets, preferential parking arrangements for employees who car share, and dedicated secure cycle parking and shower facilities. A key challenge for staff travel, however, is that public transport services may be limited or not run 24 hours a day. Overall, however, these only represent small improvements in environmental performance, and so they must be considered as part of a package of measures which collectively seek to reduce emissions and lower air transport's environmental impact.

With regards to noise management, the principle of the 'balanced approach' is often adopted. This involves modelling and monitoring current noise levels, defining a baseline, and setting objectives for reducing future noise and community annoyance. These targets will be detailed in a Noise Action Plan (NAP). The 'balanced approach' to noise mitigation contains four elements:

1 *Reduction of noise at source*: the ongoing review and adoption of aircraft noise certification standards ensures that they reflect the current state of aircraft technology. These standards reflect the combined performance of the aircraft (airframe and engines), as detailed in Annex 16, Volume 1 of the Chicago Convention. As this element is not within the direct control of airports, it is important that they work collaboratively with other service providers to help promote and support research and technology programmes which aim to reduce noise at source.

2 *Land use planning and management*: prevents incompatible development in noise-sensitive areas. This includes land use planning elements (such as 'zoning' to identify sensitive areas), mitigation measures (including the use of soundproofing schemes and installation of double glazing in affected properties), and financial aspects, including charges and incentives.

3 *Noise abatement procedures*: specific alterations to aircraft operations in-flight and on the ground to minimise the number of people affected by noise. For example, runways may be utilised in such a way so that the direction of take-off or landing enables aircraft to avoid noise-sensitive areas during the noisiest initial departure and final approach phases of the flight. Once an aircraft is airborne, flight-track monitoring of arrival/departure profiles is used to develop and enforce noise preferential routes (NPRs) which direct aircraft away from densely populated areas to minimise noise

Auxiliary power units (APUs): small jet engines fitted to some aircraft that provide electrical power when the aircraft is on the ground and its engines are switched off.

disturbance. Penalties or fines are administered to airlines that do not comply with the NPRs. Proceeds from these fines may then be used to fund local community projects. Noise-monitoring equipment will also be used to calculate noise contours and help identify flights which breach noise limits.

4 *Operating restrictions*: noise-related restrictions that limit access to or reduce the operational capacity of an airport. Restrictions can be aircraft-specific (based on the aircraft's individual noise performance), operate at certain times of the day (for example, banning movements during the night or by restricting operations from particular runways), or be progressive in nature to ensure a gradual reduction in noise over a longer period of time. Such restrictions can be imposed by national government or by the airport operator itself when devised in conjunction with local communities.

Market-based measures

Market-based approaches utilise economic instruments to either incentivise or discourage certain actions. Most measures follow the 'polluter pays principle', where polluters are charged for the environmental impact of their activities. The introduction of fuel taxation has long been suggested as a means of addressing aviation's environmental externalities. Aviation fuel that is used for international flights is exempt from tax, and only a few countries (including the US) impose taxes on fuel for domestic flights. This situation is a legacy of the 1944 Chicago Convention (➤Chapter 1), which enshrined fuel tax exemption for air travel into international law. While there have been efforts to address this situation, significant obstacles, relating to the creation of the necessary legal framework, the renegotiation of existing bilateral air service agreements, and the need to ensure that taxes are applied universally and uniformly, remain.

An alternative to taxing fuel is direct emissions charging. Here, polluters are charged according to the quantity of emissions they generate. Tradable permits can be used to create a market for carbon by assigning monetary values to quantities of CO_2. Such schemes provide financial incentives for participants to reduce their environmental impacts. Essentially, each 'permit' represents a license (or allowance) for the polluter to emit a certain quantity of CO_2. The total number of permits in any market is capped at an agreed level, and permits are then distributed among the various polluters, either freely or via auction. At the start of trading, polluters are able to buy or sell permits, depending on whether they are polluting above or below their allowance. This creates a commercial incentive for large polluters to improve their environmental performance and 'free up' permits which they can then sell. If a company is unable to submit sufficient permits at the end of each year to cover its emissions, then a fine is imposed. This type of tradable permit scheme is also described as a 'cap-and-trade' system. The success of such schemes depends on the ability to maintain a sufficiently high price for permits to provide incentives for companies to undertake environmental mitigation measures and invest in lower carbon technology to reduce their emissions. Case Study 18.5 details the application of a regional cap-and-trade system to commercial aviation.

CASE STUDY 18.5

EU ETS

The EU ETS (Emissions Trading System) is an example of a tradable permit scheme. Established in 2005, it is currently the largest emissions trading scheme of its type in the world. The EU ETS has the potential to cover many sectors but, to date, the focus has been on static site emissions which can be measured. Oil refineries, power plants, steel works, production of aluminium, metals, iron, cement, paper, glass, and cardboard have been included with participation being mandatory. This covers 45 per cent of EU GHG emissions.

From 2012, commercial aviation was included within the programme. Initially, the scope of the ETS covered all flights arriving at, and departing from, airports in the European Economic Area (EEA), which comprises EU Member States plus Norway, Iceland, and Liechtenstein. Initially, allowances were allocated free, but this was criticised on the basis that it represented a windfall profit to industry, giving them little incentive to innovate. In terms of aviation, currently, 15 per cent of allowances are auctioned and 85 per cent are allocated free of charge to the aircraft operators. It has been estimated that the system has contributed to reducing the carbon footprint of European airlines by more than 17 million tonnes per year (Europa.eu, 2019).

In recognition of the need for a truly *global* market-based measure to help address aviation emissions, in 2016 ICAO announced the development and implementation of CORSIA (Carbon Offsetting and Reduction Scheme for International Aviation). CORSIA aims to help address any increase in CO_2 emissions from international aviation above 2020 levels. Unlike the EU-ETS, which is a cap-and-trade system, CORSIA is a carbon offsetting scheme, whereby a company (in this case, an airline) compensates for its carbon emissions above the Carbon Neutral Growth (CNG) 2020 baseline by purchasing carbon credits from the carbon market. These credits (each one equivalent to 1 tonne of CO_2) are generated by projects (such as afforestation schemes or renewable energy projects) that remove an equivalent quantity of CO_2 from the atmosphere. While participation in CORSIA is voluntary for the pilot period (2021–2023) and the first phase (2024–2026) of the programme, participation will be compulsory for the majority of operators on international routes from the start of the second phase (2027–2035). Although carbon offsetting does not explicitly require airline operators to reduce their emissions, it does provide an incentive to do so by reducing the need to purchase credits and is recognised as an effective option for sectors where the costs and timescales of abatement can be high (ICAO, 2019).

Stop and think

What are the relative merits of introducing a carbon offsetting or a cap-and-trade system for reducing CO_2 emissions from aircraft?

!

Changes in consumer behaviour

Although interventions in the market may go some way towards addressing aviation's environmental impacts, growing awareness of aviation's impact on the climate is affecting a change in society's perception of air travel. Although passengers have had the opportunity to purchase carbon 'offsets' since the 1990s in an attempt to compensate the CO_2 emissions their flight creates, it is only in recent years that flight shaming (or *Flygskam* in Swedish) has emerged as a challenge to the business as usual scenario. Flight shaming has raised consumer awareness of the environmental impacts of flying and the availability of alternative transport modes, including highspeed rail, while climate protesters have specifically targeted airports and aircraft.

Such changes in consumer attitudes are driving the development of new initiatives, including proposals to create an individual flight allowance, to tax frequent flyers, and/or to abolish frequent flyer programmes to remove incentives for members to take additional flights merely to maintain or enhance their frequent flyer status. If introduced, such policies would not be the first time that governments have directly taxed passengers. Australia introduced a departure tax in 1978, the UK introduced Air Passenger Duty in 1994, while in 2019 the French Government announced plans to introduce an airline passenger tax with the resulting revenue being used to fund public transport improvements.

Such initiatives mirror the views that are held by many in society who stress the harmful impacts of flying and appeal to a collective consciousness to reduce air travel on environmental grounds. While it is unlikely that flying will ever entirely be cast as a social pariah on account of its social and economic benefits, the phenomenon of flight shaming, widespread climate change protests, and other direct action against air transport hints strongly to a future in which an informed, empowered consumer will hold service providers accountable with regards to their environmental performance.

18.5 The impact of a changing climate on aviation

So far, this chapter has focused on the impacts of aviation on the environment. However, it is important to recognise that a changing climate is affecting (and will continue to affect) air transport operations both today and into the future, and these challenges will need to be managed. A changing climate is affecting (or will affect) aviation in the following ways:

- Altering patterns of tourist demand by making some destinations more or less attractive.

- Rising sea levels will render low-lying airports inoperable.

- Higher temperatures will damage airfield infrastructure and degrade the take-off performance of aircraft.

- Changes to the jet stream over the North Atlantic will increase the frequency and severity of turbulence (and hence increase the risk of injuries to passenger and crew) and change journey times on routes between Europe and North America.

- Changes in surface temperatures will lead to new energy demands in terminal buildings for heating and cooling.

Given these challenges, there is a need for the aviation sector to undertake climate change risk assessments and develop adaptation strategies.

Key points

- While aviation generates economic and social benefits, it also imposes environmental externalities, including those related to climate change, local air quality, noise, and energy, water, and waste.

- Aircraft and engine technology continue to evolve, but important questions around the extent to which technological solutions can fully address environmental issues remain.

- Air transport both affects and is directly affected by the physical environment in which it operates.

- Future management interventions will need to consider the full range of possible options for reducing aviation's environmental impact – these may include, but are not limited to, additional market-based approaches and consumer behavioural change.

References

ATAG (Air Transport Action Group). (2018). *Aviation: Benefits beyond borders.* Available at: www.atag.org/our-publications/latest-publications.html

Crippa, M., Oreggioni, G., Guizzardi, D., Muntean, M., Schaaf, E., Lo Vullo, E., Solazzo, E., Monforti-Ferrario, F., Olivier, J. G. J. and Vignati, E. (2019). *Fossil CO_2 and GHG emissions of all world countries 2019 report.* Luxembourg: Publications Office of the European Union.

Department for Transport [DfT]. (2018). *Transport Statistics Great Britain 2018*, London: DfT.

EEA (European Environment Agency), EASA (European Union Aviation Safety Agency), and EUROCONTROL. (2019). *European aviation environment report, 2019.* Available at: www.easa.europa.eu/eaer/

Europa.eu. (2019). EU *Emissions trading system.* Available at: https://ec.europa.eu/clima/policies/ets_en

Fleming, G. G. and de Lépinay. I (2019). *Environmental trends in aviation to 2050.* Available at: www.icao.int/environmental-protection/Documents/EnvironmentalReports/2019/ENVReport2019_pg17-23.pdf

IEA. (2019). *Are aviation biofuels ready for take off?* Available at: www.iea.org/newsroom/news/2019/march/are-aviation-biofuels-ready-for-take-off.html

IATA. (2019a). *World air transport statistics*, Plus Edition. Montreal: International Air Transport Association.

IATA. (2019b). *Climate change.* Available at: www.iata.org/policy/environment/Pages/climate-change. aspx

IATA. (2019c). *Cabin waste factsheet,* June. Available at: www.iata.org/pressroom/facts_figures/fact_ sheets/Documents/fact-sheet-cabin-waste.pdf

ICAO. (2019). *Destination green: The next chapter 2019 environmental report, aviation and the environment.* Available at: www.icao.int/environmental-protection/Pages/envrep2019.aspx

CHAPTER 19

Human resource management and industrial relations

Geraint Harvey and Peter Turnbull

LEARNING OBJECTIVES

- To understand the importance of effective human resource management (HRM) and industrial relations (IR) strategies in the airline sector.

- To appreciate why people management is a difficult task for airlines.

- To be aware of key developments in the industry over the last 20 years, specifically legislative change and industry crises, and evaluate their impacts on the nature of work within the airline sector.

- To assess the prevailing model of employment relations in European airlines.

- To evaluate the sustainability of different systems of employment relations in civil aviation.

19.0 Introduction

The global air transport industry directly employs 10.2 million people and supports 65.5 million jobs overall (ATAG, 2018). Civil aviation is a service industry, and many airline employees, including cabin crew and ground staff, are front-line service sector workers. Effective people management is crucial to the success of airlines, irrespective of whether they adhere to a full service or a low cost business model (➤Chapter 8). Full service network carriers (FSNCs) that compete on service quality require a high level of customer service from their staff, whereas operators that compete on price require high levels of productivity and efficiency from their staff (see Section 19.1). Consequently, airlines have experimented with a range of people management strategies. Some of these have been enduringly successful and have enabled certain carriers, most notably Southwest Airlines in the US, to secure and maintain competitive advantage. Other airlines that have adopted a different approach to people management have been less successful.

This chapter describes four interconnected factors inherent to airline operations that highlight the importance of effective people management:

1 the cyclical nature of demand for air transport;

2 the perishability of the airline product;

3 the proportion and pliancy (flexibility) of labour costs;

4 the importance of employee performance and productivity.

Critical events that have impacted on the industry and affected people management over recent decades are then identified. Recent developments in civil aviation and the impact of the prevailing 'low road' trend in employment relations on people management are examined, before the implications of the alternative 'high road' system of employment relations pursued by Southwest Airlines and other carriers are considered.

19.1 People management in the airline industry

People management is important since employees enable an airline to deliver its service. People are at the heart of airline differentiation strategies – staff provide high levels of customer service at FSNCs, and staff are also central to the low cost model because high staff productivity at low cost carriers (LCCs) reduces unit costs and enables these savings to be passed on to consumers in the form of lower fares. However, employment relations strategies are also complicated by factors that are peculiar to, or at least more potent in, the civil aviation industry.

Cyclical demand and its relationship to HRM

Demand for air transportation (expressed as revenue passenger kilometres (RPKs)) is cyclical (Figure 19.1) and is linked to fluctuations in economic growth (expressed by gross domestic product (GDP)), with demand increasing or decreasing as GDP grows or contracts, but at a much faster rate (►Chapter 2). Business class travel is especially sensitive to economic fluctuations because firms are less inclined to spend scarce financial resources on the premium charged for service and comfort by FSNCs. This has a disproportionate impact on full service airlines' revenue and profitability.

The cyclical nature of air transport demand can lead to the expectations of management and labour being 'out of sync' with one another with respect to current and future market conditions. When an airline experiences a downturn in demand, costs will be controlled more tightly and employees may be expected to make greater concessions such as accepting a pay freeze, pay cuts or the suspension of staff travel allowances. However, when there is an upturn in demand, airline managers are typically cautious and, anticipating the next downturn in an increasingly competitive environment, are unlikely to reciprocate (at least in the short term) with improved terms and conditions of employment. In contrast, employees expect concessions to be lifted and improvements in terms and conditions to reflect the renewed prosperity of the airline, which seems evident to front-line staff who handle more passengers (Case Study 19.1). This mismatch in perception is highly problematic, particularly at the peak of the business cycle when employee expectations are still rising but airline management foresee or face falling demand or a decline in advance bookings.

CASE STUDY 19.1

INDUSTRIAL ACTION AT BRITISH AIRWAYS, 2019

Industrial action undertaken by British Airways pilots in September 2019 resulted in the cancellation of 2,000 flights which impacted on around 300,000 passengers and cost £121 million. The industrial action occurred after the British Air Line Pilots Association (BALPA), the trade union representing pilots at the airline, rejected the offer of a pay rise of 11.9 per cent over three years. The trade union argued that the pay rise did not acknowledge the role pilots have played in the success of the airline. International Airlines Group, the parent company of British Airways, made a profit of £2.6 billion in 2018 with British Airways contributing around 80 per cent.

The LCCs that emerged as a result of deregulation of the US domestic airline industry in the late 1970s and the single European aviation market in the 1990s (►Chapter 3) display a much greater seasonal (cyclical) variation in demand, which they seek to accommodate through Human Resources (HR) and employment policies. For LCCs, passengers visiting friends and relatives and leisure travellers remain important customer segments, despite the

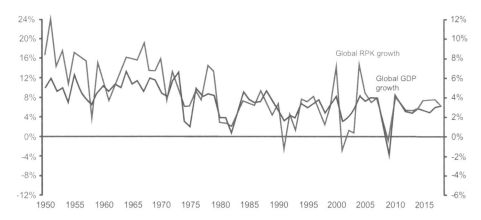

Figure 19.1 The cyclical demand for air transport

efforts of some more established low cost operators to attract business customers (➤Chapter 10). Consequently, LCCs experience a spike in demand in the summer months and encounter a more pronounced trough in the winter in comparison to most full service airlines. As a result, LCCs employ many flight and cabin crew on short-term seasonal or temporary contracts who may be recruited through an agency rather than being directly employed by the airline, and many ground services such as ground handling, catering, fuelling and check-in are subcontracted to third parties to reduce costs (Harvey and Turnbull, 2012, 2014).

Perishability

An airline's seat inventory is a perishable commodity as carriers cannot stockpile seats on cancelled flights for use on another occasion. Flight cancellations have an immediate and direct impact on an airline's performance, and so industrial action taken by airline employees can be highly detrimental to the company's reputation and profitability. Industrial action by pilots at Ryanair in 2018 led to a 20 per cent reduction in profitability, while action by pilots at SAS in 2019 cost the airline between €6 million and €8.4 million per day. The industrial action at British Airways in September 2019 cost the airline an estimated £121 million. Even the threat of a strike can lead to a loss of revenue as passengers transfer to other airlines to avoid any possible disruption or inconvenience to their journeys. In this context, the cooperation and the consent of the workforce are paramount: disgruntled staff are bad for customer service, while disruptive staff are disastrous for the company's finances and reputation.

!

Stop and think

Why is the financial cost to airlines of industrial action so high?

Proportion and pliancy of labour costs

Traditionally, labour costs have amounted to a sizeable proportion of an airline's operating costs. For most legacy (full service) airlines, labour typically accounts for around one-third of total costs. Pilots are especially well paid (around £90,000 a year on average in the UK, which, according to the Office for National Statistics, puts them among the highest paid workers in the country). Legacy carrier cadet pilot schemes accept only a small number of applicants each year, and some pilots start their careers with training associated debts of up to £100,000 and no guarantee of employment once their training ends. In addition, some airlines expect their new first officers to pay for their type-rate training, which can increase their debts by another £20,000 or more.

Unlike fuel, landing charges, aircraft and many other capital costs, labour costs are one of the few pliable costs over which airline management exerts a level of control. This is not to suggest that savings elsewhere cannot be achieved. For example, deals can be negotiated with airports to reduce or subsidise landing charges (➤Chapter 7), bulk purchasing aircraft may attract a reduction from aircraft manufacturers and airlines might negotiate advantageous fuel hedge contracts (►Chapter 12). Nonetheless, most operating costs are (quasi-)fixed, at least in the short term. Thus, airline cost-cutting initiatives invariably focus on labour (Harvey and Turnbull, 2009). It is unsurprising, then, that the figure for labour costs as a percentage of total operating costs has diminished in recent years in the US, Europe and Asia-Pacific. LCCs are leaders on this metric, with labour costs accounting for less than 10 per cent of revenue at Vueling and Ryanair, and less than 13 per cent at easyJet, compared to almost 24 per cent at International Airlines Group (IAG), which includes British Airways, Iberia and Aer Lingus.

Performance and productivity

There is a very clear relationship between employee performance and the competitive advantage of airlines. Employees at full service (legacy) airlines that compete on the basis of service quality are responsible for personifying the brand and delivering a distinctive level of service. Employee productivity, on the other hand, is a common indicator of airline cost-effectiveness. On this metric, the difference between the FSNCs and LCCs is clear. Data from airline financial statements for 2012, for example, compiled by the CAPA Centre for Aviation, illustrates the sheer size of this particular 'performance gap': the operating profit per employee at Ryanair (€80,943) was more than ten times that of British Airways (€8,030).

An important dimension of an airline's strategy for enhancing employee performance and productivity is its relationship, or lack thereof, with trade unions that represent workers in the industry. In the UK, many airlines formally recognise **trade unions** for the purposes of collective bargaining – the process through which the terms and conditions of work are determined. However, management at some airlines have opposed trade union participation in the decision-making processes (Harvey and Turnbull, 2015).

Trade union: an independent employee organisation that negotiates with the company over the terms and conditions of its members. Its primary purpose is to represent the interests of its members who may be employed by a single airline (e.g. Southwest Airlines Pilots Association) or many airlines (e.g. BALPA).

19.2 Liberalisation, industry crises and LCCs

Labour costs and productivity became even more important following the liberalisation of the civil aviation industry and two major industry crises at the start of the new millennium. Prior to liberalisation, air transportation was organised according to bilateral agreements between governments (►Chapter 1). These agreements specified the airlines that could fly particular routes and the tariffs they could charge. Bilateral agreements effectively prevented price competition and restricted market entry. Liberalisation made for a more volatile and dynamic aviation market wherein extant airline restructuring and the employment contracts offered by the new entrant airlines invariably and detrimentally impacted on labour. Management teams at full service airlines were presented with both the motive and the opportunity to exploit the pliancy of labour costs and seek concessions from staff in order to meet the challenge of the new competitive environment. Moreover, an enduring strategic aim of the FSNCs is service quality differentiation, and so alongside initiatives to reduce the cost of labour, many legacy airlines have also sought performance increments, or intensified **emotional labour,** from their staff. In part, the new entrant airlines sought cost savings they would pass onto their customers by employing staff on inferior terms and conditions. The

Emotional labour: the management of emotions in line with commercial requirements in order to engender a positive customer experience and benefit the organisation.

productivity of staff at the new entrant LCCs gives these airlines a competitive (price) advantage and much higher profits.

!

Stop and think

Why are staff costs at LCCs typically much lower than those of FSNCs?

An important innovation that is intrinsically linked with the LCC business model, and one that has direct consequences for employment in the airline industry, is subcontracting. Stelios Haji-Ioannou, the founder of easyJet, explained that his company believes 'relationships with entrepreneurial companies out there to make a profit are more efficient than having a bunch of employees yourself' (quoted by Sull, 1999, p. 25). This innovation has been adopted and adapted to a varying extent, and in different ways, by FSNCs. British Airways, for example, faced considerable resistance from its employees when the 'virtual airline' model was proposed in the mid-1990s as part of the Business Efficiency Programme, whereby the airline would retain only central functions and operate aircraft supplied and staffed under wet lease arrangements with other airlines (➤Chapter 12). Following the proposals, members of the British Airline Pilots' Association (BALPA) threatened to strike in 1996, while their cabin crew colleagues, who were members of the British Air Stewards and Stewardesses Association (BASSA), took industrial action in the following year. The airline had successfully detached elements of its non-core business on franchise agreements with other firms (not necessarily airlines), enabling the airline to operate with 'much lower costs, minimal investment, and fewer objections from the competition authorities' (Blyton et al., 1998, p. 11). The airline also succeeded in a degree of disintegration, which is at the heart of the virtual airline model, by outsourcing activities such as catering, vehicle management and maintenance and ticket services.

Whereas there was considerable growth in demand for air transport in the early 1990s, demand decreased significantly from 1995 onwards and had not recovered when the 11 September 2001 terrorist attacks on the US exacerbated what had been a very difficult period for many airlines (see Figure 19.1). Because of 9/11, airlines introduced significant cost-reduction measures that prompted accusations of opportunism from many aviation trade unions (Harvey and Turnbull, 2012). However, according to the director general and CEO of IATA at the time, the airline industry was ill-prepared to deal with even a mild economic downturn. Post-crisis cost reductions were applied to labour.

Analysis of airlines' response to the crisis revealed that many airlines across the globe, especially in liberal market economies such as the US and the UK, moved quickly to reduce labour costs by offering voluntary redundancy to (and imposing compulsory redundancy on) staff, alongside voluntary and compulsory furlough (the requirement that staff take unpaid leave). In Europe and Asia-Pacific, cost-reduction initiatives also impacted severely on junior and temporary workers, as there was widespread non-renewal of temporary contracts and probationary staff not being transferred onto full-time contracts. These measures, taken in response to the crisis, whether they reflected necessity or opportunism, profoundly impacted employment relations at many airlines. The response to the crisis served to increase tensions between labour and management, lower trust between both parties and provide the backdrop for future conflict. A dispute that resulted in strike action by ground services workers at Sabena, the former Belgian flag carrier, contributed to the airline's demise in 2001.

In the wake of 9/11, the extent of the competitive threat facing FSNCs from the LCCs was becoming clear. The response of many legacy airlines to the emergence of the LCCs in the 1990s might be described as 'studied neglect', as the new entrant operators rarely competed head on with the legacy airlines. Where competition existed (e.g. on routes between London Stansted, Luton, Gatwick and Heathrow airports and those serving Barcelona, namely Barcelona, Girona and Reus), the LCCs initially generated new markets rather than cannibalised those of the legacy airlines (Harvey and Turnbull, 2014). At the turn of the millennium, LCCs accounted for only 5 per cent of the intra-European market, but they survived the economic crisis of 2001–2 far better than their legacy counterparts. In the last quarter of 2001, the passenger traffic carried by the European LCCs easyJet, Ryanair and Go increased by around 30 per cent.

By 2008, LCCs accounted for almost 30 per cent of the US domestic market and around 40 per cent of the intra-European market. It is in this context of increasing LCC market share that civil aviation was once again impacted upon by the global financial crisis that eclipsed the problems encountered after 9/11. Whereas airline revenue fell by 7 per cent in the crisis that followed 9/11, revenues fell by 15 per cent in 2009, with the operating losses of the world's top 150 airlines totalling US$15 billion (compared with profits of US$29 billion in 2007). Several airlines ceased trading (such as Silverjet, XL Airways and Zoom Airlines), and many more responded with cost-reduction strategies that once again directly impacted on labour, with staff-reduction programmes alongside leaves of absence (furloughs) and a reduction in training. Data from studies conducted in 2001 and 2009 suggests an increased incidence of redundancy (voluntary and compulsory) in the latter period, despite the opposition of trade unions to compulsory redundancy (Harvey and Turnbull, 2010).

The financial crisis impacted on the success of the principal European LCCs in terms of passengers carried, but these airlines have recovered quickly. Immediately prior to the 2007–8 financial crisis, all the largest (top ten) network airlines were profitable, while nine of the top ten LCCs were making money. In 2008, seven of the top ten network airlines lost money compared to just three of the top ten LCCs. A year later, nine of the top ten network airlines were in the red compared to only two of the leading LCCs.

Stop and think

Why are LCCs more resilient to financial downturns than FSNCs?

19.3 Contemporary people management in civil aviation

By 2018, LCCs carried more than 40 per cent of the available seats on flights within Europe (www.centreforaviation.com), with seat capacity among LCCs growing by 78 per cent since

2009 (compared with a 31 per cent increase among FSNCs). In a single European aviation market there are far greater opportunities for 'social dumping', a strategy geared towards the lowering of wage or social standards for the sake of enhanced competitiveness, because airlines can readily take advantage of the competition between workers in different geographical regions. As previously noted, the low cost model is synonymous with various forms of subcontracting and the increased use of agency or temporary workers, whereby direct employees are replaced with 'self-employed' workers and other staff hired on fixed-term or seasonal employment contracts. Such cost-cutting does not always constitute 'fair competition'. Ryanair, for example, was found guilty by French courts in 2014 for paying less than 11 per cent of the requisite 45 per cent social security cost for its staff based in the country. In 2017 and 2018, the airline faced investigations of false self-employment by the Belgian and UK authorities (see, for example, Davies, 2017).

However, aside from increasing numbers of people in the industry working for the LCCs, the success of these airlines has further impacted on civil aviation employment in two main ways. The increased competitive pressure exerted by LCCs on FSNCs renewed the latter's efforts to replicate elements of the low cost model. Whereas abortive attempts were made by several airlines in the late 1990s to replicate the low cost model via a carrier-within-a-carrier subsidiary such as British Airways' Go and KLM's Buzz (➤Chapter 8), more recent ventures by Lufthansa (Eurowings, formerly Germanwings) and KLM–Air France (Transavia) have been more successful, especially in terms of reducing labour costs. Cabin crew at Eurowings, for example, are paid around 40 per cent less than their colleagues in the Lufthansa mainline operation and experience a much slower progression up the pay scale. British Airways has pioneered an approach whereby a new workforce has been created inside the airline with new staff hired on inferior terms and conditions of employment. Alongside its Euro and Worldwide Fleets, the airline now has a third, Mixed Fleet. Unlike the physical separation of Germanwings from Lufthansa mainline, British Airways' Mixed Fleet operates both short-haul European and long-haul intercontinental flights from London Heathrow.

! Stop and think

What are the advantages and disadvantages of FSNCs hiring staff for their low cost subsidiaries on much lower terms and conditions of employment?

The second development is a consequence of the diminishing returns from the low cost model. This is manifest both in terms of the introduction of longer and 'thinner' (lower demand) routes as the LCCs seek out new markets that are more costly to service and the (inevitable) limits to continually cutting labour costs: at some point the low motivation of poorly paid and insecure staff will result in a decline in (even basic) service quality that will outweigh any savings from lower unit labour costs. It comes as no surprise, then, that LCCs such as easyJet now differentiate their service in terms of the tariff, with FLEXI fares that include allocated seating, 'Speedy Boarding' and one piece of hold luggage. Consequently, if the options for continuous cost reductions diminish, the only financial alternative is to grow revenue. This can be achieved by 'adding value' through offering ancillary products and services such as travel insurance, car hire, hotel accommodation, surface transport tickets, on-board and online gambling and in-flight sales and/or targeting different passenger groups, especially those with more disposable income (➤Chapter 10). In the financial year ending March 2018, Ryanair earned over £2 billion (28 per cent of total revenue) from ancillary revenue, while IATA predicted a figure of $92.9 billion in ancillary revenue worldwide (airlines.iata.org). Ryanair faced criticism for the way in which it pursues ancillary revenue, for example, the 'frequency with which it splits up passengers, including families with children, who choose not to pay for selected seats' (Morris, 2017).

Ryanair's dependence on ancillary revenues is demonstrated by the fact that, based on ticket revenue alone, the airline needed to sell 98 per cent of seats to break even in 2008, whereas it actually sold only 81 per cent. The company therefore returned a profit on the back of ancillary revenues. By unbundling the different components of air travel, LCCs not only turn the flight into a commodity for the passenger, with payment for all the different elements of the service (including seat choice, checked-in baggage and in-flight food and drinks), they also change the nature and expectations of work for staff. Indeed, a significant component of pay for cabin crew is often now based on in-flight sales performance. Most LCCs use some form of variable pay for cabin crew, which can sometimes comprise more than half the employee's monthly pay.

These developments explain why the business strategies of LCCs are evolving by, for example, facilitating transfers, entering alliances and acquiring other airlines, and why the experience of work for aircrew will differ not only between legacy and LCCs but also between different legacy and low cost operators. For example, LCCs such as easyJet, with a denser route network and access to more and higher value passengers at primary airports, will have different expectations of staff and a more stable roster throughout the year with less variation between summer and winter schedules.

While easyJet and Vueling target higher value passengers and primary airports, the self-styled ultra-low cost carriers (see ➤Chapter 8) will no doubt continue to reduce labour costs, and staff will find themselves working right up to the maximum flight and duty time during the busy summer schedule, with enforced lay-offs or unpaid leave becoming the norm during the winter when aircraft are grounded. Ryanair, for example, now 'flex' the fleet between winter and summer schedules and typically ground between 60 and 80 aircraft each winter, principally because the carrier no longer makes a profit during the winter and relies on summer profits to offset winter losses.

It is clear that the continued success of the LCCs, through a strategy of increasingly direct competition with legacy airlines at primary airports for the same passenger groups, will also impact on staff at the legacy airlines. easyJet already poses a direct competitive challenge to many legacy airlines as the company has invested heavily in frequent services to and from primary airports while maintaining a low cost operating base. In some EU Member States, easyJet is now the benchmark used by management calling for a reduction in legacy labour costs, but in other Member States it is an employer of choice for many aspiring cabin crew, including many staff who work for British Airways Mixed Fleet. Direct competition from Ryanair is rather more challenging. When legacy airlines with a much higher (legacy) cost base face 'social dumping' by an ultra-low cost carrier, the pressure on revenue and staff costs can be considerable.

The churn created by low cost competition for legacy airline staff is not confined to the low cost version of the main brand (e.g. staff employed by British Airways Mixed Fleet, Iberia

Express, Eurowings and Transavia). A combination of more fuel-efficient aircraft and open skies agreements with neighbouring countries has enabled LCCs to extend the geographic reach of their route network to many of the attractive and lucrative 'long-haul' destinations traditionally served by legacy airlines (e.g. easyJet offers flights to Egypt, Jordan, Turkey and Morocco). From a multibase network, LCCs can retain the cost advantages of their original business model on these routes.

New market opportunities are being created through the negotiation of open skies agreements with non-EU countries, most notably the US. With the commission of a European sovereign state (an Air Operator's Certificate), European LCCs are now able to adopt and adapt the maritime practice of Flags of Convenience and Crews of Convenience as a way of redefining employment relationships, exerting control over labour and extracting surplus value. The clearest example of this strategy is the creation of Norwegian Air International (NAI), a subsidiary of Norwegian Air Shuttle (NAS) (Case Study 19.2).

CASE STUDY 19.2

NORWEGIAN AIR INTERNATIONAL (NAI)

Norwegian Air Shuttle (NAS) is one of Europe's largest LCCs, flying around 37 million passengers in 2018 to more than 150 destinations. To completely break all ties between labour, location and (operating) licence, the airline's transnational subsidiary, NAI, acquired an Irish Air Operator's Certificate (AOC), even though none of its operating bases are in Ireland. Irish registration is simply a convenient flag as NAS shifts the sovereign regulatory regime under which social relations take place, enabling NAI to escape from national (Nordic) class compromises and exploit the EU–US open skies agreement.

NAS claims that the company's application for an Irish AOC was motivated mainly by access to existing and future traffic rights to and from the EU (EU open skies agreements with Israel and Canada) as well as Ireland's adoption of the Cape Town Convention on International Interests in Mobile Equipment, a treaty designed to standardise transactions involving movable assets and property, including aircraft. In contrast, trade unions have pointed out that NAI's international base strategy is estimated to save around 50 per cent on salary costs, with Thai crews paid less than the minimum wage in Norway and indeed most other countries in Europe. With the entry of a LCC into the transatlantic market, the challenge for organised labour on both sides of the ocean is clear:

NAS is using the unique nature of EU aviation laws to effectively shop around for the labour laws and regulations that best suit its bottom line.

(Transport Trades Department, AFL-CIO, 29 October 2013)

NAS began operations between the UK and US in June 2017. In May 2018 a Federal Appeals Panel in the US upheld the US Department for Transportation approval for Norwegian to operate in the US. Norwegian's response to claims made by trade unions representing US workers that Norwegian were competing unfairly using crews on lower wages than other transatlantic airlines was to cite the purchase of Boeing 737 MAX 8 aircraft, worth around $14 billion, and the hiring of US crews. The complex structure of the airline group – with several AOCs – and the transfer of staff from one AOC to another facilitating changes to terms and conditions has resulted in further labour disputes, with cabin crew at Charles de Gaulle Airport represented by the French trade union UNAC taking strike action against the airline in April 2019.

19.4 Diversity, gender and work in civil aviation

There are powerful economic and social legitimacy arguments in favour of ensuring workforce diversity in civil aviation. To rule out an applicant because of their gender, age, ethnicity, sexual orientation or any other protected category is morally wrong and potentially damaging as the firm foregoes valuable skills and risks a customer boycott. This section focuses specifically on gender.

Over the years there has been a very clear gender dimension to occupations in civil aviation. Changing attitudes towards the nature of commercial flights and the increasing importance of service led to a feminisation of cabin crew jobs. The value of emotional labour and aesthetic labour both created opportunities for women in the industry and served also to undermine female workers' dignity and professionalism (Turnbull, 2013). Employment opportunities for women remain largely confined to customer service roles, such as cabin crew, and recent research indicates a challenging work environment for women in European airlines (Harvey, Finniear and Greedharry 2019). There have been several high profile female airline leaders in the US and Europe, for example Colleen Barrett (COO of Southwest Airlines, 2001–8), Barbara Cassani (CEO of Go, 1998–2001), Carolyn McCall (CEO of easyJet, 2010–17) and Anne Rigail (who became CEO of Air France in 2018). But this paucity of examples indicates the scarcity of women in airline senior management.

The flightdeck has also historically been dominated by men (McCarthy, Budd and Ison, 2015) as have other well-paid roles such as aircraft maintenance, while women seeking to enter male-dominated professions (airline pilots, for example) have faced harassment and sexism (McCarthy, Budd and Ison, 2015) and ongoing obstacles associated with 'role stereotyping' (Germain, Herzog and Hamilton, 2012).

Concerted efforts have been made by a small number of airlines to increase the gender diversity of flightcrew. Most notably, easyJet has focused particular efforts on empowering female staff, actively encouraging women to become pilots through its Amy Johnson initiative. The initiative has been highly successful since its launch in 2015, with women constituting 15 per cent of new entrant pilots in 2018 compared to 5 per cent in 2015 (compared to only 4 per cent total female pilots in the UK overall). easyJet's current target is 20 per cent female new entrant co-pilots by 2020. Activities associated with the Amy Johnson scheme include visits by pilots to schools, over 100 youth and aeronautical organisations; sponsoring an Aviation Badge for Brownies; and continuing to highlight female easyJet pilots in the media. The airline also partially covers the type-rating training costs for a small number of new pilots (around 20 per annum), at a cost of around £22,000 per pilot, in order to increase the diversity of its flightcrew and take advantage of a wider talent pool.

19.5 An alternative approach to people management

This chapter has focused primarily on the ways in which many European airlines have taken a 'low road' approach to employment relations based on eroding the terms and conditions of work for staff. However, it is important to reiterate that people are crucial to the success (or otherwise) of airlines. The recent innovations by European FSNCs to operate what staff regard as 'main-line services' via a low-cost subsidiary led to industrial action and significant costs at both Lufthansa and Air France. In July 2017, British Airways faced further industrial action by Mixed Fleet cabin crew and was forced to wet-lease nine aircraft and crew from Qatar Airways (a oneworld alliance partner) at significant cost.

Whereas most LCCs have copied the original low-cost business (operating) model developed by Southwest Airlines (SWA), to a greater or lesser extent, very few have emulated the airline's 'high road' approach to HRM and industrial relations, despite the fact that the company's people strategy is at the heart of its sustained competitive advantage. SWA is now the largest US domestic carrier, transporting more domestic passengers than any other US airline (it operates 4,000 departures per day to 99 destinations across 40 states of the US and 10 near-international countries). SWA has recorded over 40 years of consecutive profitability and posted a net income of $2.4 billion in 2018. This success has been achieved in no small part due to the 'Fun-LUVing attitude' and 'warrior spirit' of its staff, who are keen to demonstrate their 'servant's heart' to provide passengers with a novel flight experience.

The employment relations system adopted by the airline is exemplified by the former CEO, Herb Kelleher, who encapsulated the airline's approach towards staff in the following statement: 'You put your employees first. If you truly treat your employees that way, they will treat your customers well, your customers will come back, and that's what makes your shareholders happy' (Herb Kelleher, quoted by McDermott, Conway, Rousseau and Flood, 2013, p. 306). Treating staff well includes industry-leading pay and benefits. In 2018, SWA recorded over 2 million hours of training (over 34 hours per employee), with around 820,000 hours on safety and security, and the company's training programmes routinely cover environmental stewardship and sustainability. The airline's 'University for People' provides training and career development to help employees 'learn and grow'.

SWA is the most highly unionised airline in America – union density currently stands at over 80 per cent – and unions are treated as 'business partners', not 'third parties'. To illustrate how opportunities to participate in decision-making (such as on pay and benefits) can directly enhance the performance of the organisation, consider the process of collective bargaining and how this might affect customer service (such as delays caused by strikes or other forms of industrial action) or passengers' perceptions of the reliability of a particular airline (e.g. adversarial contract negotiations reported in the media that might lead to future flight cancellations if the parties cannot reach an amicable settlement). SWA leads the way in timely contract negotiations in the US through its partnership approach with trade unions. In its 40-year history the airline has only ever experienced one strike. US industry data indicates that efforts by airlines to avoid unions are not likely to produce a sustained improvement in either service quality or financial performance.

! Stop and think

Outline the benefits of airline operators treating unions as business partners.

Key points

- Airlines, irrespective of their business strategy, are reliant on the performance of their staff and, as such, employment relations play a crucial role in the success of airlines.

- People management is a task made difficult by the peculiarities of civil aviation, in particular the cyclical demand for air transport and the perishability of the 'product'.

- Market deregulation and liberalisation combined with economic crises have highlighted the importance of HRM and industrial relations. In the highly competitive European airline market, the business strategies and associated HRM and industrial relations polices of the LCCs have proven most successful in terms of profitability, creating challenges for FSNCs who have struggled to reduce their legacy (labour) cost. Here too, therefore, airline management have focused on labour productivity as the key to a sustainable competitive future.

References and further reading

ATAG. (2018). *Aviation benefits beyond borders*. Geneva: ATAG.

Bernaciak, M. (2012). *Social dumping: Political catchphrase or threat to labour standards?* Working Paper 2012.06. Brussels: European Trade Union Institute.

Blyton, P., Martinez Lucio, M., McGurk, J. and Turnbull, P. (1998). *Contesting globalisation: Airline restructuring, labour flexibility and trade union strategies*. London: International Transport Workers' Federation.

Davey, C. L. and Davidson, M. J. (2000). The right of passage? The experiences of female pilots in commercial aviation. *Feminism & Psychology*, 10(2), pp. 195–225.

Davies, R. (2017). Ryanair pilots face HMRC investigation over airline's employment structures, *The Guardian*, 3 October.

Foster, P. (2011). Diversity recruiting in aviation maintenance. In E. A. Hoppe, ed., *Ethical issues in aviation*. London: Routledge, pp. 155–168.

Germain, M. L., Herzog, M. J. R., and Hamilton, P. R. (2012). Women employed in male-dominated industries: Lessons learned from female aircraft pilots, pilots-in-training and mixed-gender flight instructors. *Human Resource Development International*, 15(4), pp. 435–453.

Gittell, J. H., Von Nordenflycht, A. and Kochan, T. (2004). Mutual gains or zero sum? Labor relations and firm performance in the airline industry. *Industrial and Labor Relations Review*, 57(2), pp. 163–180.

Gunnigle, P. and O'Sullivan, M. (2009). Bearing all the hallmarks of oppression: Union avoidance in Europe's largest low cost airline. *Labor Studies Journal*, 34(2), pp. 252–270.

Harvey, G., Finniear, J. and Greedharry, M. (2019). Women in aviation: A study of insecurity, *Research in Transportation Business and Management*, 31, 100366.

Harvey, G. and Turnbull, P. (2009). *The impact of the financial crisis on labour in the civil aviation industry*. Geneva: International Labour Office.

Harvey, G. and Turnbull, P. (2010). On the go: Piloting high road employment practices in the low cost airline industry. *International Journal of Human Resource Management*, 21(2), pp. 230–241.

Harvey, G. and Turnbull, P. (2012). *The development of the low cost model in the European civil aviation industry*. Brussels: European Transport Workers' Federation.

Harvey. G, and Turnbull. P. (2014). *Evolution of the labour market in the airline industry due to the development of the low fares airlines (LFAs)*. Brussels: European Transport Workers' Federation.

Harvey, G. and Turnbull, P. (2015). Can labor arrest the sky pirates? International trade unionism in the European civil aviation industry. *Labor History*, 56(3), pp. 308–326.

IATA. (2018). *Ancillary revenues expected to hit $92.9bn*, December 2018 Available at: www.airlines.iata.org/news/ancillary-revenues-expected-to-hit-929bn

McCarthy, F., Budd, L. and Ison, S. (2015). Gender on the flightdeck: Experiences of women commercial airline pilots in the UK. *Journal of Air Transport Management*, 47, pp. 32–38.

McDermott, A. M., Conway, E., Rousseau, D. M. and Flood, P. C. (2013). Promoting effective psychological contracts through leadership: The missing link between HR strategy and performance. *Human Resource Management*, 52, pp. 289–310.

Mills, J. H. (2002). Employment practices and the gendering of Air Canada's culture during its Trans Canada Airlines days. *Culture and Organization*, 8(2), pp. 117–128.

Morris, H. (2017). Ryanair faces more claims it is separating passengers who refuse to pay for their seats. *The Telegraph*, 13 June.

Neal-Smith, S. and Cockburn, T. (2009). Cultural sexism in the UK airline industry. *Gender in Management: An International Journal*, 24(1), pp. 32–45.

Southwest Airlines. (2018). *Southwest citizenship: Without a heart, it's just a machine*. Available at: www.southwest.com/html/southwest-difference/southwest-citizenship

Sull, D. (1999). Case study: easyJet's $500 million gamble. *European Management Journal*, 17(1), pp. 20–38.

Turnbull, P. (2013). *Promoting the employment of women in the transport sector: Obstacles and policy options*. Working Paper 298, International Labour Organization: Geneva.

CHAPTER 20

Air transport marketing communications

Nigel Halpern

LEARNING OBJECTIVES

- To identify the principles of air transport marketing with an emphasis on marketing communications.

- To recognise how marketing communications have been influenced by the changing media and messages of marketing copy.

- To understand the basic principles of engagement marketing.

- To examine a range of initiatives employed by airlines and airports to engage consumers.

- To assess possible future trends in air transport marketing.

20.0 Introduction

As previous chapters have shown, the air transport market is characterised by sustained long-term growth but also by saturation in some markets (➤Chapter 2). Deregulation and liberalisation have changed the competitive environment (➤Chapter 3) and led to a transfer of ownership and control of airlines, and to some extent airports, from the public to the private sector (➤Chapters 6 and 12). These factors have contributed to a more open and contested commercial market where airlines and airports are under increasing pressure to grow their business through marketing.

The aim of this chapter is to discuss the key issues relating to air transport marketing. The focus is on how both airlines and now, increasingly, airports market themselves to passengers. Air cargo is also important. However, air cargo is sent by individuals or organisations that, unlike passengers, rarely come into contact with an airline or an airport. Instead, they tend to deal with forwarders or integrators

(➤Chapter 17). As a result, approaches to marketing air cargo differ from passenger marketing.

This chapter considers the principles of air transport marketing, including services marketing, marketing communications, and the changing media and messages of marketing copy. Principles of engagement marketing including a range of marketing initiatives that can be used by airlines and airports to engage particular consumers are also examined.

20.1 Principles of air transport marketing

Services marketing and marketing communications

The UK Chartered Institute of Marketing defines marketing as 'the management process responsible for identifying, anticipating and satisfying customer requirements profitably'. Marketing is a management process that has clearly defined objectives and outcomes. The objectives include flexible long-term efforts to understand and anticipate customer requirements and short-term efforts to satisfy those requirements with profitability the desired and intended outcome. However, marketing plays a different role for aviation service providers than it does for companies dealing purely with goods (see Table 20.1). In particular, services marketing focuses on exchanging offerings of value with customers rather than simply seeking profits. It also seeks to develop and maintain close relationships, reinforce brand identity, encourage loyalty and develop tangible cues that provide evidence of the benefits available.

Marketing communications are a specific type of marketing practice that is fundamental to a company's marketing efforts. It is concerned with the media (the tools that are used to store and deliver information) and the messages used to communicate with target markets and is therefore associated with the promotional mix – a subset of the **marketing mix** that

> **Marketing mix:** the combination of tactical and controllable decisions made by companies about their product/service, price, promotion, place, processes, people and physical environment in order to produce the desired response from target markets.

Table 20.1 Characteristics of a service and implications for marketing

Service characteristic	Implications for marketing
Inseparable. Service is often produced and consumed at the same time and through interaction between providers and consumers.	Important to develop and maintain close relationships with consumers because interaction determines the service outcome.
No transfer of ownership. Consumers rarely gain personal ownership of the service they pay for.	Important to reinforce brand identity and encourage loyalty.
Intangible. Generally has no substance – cannot be seen, tasted or touched.	Important to develop tangible cues for the benefits available such as quality of service.
Heterogeneous. Quality depends on where, when and how they are provided and by whom.	Important to invest in quality, including staff training and management systems.
Perishable. Cannot be stored for later sale or use.	Important to anticipate future demand and use the marketing mix to influence and respond to change.

consists of promotional activities that are combined and used by companies to communicate with target markets and produce the desired response from them (see Table 20.2).

Table 20.2 Traditional activities included in the promotional mix

Activity	Main strengths	Main weaknesses
Advertising. Mass communication via radio, television, print or display.	• Reaches a large audience • Repeatable • Expressive	• Not personal • General messages • Lacks persuasiveness • Limited control over who views it • Expensive
Direct marketing. Direct communication via post or telephone.	• Quick and effective • Can be targeted • Can customise messages • Develops personal relationships	• Requires a mailing list • Expense of printing and postage
Personal selling. Face-to-face communications at company offices or networking events.	• Personal contact • Targeted and persuasive • Can provide detailed information and answers • Immediate feedback	• Time-consuming and costly • Long-term commitment • Effectiveness determined by the salesperson
Sales promotion. Short-term incentives such as competitions, samples or discounts.	• Grabs attention • Stimulates demand • Initially effective	• Limited long-term effect • Does not always reflect genuine value • Can lead to **price wars**
Public relations/publicity. Indirect communications, often via a third party such as the press.	• Includes free or paid efforts • Often viewed as credible	• Difficult to generate • Limited control over content • Hype may not be met

Price war: a situation in which companies reduce the price of their products or services to undercut those of their competitors so as to increase their market share.

Stop and think

To what extent does marketing differ from selling?

The main objectives for marketing communications are to create awareness, inform target markets about the brand, influence attitudes and buyer behaviour, and encourage preference, repeat business and loyalty. The focus is therefore generally on the brand. This compares to related concepts such as corporate communications (which focus on the company), crisis communications (which focus on dealing with disruptive events or crises), and customer services (which focus on enhancing customer satisfaction). However, there is a great deal of overlap between the different concepts because no matter how effective a company is with marketing communications, the success or failure of a brand can still be influenced by the company's ability to communicate, and problems can arise if brand promises are not met (see Case Study 20.1).

CASE STUDY 20.1

THE CASE OF SCOOT

YouGov's BrandIndex asks consumers if they would recommend a particular brand to a friend or colleague or tell them to avoid the brand. The score ranges from +100 to −100 and is calculated by subtracting negative feedback from positive feedback. A score of zero means there is an equal amount of positive and negative feedback. The score for Singaporean low cost airline Scoot was +25 among consumers in Singapore at the end of October 2018. On 9 November 2018, one of the airline's flights from Berlin was delayed by ten hours on departure from Singapore Changi Airport. A few weeks later, on 25 November 2018, a flight from Singapore to Bangkok was delayed for seven hours. On 20 December 2018, a Singapore-bound flight from Athens was delayed by 56 hours. Just ten days later, on 30 December 2018, a Singapore-bound flight from Taipei was delayed for two days, affecting passengers' plans for New Year's Eve. The delayed flights and the airline's handling of communications with passengers were heavily criticised by the press and on social media. By January 2019, the airline's Recommend score had fallen below zero, meaning that more consumers recommended avoiding the brand than using it.

Brand: the name, term, logo, symbol, design, style and other tangible and intangible features that stand for the qualities of the product or service and help to distinguish it from others and encourage customer preference, purchasing behaviour and loyalty.

Managing marketing communications is a complex process because each source of information – whether it is a department or an individual within an airline or airport – has the potential to use different forms of media to communicate their own, and possibly conflicting, messages about the **brand**. It is therefore important for airlines and airports to integrate communications throughout the organisation to ensure that consistent and compelling messages are communicated via appropriate media. In order to develop effective communications, it is important to have a clear understanding of the target market, to identify the response sought, to construct a message with effective content and structure, to select relevant media and to research market awareness, satisfaction and response to communications.

Effective marketing communications can be time-consuming and costly. Airlines and airports will always need to make trade-offs between what they would like to do and what they can afford to do. Effective communication is further challenged by the diverse range of customers who are served by airlines and airports, because they need to communicate not only with passengers that consume their products and services in different markets, languages and cultural contexts but also with other stakeholders such as travel agents, tour operators and employees. Airports have a particularly wide range of customers that also includes airlines, tenants and concessionaires, and visitors to the airport (➤Chapter 7). In addition, air transport tends to be a derived demand (➤Chapter 2). This means that airlines and airports need to be aware of the marketing objectives of different stakeholders, such as tourism boards, and may benefit from entering into collaborations that facilitate the pooling of resources and the development of an integrated approach to marketing.

Changing media and messages of marketing copy

Changes in the business environment are affecting the marketing communications landscape and encouraging a shift away from traditional marketing activities that, with the exception of personal selling, are largely mass media-orientated and intended to reach a large and general audience.

One key change has been the fragmentation of mass markets as air travellers have become more diverse, knowledgeable and discerning. This has been encouraged by the growth and expansion of air travel as a growing number of people fly more frequently and air travel becomes accessible to more regions and sections of society. This reduces the effectiveness of mass media communications because the all-inclusive approach does not work when markets are fragmented. As a result, airlines and airports need to develop more targeted communications that encourage closer relationships with consumers in more narrowly defined markets. This is problematic because airlines and airports cater to a wide range of passenger segments (➤Chapter 10). Some airlines address this by offering differentiated products and services, while others operate multiple brands in order to provide passengers with distinct and separate choices. IAG, for instance, operates several brands for different passenger segments such as Aer Lingus, British Airways and Iberia (as network carriers) and LEVEL and Vueling (as low cost carriers). However, companies sometimes fail to achieve the necessary levels of differentiation because customers still associate individual brands with a single company.

Perhaps the greatest change has come from the development of new technologies that allow airlines and airports to collect detailed information about passengers and target more narrowly defined markets with specific messages via, for instance, digital media. The marketing **media mix** has therefore become broader and more complicated, and while digital media is often associated with relatively new forms of communication such as the internet, mobile technologies and social media, it is important to note that it increasingly encompasses all media channels due to the introduction of digital radio, television, print and display.

The internet has encouraged a shift away from mass marketing to transactional marketing, where the focus has changed from creating general messages for a wide audience to encouraging click-through behaviour where consumers can follow a link from an advert on a website to make a transaction. Low cost carriers were innovators, as they focused on the internet for advertising and **distribution**. However, the transition from static to interactive websites, commonly referred to as the transition from Web 1.0 to Web 2.0, and recent developments in mobile technologies have encouraged a new approach to communications. This has been further influenced by developments in **social media** that allow users to create, share and exchange content in a virtual community or network. Marketing communications have therefore shifted away from communicating to customers and focusing on **transaction marketing** to engaging with them through **relationship marketing** in order to develop more personal, meaningful and long-term relationships.

Media mix: the combination of channels used to meet marketing objectives. This includes traditional (radio, television, print or display) and digital media (the internet, mobile applications and social media).

Distribution: the means by which goods or services reach the consumer.

Social media: websites and mobile applications that enable users to create, upload and share text, photo, audio and video content on the internet and engage in instantaneous social networking.

Transaction marketing: marketing that focuses on the efficiency and volume of sales.

Relationship marketing: marketing that focuses on customer retention and satisfaction.

Touch point: a point of human or physical contact a customer has with a brand.

Developments in technology have given airlines and airports more opportunities to interact and quickly share information with people, directly or indirectly. The nature of such interactions can often be more persuasive than traditional forms of communication, and the potential reach from people sharing information with their own communities and networks is significant. It also allows airlines and airports to engage with consumers across the entire travel chain and via a much broader range of potential customer **touch points**.

New technologies have had a particular impact on airports because they have given airports the opportunity to engage directly with passengers and gather intelligence on their preferences and buying behaviour. Traditionally, passengers were viewed as airline customers, while airlines were an airport's customer (➤Chapter 7). Airports are still focused on marketing to airlines. However, they also now directly target passengers.

Traditional activities (listed in Table 20.2) will continue to be important. However, they need to evolve and coexist with new forms of communication in a way that allows companies to target more narrowly defined markets with specific messages rather than targeting a wide audience with general messages. It is also important for companies to develop marketing tactics that include online and offline spaces (see Case Study 20.2).

Copywriting: producing copy (written material) that can be used for marketing purposes to influence customer opinions and/or purchasing behaviour.

As companies adapt to a broader and more complicated media mix, they need to consider the messages of the marketing copy that they use because digital media has expanded the range of **copywriting** opportunities to include web content, emails, blogs, posts, tweets and other forms of electronic communication. Writing for digital media differs from traditional media because the content needs to be more concise (there is a 280-character limit on Twitter) and subsequent dialogue or the co-creation of messages is likely. There is also a certain style, and a growing list of online acronyms and terms, that copywriters need to be familiar with.

Regardless of whether marketing copy is written for online or offline spaces, companies need to plan carefully in order to avoid negative attention. This was the case for Air Asia in 2019 when they advertised on buses in Brisbane with the words 'Get off in Thailand'. In several English-speaking countries including Australia, to 'get off' is used not only to describe the process of disembarking an aircraft but also as a sexual euphemism. Whether it was the intention of the airline or not, it quickly faced complaints and criticisms for promoting sex tourism, and the advertising campaign was ultimately withdrawn.

CASE STUDY 20.2

DALLAS FORT WORTH INTERNATIONAL AIRPORT'S HOLIDAY PARKING CAMPAIGN

In December 2015, Dallas Fort Worth International Airport (DFW) launched a campaign to attract new customers to trial their premium parking product, Terminal Parking, which featured close-to-gate and covered parking facilities. The campaign was part of a wider 'Find Joy' this holiday season campaign that targeted the 3.1 million customers expected to pass through the airport during the 2015–16 holiday season. The campaign included special offers on parking, for instance, with a 'Park 3, Get 3' promotion that provided three days' free parking to customers parking for at least three consecutive days.

To generate awareness among the target audience, DFW used:

- Search engine optimisation to optimise paid and organic placement of advertisements online (e.g. using Google AdWords) when customers searched for parking options at DFW.
- Various sizes of animated online banner advertisements placed through general market and travel platforms.
- A homepage banner and feature article on the airport's own website, social media sites (Facebook, Instagram and Twitter) and e-newsletter.
- A dedicated microsite for the wider 'Find Joy' campaign featuring the message 'Download the DFW mobile app to enjoy the gift of stress-free parking'.
- Media outreach focused on local and national placements in print and online.
- Radio advertising consisting of 10 to 15 spots on 17 stations within the DFW catchment area.

A fair amount of media coverage was received during the campaign period, including four print articles, 13 online articles and four television broadcasts with a total value of US$515,000. Over 50,000 online impressions (the number of times an advertisement is delivered online) resulted in almost 3,000 clicks on the advertisement, and there was an estimated radio audience of almost 1.5 million. Parking revenue for December 2015 was US$482,000 higher than in the same month of the previous year. Prior to the campaign, parking was trending away from the airport terminal to remote and express surface facilities. The campaign helped to reverse this trend and accelerated on-airport parking growth and occupancy rates.

20.2 Principles of engagement marketing

It is no longer enough to simply communicate to consumers. Instead, it is necessary to interact with and engage them. The concept of **engagement marketing** has become increasingly important for marketing communications. It represents a shift from mass to niche communications. Decisions therefore need to be based on a detailed knowledge and understanding of the target market, and the objectives for engagement marketing should be clear and measurable. With the exception of personal selling, many traditional campaigns were characterised by disconnected, point-in-time communications. Engagement marketing offers a more continuous approach that seeks to nurture long-term relationships. It should seek not only to engage consumers but also to involve them in the production and co-creation of marketing programmes and encourage them to actively connect with the brand.

Initiatives for engagement may be delivered outside or inside an aviation setting. Outside initiatives may provide live interactions that help to develop **brand advocates**, while initiatives inside tend to drive purchases and satisfaction at the point of consumption. Traditional initiatives based on the activities listed in Table 20.2 are still used by airlines and airports, especially when the objective is to reach a mass audience. However, the more innovative and engaging campaigns also tend to incorporate new technologies. For example, in late 2017 American Airlines covered 85 taxis in London with the airline's branding (see Figure 20.1). The taxis were mainly based around Canary Wharf and The City, where potential business class travellers who fly frequently between London and the US are

Engagement marketing: marketing used by companies to engage consumers, involve them in the production and co-creation of marketing programmes, and encourage them to actively connect with the brand to shape the way in which it is developed and form a long-term relationship with it.

Brand advocate: a person who voluntarily promotes a brand or product through word-of-mouth or online.

Geofencing: uses GPS or other similar technologies to create a virtual geographic boundary around something. Software can then be used to trigger a message to a mobile device when it enters or leaves that geographic boundary.

Marketing buzz: where consumers amplify and/or alter an original marketing message by generating a sense of anticipation, energy or excitement.

concentrated. **Geofencing** technology on board the branded taxis was then used to send highly targeted digital advertisements to people's mobile devices. Additional outdoor advertising was used on office digital panels, the Canary Wharf shopping centre and other outdoor spaces, along with online advertising. The campaign reached an estimated 4.1 million people, and geofencing as a moving object for marketing purposes was a world first, so the campaign received a significant amount of press coverage on both sides of the Atlantic, and purchase intent grew by 34 per cent.

Common engagement initiatives used by airlines and airports are shown in Table 20.3. Initiatives can be delivered offline and online – preferably using an integrated, multichannel approach that may, for instance, deliver an offline initiative that provides a real interaction which is then supported and strengthened by online initiatives that create a **marketing buzz**, or vice versa.

Figure 20.1   American Airlines branded and geofenced taxi.
Source: American Airlines

Table 20.3 Common engagement marketing initiatives

Mainly offline – experiential	*Mainly online – digital*
• Mobile marketing	• Online advertising
• Street marketing	• Email marketing
• Marketing through amenities	• Mobile technologies
• Events and micro-events	• Social media
• Thematic marketing	• Crowdsourcing

Experiential initiatives

Experiential initiatives provide a physical or emotional experience of the brand through direct exposure to and interaction with it (see Case Study 20.3). Mobile marketing takes the brand to the consumer and provides opportunities for interaction and real experiences. These events may occur outside of an aviation setting and use custom-branded vehicles to draw attention to the offering. The vehicles act as mobile displays but can also stop and provide brand experiences, for instance in town centres, in car parks or at events. In 2017, United Airlines themed food trucks travelled around New York and San Francisco offering complimentary themed desserts from their international destinations, which included *Rugelach* (a filled pastry) and black and white cookies from Israel, *Dulce de leche* (coconut cream and pineapple donuts) from Latin America, *Sakura* cherry donuts and *Yuzu* lemon zest donuts from Japan, and traditional almond horseshoes and apple crumb cake from Germany. It was part of their 'Taste the World' promotion that highlighted destinations served by the airline.

Street marketing is similar to mobile marketing in that companies tend to use spaces in a non-aviation setting to draw attention to their offering, for instance, by providing competitions or real brand experiences. In addition to the taxi geofencing campaign mentioned earlier in this chapter, American Airlines launched a street marketing campaign in 2017 to engage target passenger markets in the Canary Wharf area of London. The airline set up a bespoke Transatlantic Sports Lounge at the Canary Wharf Plaza where commuters could experience a taste of American culture. Visitors to the lounge were greeted by American Airlines flight attendants. They could then work or relax in private booths, enjoy complimentary canapes and drinks and take part in sporting activities and competitions such as throwing an American football for the opportunity to win flights to US destinations served by the airline.

Events are often used to capture the public's imagination. They may be one-off or recurring events that take place inside or outside an aviation setting. They sometimes include an element of surprise and creativity such as a 'flash mob' and may include an element of generosity; for instance, buying gifts for passengers or making charitable donations that demonstrate an investment in shared values with the target market. Some events are recorded on camera and go viral (termed **viral marketing**) when posted on social media sites such as YouTube. Smaller micro-events at an airport or on board an aircraft are also used to surprise passengers. Examples include Virgin Australia's Fashion Runway in the Sky – a fashion show on board a flight from Sydney to Melbourne in 2018 to celebrate the airline's partnership with the Virgin Australia Melbourne Fashion Festival.

Similarly to street marketing, airlines and airports may engage consumers by marketing through amenities such as a bus shelter. Hong Kong International Airport did this for a three-week campaign in 2018. The campaign aimed to raise awareness for the airport's mobile application 'HKG My Flight' and their smart luggage tag initiative 'MyTag', which offers passengers who have downloaded the application an opportunity to receive notifications as to when their luggage is due to arrive at the airport's baggage reclaim

Viral marketing: using social networking sites or other technologies to pursue marketing objectives such as increased brand awareness through a process of self-replication. This may be achieved when a message (in any form such as text, an image or a video) that is placed by a company or individual, for instance, on a social networking site, website or email, is spread rapidly to a wider audience as a result of being copied and passed on by others.

carousel. To do this, the airport created a mini baggage reclaim hall to showcase the mobile application and smart luggage tag functions at a bus shelter on Nathan Road in Tsim Sha Tsui – a busy downtown shelter that serves over 20 bus routes to all parts of Hong Kong. A television was installed to play a videogame to win free MyTag redemption cards. Complimentary MyTag redemption cards were also given to people that downloaded the HKG My Flight application to their mobile device, with complimentary Wi-Fi provided so that people could download it. Facebook and a collaboration with *U Magazine* (a top travel magazine in Hong Kong) helped to promote the campaign, which resulted in 13,000 downloads of the HKG My Flight application over the three-week period (the number of downloads had stagnated prior to the campaign) and over 4,000 MyTags were redeemed within three months after the campaign.

Another example of a marketing through amenities campaign was launched by Norwegian in 2018 in Helsinki to encourage Finnish residents to fly to Mediterranean destinations from Helsinki. Digital displays at bus stops in the city compared real-time weather data for Helsinki and chosen destinations such as Tenerife – hoping that the warmer and sunnier climate would encourage people to book flights with them. Sales for advertised destinations increased by up to 20 per cent during the campaign period.

As Case Study 20.3 shows, thematic initiatives are often used. These may be concept-based and time-limited but offer an effective and attractive way of engaging people. Several airlines have used themed aircraft to celebrate anniversaries. For instance, in 2019 British Airways celebrated its centenary by painting four of its aircraft with special heritage liveries from its past. Thematic initiatives may also be based on a subject such as a famous film or character. For instance, Brussels Airlines operates six aircraft themed in the style of Belgian cultural icons including cartoon characters the Smurfs, the festival Tomorrowland, comic hero Tintin, the Belgian artist Rene Magritte, the national football team of Belgium (known as the Red Devils) and Renaissance painter Pieter Bruegel the Elder. The aircraft liveries and themed interiors will be time-limited, with the airline expecting to have phased all of the current ones out by 2024. The Bruegel-themed aircraft was added in 2019 (see Figure 20.2) and is a collaboration between the airline, the tourism board Visit Flanders and Bozar – the Centre for Fine Arts. The themed aircraft was part of a wider project to promote Flemish Masters in 2019 and 2020 with 2019 being the year of Bruegel. To help attract tourists to Belgium, the project also includes a 'Hi Belgium Pass' that allows tourists to fly from 48 European cities to Brussels, take unlimited train journeys to Belgian cities and visit various cultural attractions for free including Flemish Masters exhibitions and other collections throughout Flanders and Brussels. Visit Flanders has also introduced a Bruegel art installation at Brussels Airport to help immerse tourists in the Bruegel experience upon arrival.

CASE STUDY 20.3

WESTJET'S CHRISTMAS CAMPAIGNS

December 2018 saw Canadian low cost airline WestJet deliver its seventh successive Christmas campaign. Their first one in 2012 surprised passengers at Calgary International Airport waiting to board a flight to Toronto with a Christmas flash mob of carol singers, elves, snowmen, fairies, reindeer, Santa and gifts. In 2013, they surprised more than 250 passengers on two flights to Calgary International Airport. Passengers were asked to tell Santa via a video screen what gifts they wished for that Christmas before departing for Calgary. The gifts were then waiting for them as they arrived in the baggage hall. In 2014, they placed a sled in the town centre of Nuevo Renacer in the Dominican Republic – one of the destinations WestJet serves in the country. The sled included a video screen where residents could tell Santa what they wished for that Christmas. The next day, residents were invited to a Christmas-themed party on a beach where the gifts they had wished for were waiting for them. WestJet also unveiled a small playground in the town centre. In 2015, thousands of WestJet employees hit the streets and airports in over 90 destinations served by the airline to carry out 12,000 random acts of kindness in a 24-hour period. In 2016, the airline invited residents of Fort McMurray (the location of a major wildfire in May 2016 that destroyed approximately 2,400 homes and buildings) to a Christmas party where they could take their minds off the wildfire and create new family memories. In 2017, the airline met with children from the Boys and Girls Club of Canada to determine what Christmas means to them. It then brought their ideas to life via 12 stunts over 12 consecutive days across the WestJet network.

In their 2018 campaign called 'Christmas Miracles: Uniting through Traditions', the airline wanted to emphasise how its fleet expansion will make it a truly global carrier that is able to connect loved ones all around the world. The campaign celebrated Christmas traditions around the world including those in Australia, Japan, Haiti and the UK. It featured a Christmas advent calendar that released a new video each day that featured the airline's Blue Santa going to a different global destination to showcase the traditions of that country or city, and also to surprise people along the way with reunion experiences and the opportunity for people to create their own reunion experiences through an interactive global map. It also incorporated a reunion experience contest to win flights for two to anywhere that the airline flies to. For the 2018 campaign alone, WestJet shared 25 videos on their Facebook site and designated content hub, resulting in 438 million earned media impressions, over 8 million video views, and a 19 per cent increase in organic views compared to the previous year. Overall, the campaign generated CAN$11 million in direct sales, which was a 27 per cent increase from the previous year.

Stop and think

!

Why do airlines and airports engage in experiential and thematic initiatives, and to what extent do you consider them to be effective marketing tools?

Figure 20.2   Brussels Airlines Bruegel-themed aircraft.
Source: Brussels Airlines

Storytelling: initiatives used by companies to convey illustrative and memorable events in words and/or images in order to create strong emotional bonds with consumers and enhance their loyalty to the brand.

Another initiative associated with thematic concepts or subjects is the art of **storytelling**. Storytelling has always played an important role in society as a means of entertainment, education and the preservation of cultures. It has also become recognised for the role it can play in marketing communications. Stories provide an effective way of communicating messages about a brand and can create strong emotional bonds with consumers that enhance their loyalty to the brand. When told in the correct way, stories can be a good way of simplifying and communicating complex messages in a memorable way. The stories will be particularly compelling when conveyed by independent voices – thereby building trust through transparency and authenticity. This compares to traditional ways of communicating messages that may evoke counter arguments or be quickly forgotten. Of course, there are some potential weaknesses. Sequencing and progression of the story will be important, especially if it is part of a multiple narrative, and care should be taken to make sure that the story does not distract people from core messages about the brand. Complex stories will require too much effort to digest, and people will quickly lose interest. In addition, people may be wary of stories that are told from a single viewpoint.

Storytelling has been a key feature of several Southwest Airlines campaigns. In 2017, they ran a '175 Stories' campaign that profiled reasons for flying from 175 of their customers – each story was featured on a microsite for the campaign and linked to a seat number on a virtual Boeing 737 aircraft that the airline was due to add to its fleet. The campaign ran on television, digital and social media, radio and outdoor advertising and in airports and other locations, and included a '#175Stories' tag and a contest to win free flights with the airline. In 2018, the airline appointed several loyal customers including a former professional football player, a radio personality and a barefoot water skier as social media ambassadors as part of their 'Southwest Storytellers' programme, while in 2019 the airline launched a storyteller contest for social influencers to win up to 12 round trips with the airline in return for stories about their travels.

Similar initiatives have been used by airports. For instance, Washington Dulles International Airport hosted 23 social media influencers as part of their 'Passport to #Dulles2018' campaign for a day to experience first-class travel at the airport. The influencers were also introduced to new products and services planned at the airport for 2018. The influencers subsequently shared photos and comments online, for instance, about a new online booking system for the airport, a curbside valet and new food and drink outlets. Several airports such as London Heathrow, Vancouver and Helsinki have used resident storytellers who live in the airport for a time-limited period and share their experience through storytelling. For instance, Helsinki Airport, which has the three-letter code HEL, launched a '#LIFEINHEL' campaign in 2017 where a Chinese man named Ryan Zhu lived in the airport for 30 days. His social media postings and media coverage of the stay were estimated to have reached 2.2 billion people. The airport also used feedback from Ryan to try to improve the passenger experience, for instance, addressing issues relating to terminal space, language barriers, dining options and payment systems for Chinese passengers.

Stop and think

Why might airlines and airports want to personalise their marketing messages, and what are the potential disadvantages of this approach?

Digital initiatives

Online advertising initiatives that make use of targeted and **remarketing** tools to target users with personal messages are now commonly used by airlines and airports, and **flow advertising** is also used to adjust elements of a company's online advertising in real time based on signals from the consumer, such as the device that they are using, the amount of time that they spend watching an advertisement or the search that they perform after viewing an advertisement. This is a more dynamic type of sequential advertising, which is when a campaign is delivered over a series of pre-determined stages. Web 3.0, which is described as the next stage in the evolution of the internet from interactive to **semantic websites**, may also be used in the future. This will enable airlines and airports to be more effective in responding to complex web-based searches such as 'find me a flight from London to New York with my favourite airline'.

The rapid growth of emails as a form of communication facilitated considerable interest in email marketing. This form of direct marketing shares many of the strengths and weaknesses listed in Table 20.2. The idea is that companies customise messages in electronic form that can be sent by email to a target market. Individuals can then be encouraged to share the messages by forwarding the email on to their own contacts. Approximately 235 billion emails are sent and received each day worldwide, so marketers need to find ways to get their email seen and may benefit from combining their email campaign with other activities, for instance, with direct marketing by post and/or the use of innovative mobile technologies and social media.

Remarketing: reintroducing a product or service to existing customers to remind them about the brand.

Flow advertising: adjusting the way in which various elements of a marketing campaign are delivered in real time based on signals from consumers.

Semantic websites: intelligent and intuitive websites that are better able to serve user needs.

The vast majority of air travellers now use mobile devices before or during their journey, and the number of airlines and airports that have developed dedicated and branded mobile applications has increased dramatically. Common services on airline mobile applications include flight search and booking, mobile check-in and boarding passes, flight status and loyalty programmes. For airports, common services include flight status and airport information on wayfinding, public transport links and support for passengers with reduced mobility. However, mobile applications are also increasingly used to advertise sales promotions, distribute non-aviation-related products and services (such as duty free purchases or the pre-booking of airport car parking) and provide links to social media.

Mobile technologies are used to support a wide range of initiatives such as American Airlines' geofencing campaign and Hong Kong International Airport's HKG My Flight and MyTag campaign that have already been mentioned in this chapter. Another interesting initiative is Delta's 'Dating Destination Wall' that was launched in partnership with social search and dating site Tinder in 2017. It was based on selfies that involve people taking a self-portrait photograph using a smartphone or tablet device and then sharing it on social media. The wall was located on Wythe Avenue in Brooklyn, New York, and featured painted images of nine exciting destinations served by the airline. Singles could then take a selfie with the destination in the background to make them look like international jet-setters when adding it to their dating profile on Tinder. Users were also encouraged to post their images to Instagram with the hashtag #DeltaDatingWall, therefore extending the reach of the campaign and inspiring people to fly to the destinations with Delta.

Growth in the use of social media during the last decade or so has been considerable and has provided new instantaneous online engagement opportunities for airlines and airports. According to the respective sites, the number of active monthly users towards the end of 2019 was approximately 2.4 billion on Facebook, 1.9 billion on YouTube, 1.6 billion on WhatsApp, 1.3 billion on Facebook Messenger, 1.1 billion on WeChat and 1 billion on Instagram. The number of airlines and airports that use social media has also grown rapidly, and many airlines and airports now have huge followings on their social media accounts. For example, at the end of April 2019, Qatar Airways had 14 million followers on Facebook, 2.5 million followers on Instagram, 1.4 million followers on Twitter, 722,000 followers on LinkedIn, 150,000 subscribers on YouTube and 18,000 followers on Pinterest. Singapore Changi Airport had 3.8 million followers on Facebook, 262,000 followers on Instagram, 98,000 followers on Twitter, 77,000 followers on LinkedIn and 26,000 subscribers on YouTube.

The relative ease and speed with which social media campaigns can be launched means that many companies initiate social media campaigns without setting clear and measurable objectives or using **marketing analytics** to assess and optimise their performance. Companies often find that it takes a great deal of time and resources to maintain and manage social media efforts effectively, especially given the vast range of platforms that can be used. A particular challenge with using social media is that the quality of content can vary, and it can be misused or abused and attract negative comments from users. Negative comments can spread quickly and potentially damage a brand.

Marketing analytics: assess the performance of marketing efforts so that they can be adjusted to provide the optimal return on investment. Analytics assess and optimise efforts delivered via offline spaces such as television, radio, printed and outdoor media and online spaces such as websites and social networking sites. The latter include on-site analytics that relate to a company's own sites and off-site analytics that relate to what is happening on other sites and in online spaces in general.

Stop and think

What challenges does social media use pose for airlines and airports, and how can social media be used to transform negative customer experiences into positive ones?

Crowdsourcing is increasingly used by airlines and airports as a tool for engagement. It is listed under digital initiatives because developments in technology have made it easier for companies to create online crowdsourcing initiatives. However, it can be conducted online or offline, or via a hybrid approach that combines both. The aim is to solicit ideas and opinions from a large audience, and then to use that intelligence to make business decisions. The aim is to develop loyalty to the brand through a process that engages people and encourages them to associate with it. It should complement, rather than replace, more traditional feedback mechanisms. Many airlines and airports have introduced crowdsourcing initiatives in recent years. For instance, Copenhagen Airport has a platform called CPH Ideas, where users can suggest ideas for the airport of the future. Users can then vote on whether they like the idea or not, and the most popular suggestions are then considered for implementation. However, crowdsourcing initiatives such as this are open to anyone and may therefore result in a large volume of low quality suggestions being submitted. Air France recognised this by introducing a targeted crowdsourcing approach by selecting students and recent graduates in design, fine art, graphic art and industrial design from high-calibre universities in France, Belgium, the Netherlands, Italy, Switzerland and the UK. This limited the number of participants and resulted in a more targeted, knowledgeable and manageable pool of participants, and effective innovations such as the design of new tableware for the in-flight catering service.

Companies need to track and assess the effectiveness of any marketing initiatives, and a particular need to do so for digital initiatives comes in part from concerns that consumers' responses to brands are not always associated with the level of engagement that is experienced. This is especially the case with social media because, while campaigns may go viral, boosting the number of followers on the company's social media sites and extending exposure to their brand, they do not necessarily result in a desired response. Airlines and airports therefore need to try to build real-world relationships and not just relationships on social media, which is why it will be necessary to focus on integrated, multichannel initiatives in the future that are delivered across both online and offline spaces.

> **Crowdsourcing:** the process of obtaining intelligence by soliciting ideas and opinions from a large audience, and then using that intelligence to make business decisions, for instance, on new or improved products. It should also seek to engage people in a way that encourages them to associate with the brand and develop loyalty towards it.

Stop and think

How can airlines and airports turn social media into a revenue stream?

Key points

- Changing consumer trends and new technologies are influencing how airlines and airports market themselves to consumers.

- Marketing emphasis has switched from communicating to customers to actively engaging with them in order to develop more personal, meaningful and long-term relationships.

- Traditional, experiential and digital initiatives can be used.

- Markets are increasingly crowded with messages, and airlines and airports need to find innovative ways to break through the marketing clutter and be more effective in delivering their messages to the right people through the right media at the right time. To help facilitate this, they will need to have a clearer understanding of their customers and their willingness to engage in different media and provide them with offers of true value.

- The shift from traditional to digital media and from desktop to mobile internet, possibly via wearable technology, is likely to continue. This will increase opportunities for engagement across the entire travel chain.

- Airlines and airports need to focus less on the types of media that they use, and more on how they can develop creative and compelling messages to engage consumers. Messages that spark excitement, tell the brand's story or strike an emotional chord with consumers are likely to be popular.

- Social media has become widely used for marketing communications. However, relentless and potentially damaging effects of social media also mean that companies that do use it need to remain vigilant at all times. Problems can quickly go viral, and inappropriate responses or attempts to cover them up often only make things worse.

- Airlines and airports therefore need to manage such communications in a calm, transparent and timely manner, which can help to build trust and develop brand advocates.

- While airlines and airports should embrace new opportunities and experiment with new technologies, they must integrate the range of messages and media that they use and track their effectiveness against clear objectives.

- Marketing analytics help understand and influence market responses and maximise returns on marketing investment in terms of sales and profits.

References and further reading

Halpern, N. (2018). Airport marketing. In: N. Halpern and A. Graham, eds., *The Routledge companion to air transport management*, Abingdon: Routledge, pp. 220–237.

Halpern, N. and Graham, A. (2013). *Airport marketing*. Abingdon: Routledge.

Hanke, M. (2016). *Airline e-commerce: Log on: Take off*. Abingdon: Routledge.

Waguespack, B. P. (2018). Airline marketing. In: N. Halpern and A. Graham, eds., *The Routledge companion to air transport management*, Abingdon: Routledge, pp. 206–219.

CHAPTER 21

Air transport in regional, rural and remote areas

Rico Merkert

LEARNING OBJECTIVES

- To recognise the social and economic importance of aviation to remote regions.

- To understand the management challenges of providing air services in remote regions.

- To evaluate aircraft types, operational challenges, financial viability and franchising.

- To appreciate how essential air services are supported financially, procured and regulated.

- To assess future developments in this market segment.

21.0 Introduction

Air transport in remote regions is often seen as a niche market with very low margins and services that are more exotic than meaningful in terms of volume. Given their niche character and often monopoly market structure, however, some services, such as those contracted in by primary industries, including mining and mineral exploration, are highly lucrative. Even the many regular scheduled air services on low volume routes that are not commercially viable without public support (e.g. subsidies or regulation) are much more than a niche product to the communities and businesses that depend on them. To these communities, regional air services are much more than a business, they are the connection to the next island, the mainland, the regional centre and the rest of the world. Air transport to and within remote regions is a market segment that goes beyond a narrow focus on just profit margins to an appreciation of wider tangible economic and social benefits that air services can potentially deliver. Indeed, it can be argued that without aero-medical services and regular public flights (also referred to as air buses), many remote communities would become marginalised and even more isolated, leading to a risk of de-population and economic decline.

It is widely accepted that aviation plays a vital role in the regional, rural and remote (RRR) context. Local businesses, airline operators and remote airports frequently highlight the substantial economic impacts of air transport to geographically remote or isolated regions. IATA has recognised the wider economic benefits of aviation in small island developing states (SIDS), particularly its role in developing tourism. Economic impacts usually refer to employment and income generated and include four main types: direct impacts (generated by the direct construction and operation of remote aviation), indirect impacts (generated by the chain of suppliers of goods and services related to remote aviation), induced impacts (generated by employees spending their income) and catalytic impacts (generated by the role of air transport in remote regions as a driver of productivity growth and an enabler of inward investment). Research on these impacts has occurred in Europe, where in Norway it has been shown that remote airports are important catalysts for local investment (see Bråthen and Eriksen, 2018). In Canada, the US and Australia, Baker, Merkert and Kamruzzaman (2015) found a strong bidirectional relationship between regional aviation and economic growth.

This chapter will explore the management of air transport in remote regions. The provision of safe and successful operation of such services demands particular management strategies that are adapted to the unique social, commercial and natural environment of small communities, small island states and other remote regions. The role of government subsidies, public service obligation routes, franchising and aero-medical services will also be discussed.

21.1 Market segments of air services to remote regions

Studies undertaken on air transport in remote regions have focused on regular passenger flights that are not commercially viable unless subsidised by public money. While this is certainly an important market segment with a number of interesting contractual, economic and social implications, air transport in remote regions is more diverse and complex than that. There are, for instance, regular/scheduled passenger air services that are operated on denser, commercially viable routes to remote centres and tourist destinations. These air services don't require public support and are therefore an entirely different proposition for airline management. Often more viable are charter tourism and corporate flights to remote destinations, which include the recently rapidly growing fly-in, fly-out operations (FIFO) for mining and natural resource companies around the world, most notably in Australia, Canada, Russia, Africa, Brazil, Norway and the US. New South Wales and Western Australia, for example, have experienced significant development in the mining sector, and remote airports play an important role in the construction and operation of the mine as well as the economic development of these regions and the movement of skilled labour, machinery, supplies and services.

Another dimension of air transport to remote regions worthy of consideration is cargo. While most of the cargo is limited to small consignments due to the limited capacity of the aircraft in operation, mail and newspapers have always been very important aspects and revenue streams of the remote airline business model. Perishable freight can also be readily exported to world markets. Further aspects of air transport in remote regions are government and military traffic as well as surveillance, border and maritime patrols. Emergency and general medical services, such as those provided by the Royal Flying Doctors in Australia, add to the variety of remote air transport applications. Finally, air transport in remote regions is also of importance to supply chain resilience, disaster relief and humanitarian missions.

Another fact that makes consideration of air transport in remote regions multifaceted is that there is no universal definition of what constitutes a remote region. In the context of public support for scheduled public air services, the only indication of 'remoteness' appears in the documents that govern public support for remote air services in the US and Europe, but even here the definitions differ. The US Essential Air Services Program (EAS) funds only services to communities that are located at least 70 miles from the nearest hub, while the European **Public Service Obligation (PSO)** Scheme allows Member States to impose PSOs on routes to peripheral areas and airports where air services are vital for the economic and social development of the surrounding region. The EU definition includes economically underdeveloped regions, including some in France, that do not have to be at the periphery or on an island. A more specific definition of remoteness is that without air services it would not be possible to achieve a day business trip to the next largest regional centre.

What all these definitions have in common is that the region in question is highly dependent on air transport as alternative ground transport provision is not available or incurs significant travel time penalties. Whether or not these characteristics translate into sufficient demand for commercially viable transport services is another matter and not part of the definition of remoteness. Another way of defining the scope of air transport in remote regions is to focus on the airports that are located in those regions. For example, the Australian Airport Association clusters its member airports into the following:

- Tier 1 State Capital City Airports;

- Tier 2 Non-Capital International Gateway Airports;

- Tier 3 Major Regional Airports with direct interstate services;

- Tier 4 Major Regional Regular Passenger Transport airports without direct interstate services (with more than 20,000 passengers);

- Tier 5 Regional Airports without direct interstate services (with less than 20,000 passengers);

- Tier 6 Regional Airports without Regular Passenger Transport services (general aviation operations only);

- Tier 7 Remote Community Aerodromes (which exist for community service aviation and medical emergency flights).

PSO (Public Service Obligation): regulated and/or financially supported regular air services that operate under contract and with obligations set by a public authority to remote or economically underdeveloped regions that would not exist without government support.

While only Tier 1 airports are regulated, Tiers 4–6 and particularly Tier 7 airports are of interest here. That said, given that Perth is the most isolated city in the world, one could argue that despite its Tier 1 and regional hub airport status, the airport still faces challenges owing to its remote geographic location on the west coast of Australia. In general, the airports of concern here include many facilities that are classified as regional, rural and remote airports. All of these provide services that are vital to the communities they serve.

Out of the 2,000 aerodromes in Australia, only around 250 airports receive regular passenger services. The other, often much smaller, airfields and landing strips are, however, vital as they support important connecting flights, emergency services, pilot training (general aviation), maintenance, mail and freight. However, as with most other countries, only 50 per cent of regional airports in Australia are able to cover the cost of their operations. This has long-term implications for their commercial viability and hence the provision of passenger, freight and emergency services. While air transport to remote regions presents many opportunities and has significant economic and social impacts for the communities and businesses they serve, it also poses a number of particular operational and financial challenges. However, even though regular public air transport services and loss-making remote airports have received considerable political and academic attention (principally owing to the need to justify the public support they receive), the overall market for remote air services is diverse and complex, as Case Study 21.1 details.

HIGHLAND AIRWAYS

CASE STUDY 21.1

Highland Airways was a small airline based in Inverness in the north of Scotland. It was founded as Air Alba in 1991 as a flying school and diversified to become a significant operator of remote air services in the UK before it entered voluntary administration in March 2010. At the height of its operation Highland Airways employed 110 staff, had a fleet of 11 aircraft (including Cessna F406, BAe Jetstream and BN2 Islander aircraft) and operated limited scheduled as well as ad hoc charter services for leisure and corporate clients. It operated six scheduled passenger routes (most of them PSOs) in Scotland, including Stornoway–Benbecula and Oban–Inner Hebrides, as well as Cardiff–Anglesey in Wales and Lappeenranta–Helsinki in Finland. It was also involved in air freight charter services (including daily newspaper distribution to Orkney, Shetland and the Western Isles of Scotland as well as Royal Mail charters). It also operated a five-year contract for the Scottish Fisheries Protection Agency. From 1998 it was involved in seasonal contract charter flights for the oil industry, connecting Aberdeen-based oil rig workers with the rest of the UK, Ireland and Norway. It flew corporate shuttles for both Naval and Air Force supply contractors and fulfilled complex UK itineraries between small private airports.

Despite its niche market experience and a management buyout in 2007, the airline did not survive the oil price rise in 2008 as it had not hedged its fuel requirements (see Chapter 12) and did not have reactive fuel price adjustor clauses built into some of its long-term PSO and industry contracts. The airline experienced difficulties transferring from their J31 and C406 aircraft to larger aircraft. Operating different aircraft from several remote bases with limited spares and operational resilience when technical problems were experienced proved to be challenging, and the business model was ultimately unsustainable.

Stop and think

What are the key characteristics of air services to remote regions?

21.2 The management challenges of providing air services to remote regions

Providing air services to remote regions is often challenging. Low temperatures, hazardous terrain, limited infrastructure, **Short Take-Off and Landing (STOL)** and unprepared runways, low visibility, inclement weather conditions, low demand and high costs are just some of the challenges that operators face. In order to assess the market risks (some of which also apply to leasing companies, global distribution systems, travel agents and ground handlers in this market segment) it is necessary to discuss airlines and airports separately. Both airlines and airports face significant financial challenges as evidenced by the long list of airline failures (many small but also larger regional players such as Iceland's WOW ceasing operations in 2019) and RRR airport closures (including Blackpool, Filton, Manston and Plymouth airports in the UK where the land has been proposed for real estate development). While the market challenges can differ according to geographical context, the biggest concern for both airlines and airports is usually low and often seasonal volumes. For airports in particular, the seasonal or peak demand aspect of remote air services is often a challenge. Although individual airports may average low passenger volumes over prolonged periods of time, demand peaks, such as the construction of a new mine in the area, can result in serious capacity shortfalls.

For airlines, one challenge is managing the volatility and complexity that derives from unregulated air transport services in remote regions alongside their diverse product portfolio. Many airlines have to invest in expensive and specialised equipment to operate from remote airports with limited infrastructure (such as gravel runways) or low visibility, but if they lose the contract to provide that service, they cannot easily reassign these assets to other routes. These sunk costs increase the financial risks of operating the services and puts potential clients in a position of relative power from which to negotiate on price. One approach to this problem for procured regular passenger air services to remote regions is that the procuring transport authorities should own the aircraft and lease them to the winning bidder of the tender for the duration of the contract. Both the Shetland Islands Council and Transport Scotland have adopted this approach. This ensures that the risk is minimised and shifted away from the private sector. Arguably the public sector should control for that risk, as ultimately the route is owned and re-tendered by the public transport authorities and not the airlines.

As airlines serving remote regions need to maximise the utilisation of their aircraft, then even those that operate regular high-demand scheduled routes will often also fly charter, corporate and freight assignments as part of their business strategy. Although the majority of charter and many of the scheduled public air services to remote regions are undertaken by small airlines, carriers range in size from very small independent operators to large

STOL (Short Take-Off and Landing): aircraft with specific performance characteristics that enable them to operate from short runways that may be constructed from snow, ice or gravel.

international airlines. Some FSNCs (including Air France) operate to remote regions or have franchises (such as Brit Air which is owned by Air France/KLM) or subsidiaries (such as QantasLink which is owned by Qantas) that operate in those markets. Large operators, however, are the exception rather than the norm.

> ## ❗ Stop and think
>
> Why do relatively few large airlines operate dedicated regional subsidiaries?

Once small carriers have diversified into a number of niche markets, their often limited financial and management capabilities combined with their limited experience of risk management techniques may leave them exposed to sudden changes in demand and cost shocks as demonstrated by the jet fuel price spike and global financial crisis in 2008 (or the 15 per cent fuel price spike in September 2019). Most airlines serving remote regions do not have sufficiently robust balance sheets or the financial ability to engage in fuel hedging, and many of their long-term corporate or PSO contracts did not include fuel cost indices that would have protected them to some extent from rapid rises in the price of this essential commodity. As a consequence of this experience, fuel cost indices are often now included in the contracts. However, fuel hedging can only protect from sudden price increases and cannot insulate a carrier from the effect of a price rise indefinitely, which can cause problems if these price rises are mid contract. A further key issue is attracting high-calibre staff (such as engineers and experienced managers) to remote airports and regional airlines to operate in what may be perceived to be less glamorous and less well-paid positions. The various skills needed to run such an airline – including yield management, safety compliance, effective HR, financial planning and quality management practices – are often delivered in a multi-tasked, often self-taught, way.

Low or volatile demand is, however, only one part of the problem. Another, interrelated challenge is that operating aircraft in remote regions is relatively expensive when assessed on a per seat basis. Small aircraft and relatively short sectors (such as the famous 45-second scheduled flight between the islands of Westray and Papa Westray in Scotland), shorter operational windows, because thin routes can bear only low frequencies, and limited bargaining power with suppliers mean remote airlines are not able to spread their operating cost over the 500 seats found in an A380 but the 12–30 seats typically available in small regional aircraft. This results in relatively high costs per seat and per available seat kilometre, which translates into significantly higher fares. This affects consumer demand for the services and obliges regional residents who rely on the services to pay more for their tickets.

The unfavourable cost structures which result from diseconomies of scale and scope do not have to be a challenge to profitability as long as the airlines can secure sufficiently high yields. Although high yields can be obtained from the corporate, charter tourism and mining sectors, airlines operating in remote regions remain heavily dependent on regional economic development and the prosperity of a specific industry or economic sector. If one of their key corporate clients changes their operation, or if the airline loses key publicly subsidised transport contracts, the lack of alternative revenue sources means that the airline may enter administration. For this reason, airlines operating in remote regions should try to diversify their operation to geographically and operationally spread the risk. Another important factor to consider is that if business traffic and low elasticity traffic drives up yields in remote air services, local residents may be priced out of air travel, creating a type of market failure which may then prompt the need for some form of state intervention via airfare discount schemes or PSOs.

Stop and think

What are the potential disbenefits of operational diversification for airlines?

Challenges specific to airlines that provide publicly procured remote air services include the lack of access to global distribution systems (GDS) and sophisticated IT systems that can support their revenue management functions. This is despite many PSO contracts requiring the airline to have an interline agreement in place at the larger airport the PSO serves. Research by Merkert and O'Fee (2016) suggest that managers of airlines operating remote routes in Europe often find it difficult to operate in an environment where public transport authorities are not prepared to accept any risk on the demand or cost side. What is more, there is often no effective marketing or route development for remote air services, and PSO contracts often offer few incentives to grow patronage. Ideally, local transport authorities need to 'own' rather than simply administer their PSO route portfolios and include marketing and tourism campaigns that would enable the route to develop into a profitable venture that does not rely on public subsidy.

Barriers to entry, which effectively prevent them from accessing larger hub airports, is another problem remote airlines encounter. In the US and Europe particularly, continued consolidation has resulted in hubs becoming even more congested, and incumbent airlines have reassigned valuable slots to more lucrative and higher-yielding long-haul routes. For major hubs, it is much more profitable to have wide-body aircraft operating into the airport than a 19-seat airframe and, as a result, many regional and remote airlines have been priced out of hub airports. As a consequence, many remote regions have lost their direct connection to global markets, and passengers are required to connect via secondary airports which add additional time and cost to the journey. As an example, the number of domestic UK destinations served from London Heathrow fell from 18 in 1990 to 9 in 2019, and three times as many UK regional airports are served by direct flights to Amsterdam than London Heathrow. Some airports such as Sydney Kingsford Smith are regulated in such a way as to ensure that regional/remote carriers are always allocated a certain proportion of available slots, a process that is called ring fencing.

Stop and think

To what extent could ring fencing slots for regional services at major airports be considered anti-competitive?

For RRR airports, connections to large airports and regional airport networks is of equal importance. As with remote airlines, remote airports are often subject to low, seasonal or volatile traffic volumes and are invariably loss-making. Unlike the large international hubs, which now generate up to 50 per cent of their revenues from non-aeronautical activities such as car parking and duty free purchases (►Chapter 6), remote airports rely much more extensively on aeronautical revenues (either direct or indirect through departure taxes). Changeable and extreme weather conditions necessitate the provision of de- and anti-icing (or anti-flooding) facilities and snow removal equipment, which add to the operating costs. These costs have risen further in recent years as new safety, security, environmental and regulatory compliance measures have been imposed. From a market perspective, cost presents the biggest challenge to remote airports, many of which are owned and operated by local councils. In Australia, recent figures suggest that more than 50 per cent of RRR airports are unable to cover their operating costs, and this places additional pressure on local council budgets. It is hardly surprising that local councils, private airport owners and operators try to find innovative ways of minimising operating costs (see Example 21.1).

Example 21.1

Remote Towers

The biggest challenge for regional and remote airports is high fixed costs. Staff costs per departing/arriving passenger are usually high and, as they are usually under management control, most cost efficiency innovations at remote airports concern staffing. The Remote Tower concept enables air traffic control services and aerodrome flight information services to be provided to remote aerodromes from a central facility that is geographically distant from the airport. For facilities with limited demand and/or restricted operating hours, a Remote Tower offers significant cost saving potential and can confer important social and economic benefits to remote regions. Air navigation service providers around the world are very interested in this innovation as they are all trying to reduce costs without affecting safety or operational availability. This will, however, also have some change management implications. Controllers will need to adopt to new ways of working, and there may also be a need to relocate to central control areas. The concept does, however, address the ATC controller shortage that exists and the challenge of recruiting controllers to remote areas.

Remote Tower trials have been conducted in Sweden at Ängelholm Airport (with ATC services remotely provided from Malmö Airport 100 km away) and Landvetter Airport (the second largest airport in Sweden) and also at two airports in northern Norway (where aerodrome flight information services were provided by Bodø airport). Approval to operate the new technology at Sundsvall Remote Tower Centre to provide ATC for Örnsköldsvik Airport, 150 km away, has also been granted. These trials have confirmed the operational feasibility and safety of the systems as well as their ability to deliver cost savings of up to 60 per cent as staff does not need

> to be employed at individual sites. Airservices Australia have also tested a Remote Tower at Alice Springs Airport which is remotely controlled from Adelaide, over 1,500 kilometres away.

A key challenge for remote airports' cost management is the high and unavoidable sunk capital cost of airport infrastructure. Airports try therefore to save elsewhere, particularly with staff, which is their second largest cost component. Airports seek to improve labour productivity by obliging their staff to undertake a diverse number of activities. An airport baggage handler at a remote airport could, in addition to their primary role, also be responsible for firefighting or airfield operation duties. In terms of human resource management, a further key challenge for remote airports is attracting and retaining skilled staff. Many potential employees relocate to larger towns and cities in search of education and employment and leave a skills gap. With technology advancements (such as automation at airports, for example automated snow ploughs at airports in Norway) this trend may intensify in the future.

From a cost, risk management and business development perspective, another challenge facing regional airports is the fact that the majority of them are served by only one airline operator. For example, more than 70 per cent of Australia's regional airports are served by only one airline. This demonstrates the economic power that these airlines hold over remote airports. Increased airline competition would therefore be beneficial.

Stop and think

!

Consider the relative merits of the strategies remote airports can use to minimise their operating costs.

21.3 Aircraft types and operational considerations

The types of aircraft that are operated in remote regions are as diverse as the destinations they fly to as a consequence of the services' unique demand and operating profiles. In remote areas it is not uncommon for the operator to have to invest in short take-off and landing (STOL) aircraft that can operate from short runways, such as the De Havilland Canada DHC-6 Twin Otter, or, in the case of oil rigs or very mountainous terrain, helicopters. Helicopters are also used on some PSO services (such as those that serve the Lofoten Islands in Norway) but more typically for medical purposes or to connect offshore oil rigs with the mainland. Bristow has substantial fleets of helicopters based in Aberdeen, UK, and Houston, US, to serve the North Sea and Gulf of Mexico oil and gas installations.

The aircraft used in remote regions vary depending on the demand, available infrastructure and the operating environment. PSO air services in Europe use aircraft as big as Airbus A320s and Boeing 777–300ERs to tourist destinations such as Corsica/Bastia. Although

France, Italy and Portugal operate high demand routes using the PSO mechanism, PSO services are normally flown by much smaller aircraft. UK operators, for instance, typically employ aircraft smaller than 20 seats on PSO routes.

A good example of such operations is UK operator Skybus which (as of September 2019) operates a fleet of three Britten Norman Islander 8-seater aircraft and four DHC-6 De Havilland Twin Otter 19-seater aircraft to provide regular scheduled flights between the Isles of Scilly and Land's End, Newquay and Exeter airports on the UK mainland. These relatively simple and mechanically robust turboprops are not only reliable but can operate routes that are often challenging in terms of infrastructure and weather conditions. Flights to the Scottish island of Barra, for example, require aircraft to land on the beach, and scheduling is dependent on the tide (see Figure 21.1). Some of these aircraft can be fitted with specific equipment to enable them to operate as land or seaplanes (see Case Study 21.3).

Table 21.1 details the most popular aircraft in the sub-30-seat market segment. It is worth noting that the total number of aircraft presented in Table 21.1 is 3,628 which represent roughly 1 per cent of the total global fleet in operation (34,344 aircraft) and hence illustrates the niche market character of this market segment. The Caravan (208), B99-B1900 and DHC6 (Twin Otter) are the most popular aircraft. These aircraft are ideal for remote air services not only because of their size but also because they can operate in all climates, in particular the DHC6. This aircraft can be fitted with floats or skis and, as a consequence, is utilised in Norway, the UK, Antarctica and tropical islands in Asia. A key challenge for operators of these aircraft is their age and associated high maintenance costs. Many Twin Otters are around 40 years old, and the need for a replacement is becoming increasingly acute as production ceased in 1988. This has created significant challenges for airlines and remote communities who fear that they could lose their service if cost-efficient alternative

Figure 21.1 A DHC6–400 Twin Otter, owned and operated by Loganair, on the beach at Barra Airport, Scotland

Source: Photo credit: David Savile (with thanks to Highlands and Islands Airports Ltd)

Table 21.1 Sub-30-seat fleet in service (September 2019)

Manufacturer	Variant	Seats	Aircraft
Cessna Aircraft Company	CARAVAN (208)	6–14	678
Beech Aircraft Corporation	B99, B1900	15–19	560
de Havilland of Canada	DHC6	12–20	441
Beech Aircraft Corporation	B1900	16–19	439
Swearingen Corporation	METRO23-METRO III	10–19	270
British Aerospace	Jetstream 31 & 41	12–29	233
Saab	SAAB340	17–25	204
Britten-Norman	BN2 (incl. Trislander)	8–15	159
Let	L410	14–19	154
Dornier	DO228	17–20	127
Beech Aircraft Corporation	B100, B200	6–9	83
Pilatus Aircraft Ltd	PC6, PC12	6–9	61
Embraer	EMB110	12–21	58
CASA	CASA212	22	46
Harbin Aircraft Industries Group	Y12	17–19	44
Yakovlev	YAK40	16–29	35
Boeing/McDonnell Douglas	DC3T	27	16
Reims-Cessna	CARAVAN II (F406)	12	15
Antonov	AN38	26–27	5

Source: Author's analysis based on CAPA and aircraft manufacturer data, correct as of 2019.
Note: Some of these aircraft (such as the SAAB340 or DHC-6) are sometimes operated in configurations that increase their capacity above 29 seats.

airframes cannot be found. This market gap was identified by Viking Air which, from 2006, began manufacturing replacement parts for all of the out-of-production de Havilland Canada aircraft but also the fully re-designed Series 400 (of which more than 60 had been produced by the end of 2014). The benefits of sub-20 aircraft are not only operational but also commercial, as they often receive tax benefits or regulatory exemptions. For example, they do not attract Air Passenger Duty in the UK and are exempt from the EU Emission Trading System (►Chapter 18).

While the majority of remote services are provided by small aircraft, some services are operated by larger aircraft and an up-sizing (to 60–120 seats) trend is apparent in some RRR markets. Table 21.2 illustrates the most popular sub-120-seat turboprop and jet powered aircraft. It is worth noting that there is a growing shortage of 30- to 40-seat aircraft which often leaves airline management with one option, that is to up-scale to larger, more expensive equipment. This change of gauge may have implications for RRR air service economics and air service continuity. There is also growing interest in electric and other,

Table 21.2 Popular sub-120-seat aircraft in service (2019)

Turboprops	Jets
30–60 seats	**30–60 seats**
ATR42	ERJ-135/140/145 CRJ 100/200
DHC8–100/200/300	
EMB-120 Saab 340/2000	**61–90 seats**
	E170/175 CRJ 700/705/900
61–90 seats	Mitsubishi MRJ Avro RJ70/85
ATR72	Comac ARJ21–700; B717–200
DHC8–400	
	91–120 seats
>90 seats	E190/195-E2 A318 B737–200/500/600
TBD Turboprop	Fokker 100 BAe146–300 Avro RJ100
	SSJ100 CSeries CS100 CRJ 1000
	Comac ARJ21–900

non-fossil-fuel-based engine propulsion. Potentially, technological breakthroughs will be provide opportunities initially at the ultra-short haul market (9–10 seats). Several propositions have already been offered by a few manufacturers where there are also opportunities to retrofit existing fleets (BNI and Twin Otter). However, there will be implications for per seat costs and for subsidy levels.

Given the operational challenges that come with many remote routes, niche aircraft or aircraft with specific equipment will always be in demand, as Case Study 21.2 shows.

CASE STUDY 21.2

COBHAM AVIATION SERVICES, AUSTRALIA

Jet Systems Pty Ltd, trading as Cobham Aviation Services Australia, is a diversified and successful niche market player based in Adelaide, Australia. In addition to lucrative fly-in, fly-out (FIFO) and charter operations throughout Australasia, they provide freight (mainly for Australian Air Express, a joint venture of Qantas and Australian Post) and regular scheduled passenger services. Their key selling points are flexibility, safety and reliability. With their fleet of 20 Boeing 717–200 (employed on passenger services operated on behalf of QantasLink), a Bombardier Dash 8 turboprop (mainly for surveillance flights), 13 British Aerospace BAe 146 regional jets (mainly freight) and one Embraer E190, they can operate flights seating between 4 and 100 passengers. Their corporate clients can determine their own itineraries, and this attracts not only government, civilian aerial surveillance programmes and aerial tours but also big resource companies such as Santos, Chevron, Rio Tinto and BHP Billiton. For these clients, Cobham is often the only possible option. One of Cobham's 82-seat BAe 146 aircraft is equipped with a gravel kit that allows the jet to land on unprepared gravel runways close to mines or potential mining sites, which provides the airline with a distinctive competitive advantage.

Stop and think

What are the cost implications of utilising small aircraft on RRR routes, and how might they be managed?

21.4 Financial viability, franchising, public subsidies and PSO routes

On account of the substantial economic and social impact of remote air services, many jurisdictions have decided that it is in the public interest to support the airlines, the airports or (indirectly) the residents that live in remote regions; for example, through subsidised air fares, which often just inflate the overall uncapped fares and distort economic welfare. It is argued that public authorities who have an interest in the economic and social development of the (often thinly populated but still strategically important) regions in question can achieve this most effectively by directly supporting aviation that connects these regions with the world rather than through indirect economic support. Public support can include direct or indirect subsidies, marketing support, tax breaks or protection from competition on regulated monopoly routes.

Airlines usually bid in a competitive tender process for the right to operate regulated and/ or subsidised routes. The most prominent examples of these schemes are the Essential Air Services (EAS) programme and the Small Communities Air Service Development Program (SCASD) in the US, the Remote Air Services Subsidy (RASS) Scheme in Australia and the European Public Service Obligation (PSO) air service mechanism. There are 269 potential, historic and approved PSO routes in Europe of which 176 were still approved and 168 in operations in September 2018.

As Table 21.3 shows, 14 countries operate PSO routes in Europe, and the number of operators in each country varies significantly. While public subsidy can be justified on the grounds of aviation being a **merit good** in the remote aviation context, the resulting principal (transport authority)–agent (airline) contractual relationship and the potential for (mis) interpretation of policy guidelines can result in inefficiencies, inappropriate incentives and a lack of competition for contracts, as evidenced in some European countries operating under the PSO scheme.

Similar schemes of airline support can be found in Canada, Russia, India, China and New Zealand, although these often do not involve competitive tendering and are targeted more at specific airports or specific objectives. Direct financial support for remote airports exist in most countries, and the most researched one is the Route Development Fund (RDF) developed in Scotland and later replicated in other parts of the UK. Australia is another interesting example of airport support. While the RASS scheme provides more than 350 communities in remote and isolated areas of Australia with improved access through the subsidy of a regular air transport services (only six small airlines operate on those thin routes and are so far apart from each other that they don't compete), the umbrella government's Regional Aviation Access Programme (RAAP) also includes the Remote Aviation Infrastructure Fund (RAIF), the Remote Airstrip Upgrade (RAU) Programme, the Remote

Merit good: a product or service that is considered by society or government as being deserving of public finance.

Table 21.3 Number of routes and operators across the European PSO air service scheme (September 2018)

Country	Number of operated routes	Number of airlines operating
Croatia	10	2
Cyprus	1	1
Czech Republic	1	1
Estonia	3	2
Finland	2	2
France	35	12
Greece	28	3
Ireland	3	2
Italy	11	4
Lithuania	1	1
Portugal	20	3 (+1)
Spain	20	4
Sweden	11	4
United Kingdom	22	6

Source: European Commission

Aerodrome Safety Programme (RASP) and the Remote Aerodrome Inspection (RAI) Programme. In addition, there is also the Airservices Australia En-route Charges Payment Scheme that provides a subsidy to air operators providing aero-medical services to regional and remote locations, which includes reimbursement of en-route air navigation charges. On the airline side it is only the thin routes where public authorities support scheduled regular air services, while on the airport side direct and indirect public support benefits all types of air transport to remote regions.

Stop and think

What are the advantages and disadvantages of PSO routes (and non-European equivalents) to remote regions for airlines, airports, passengers and the remote regions themselves?

21.5 The future

Although jet fuel prices have not reached their peak of 2008 and 2014, it is likely that they will increase again given the scarcity of oil. It is therefore likely that in the future fuel prices will continue to present a high share of airline operating costs, and if no fuel adjustors or hedging mechanisms are available, the volatility of fuel prices will mean that air transport to remote regions will become even more risky and even less commercially viable. As many regional airlines are not in a financial position to hedge; they are particularly exposed to that risk, as evidenced by the 15 per cent jump in fuel prices in mid-September 2019. Changing (and often ageing) remote populations and increasing safety and security compliance costs will all contribute to the challenge.

The other key challenge will be a shortage of technical staff at remote airports and of pilots of aircraft operating in remote regions. While in the remote airline context training is much less of an issue, as pilots use this air time as their training ground and hour building environment, retaining them is far more problematic. In the remote airport context, staff training and retention will become ever more challenging regardless of the location. All of this points to an increasing need for public support, be it through subsidies, regulation, regional development or aviation friendly policy making. It is also likely that the aviation industry will, as so often in the past, seek to address its issues particularly with respect to productivity and environmental challenges. Innovation is the key to this, and Case Study 21.2 provided an illustration of the potential for such innovative technologies and processes. What is encouraging is that the latest research has revealed that not only passengers from regional areas but also residents of metropolitan areas value regional air services (Merkert and Beck, 2017). An interesting opportunity for increasing that appeal has been shown in better integration of regional air services with ground transport offerings (Merkert and Beck, 2020) and also in the use of technology which may contribute to improving the attractiveness of public transport or in some parts reduce the need for travelling.

Key points

- No single definition of a 'remote region' exists, and individual countries use their own definitions.

- Air transport in remote regions is of crucial importance to the communities and businesses it serves. RRR services are often considered to be a 'lifeline', as without them some remote communities could not continue to exist in their present form.

- Air transport in remote regions includes publicly supported passenger routes as well as charter services and corporate flights related to tourism and the resources sector.

- Remote air services are operated by both FSNCs and specialist niche operators using a range of fixed wing and rotary wing aircraft. Although a range of aircraft types are used, most fixed wing aircraft used on remote services have fewer than 30 seats, and some are specifically equipped to operate from short and/or unprepared runways.

- Air transport provides substantial direct, indirect, induced and catalytic benefits to remote regions, but the services pose significant operational and market challenges.

- The low demand for air services and the consequent precarious financial state of some remote airlines and airports means public support for certain routes is required.

- Managing remote airports poses unique challenges and requires particular skills. Remote airports are seeking ways to minimise operating costs, and the Remote Tower concept is one way in which greater efficiencies and cost savings may be achieved.

- Innovation, increased marketing and route/business development efforts, as well as more appropriate risk sharing between stakeholders, will aid in securing the future of air transport in remote regions and promoting the economic and social viability and vitality of communities and businesses they serve.

References and further reading

Abreu, J., Fageda, X. and Jiménez, J. L. (2018). An empirical evaluation of changes in public service obligations in Spain. *Journal of Air Transport Management*, 67, pp. 1–10.

Baker, D., Merkert, R. and Kamruzzaman, M. (2015). Regional aviation and economic growth: Cointegration and causality analysis in Australia. *Journal of Transport Geography*, 43, pp. 140–150.

Bråthen, S. and Eriksen, K. S. (2018). Regional aviation and the PSO system: Level of service and social efficiency. *Journal of Air Transport Management*, 69, pp. 248–256.

Merkert, R. and Beck, M. J. (2017). Value of travel time savings and willingness to pay for regional aviation. *Transportation Research Part A: Policy and Practice*, 96, pp. 29–42.

Merkert, R. and Beck, M. J. (2020). Can a strategy of integrated air-bus services create a value proposition for regional aviation management? *Transportation Research Part A: Policy and Practice*, 132, pp. 527–539.

Merkert, R. and Hensher, D. A. (2013). The importance of completeness and clarity in air transport contracts in remote regions in Europe and Australia. *Transportation Journal*, 52(3), pp. 365–390.

Merkert, R., Odeck, J., Bråthen, S. and Pagliari, R. (2012). A review of different benchmarking methods in the context of regional airports. *Transport Reviews*, 32(3), pp. 379–395.

Merkert, R. and O'Fee, B. (2016). Managerial perceptions of incentives for and barriers to competing for regional PSO air service contracts, *Transport Policy*, 47, pp. 22–33.

Index